T0332590

Geometry of Crystals, Polycrystals, and Phase Transformations

Geometry of Crystals, Polycrystals, and Phase Transformations

Harshad K. D. H. Bhadeshia

CRC Press
Taylor & Francis Group
Boca Raton London New York

CRC Press is an imprint of the
Taylor & Francis Group, an **informa** business

CRC Press
Taylor & Francis Group
6000 Broken Sound Parkway NW, Suite 300
Boca Raton, FL 33487-2742

Printed on acid-free paper
Version Date: 20170802

International Standard Book Number-13: 978-1-138-07078-3 (Hardback)

Visit the Taylor & Francis Web site at
http://www.taylorandfrancis.com

and the CRC Press Web site at
http://www.crcpress.com

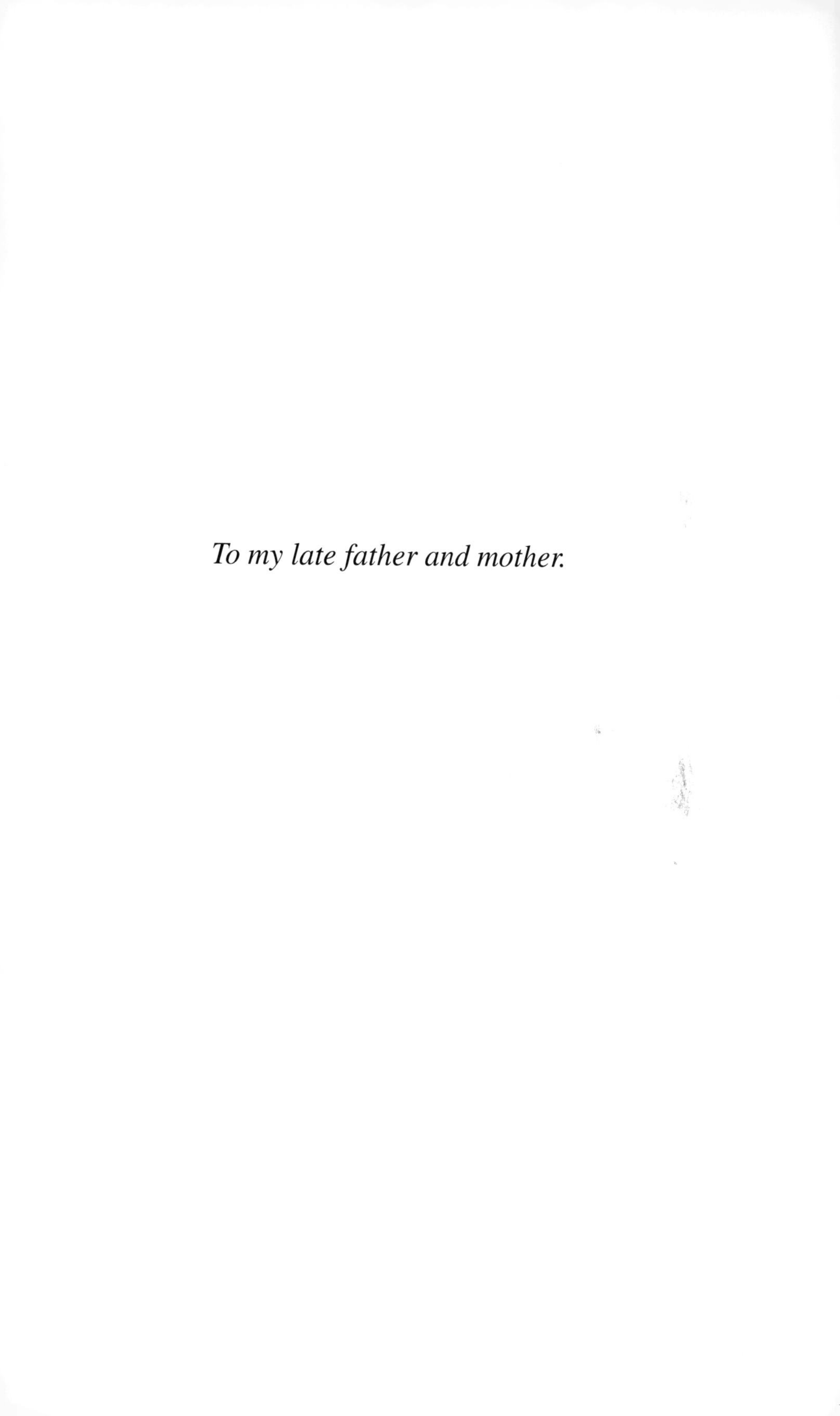

To my late father and mother.

Contents

Preface and Acknowledgments

To state the obvious, crystals contain order. Even when that is disturbed locally, the perturbations themselves may sometimes form regular patterns. This is evident in the structure of interfaces where two or more crystals meet in an apparently haphazard manner. Disorder can sometimes be ignored without compromising some of the consequences of long-range order. Crystalline solid solutions in which atoms are dispersed at random would fail the strict definition of long-range order, but they nevertheless show the characteristics of crystals when probed by X-rays. We shall attempt in this book to understand not only the elegance of individual crystals, but also of clusters of space-filling crystals and transformations between crystalline phases.

Crystallography is hardly a new subject so there are numerous books available, many of which are beacons of scholarship. So why another text on this much mooted topic? First, as Buckminster Fuller pointed out, there has been a massive expansion in human knowledge, with the process continuing at an unabating pace. New subjects spring up with notorious regularity and some of these have come to be regarded as essential first steps in the higher education curriculum. Modern undergraduate students are therefore faced with a much greater palette of distinct courses than has been the case in the past.

This book is partitioned into two, the first part of which is meant to be self-contained and deals with what I feel is essential learning for any student in the material sciences, physics, chemistry, earth sciences, and the natural sciences in general. This part, intentionally concise, is based on a set of nine lectures that I give annually to undergraduate students of the Natural Sciences Tripos in Cambridge University. It is of generic value and has just sufficient material to deliver concepts. It covers crystals, polycrystals, interfaces, and transformations that occur by the disciplined motion of atoms. I feel that most books on crystallography are too detailed to accommodate within the schedule of a contemporary undergraduate.

The second part has depth which would be appreciated most in the context of research. It is a development from a book that I taught and published in 1987 on the Geometry of Crystals. The treatment is limited to phenomena dominated by crystallography.

The book contains worked examples throughout because crystallography is a subject that thrives on practice. Video lectures and other electronic materials that can enhance the content of this book are available freely on

http://www.phase-trans.msm.cam.ac.uk/teaching.html

Crystallography has rules, established by convention, but many of these have exceptions and there sometimes are multiple conventions. Donnay in his 1943 paper proposed rules for defining the crystallographic orientation of a crystal [*American Mineral* 28 (1943) 313–328], punctuated by exceptions, special cases, and difficulties. In the Bunge convention for Euler angles, the sample frame is rotated into the crystal frame, but there is another convention where the reverse is the case. My view is that in the application of crystallography, only a limited number of conventions are important. For example, the handedness of cell axes should be consistent throughout. Similarly, although only three indices are necessary to define vectors in three dimensions, the four index system for the hexagonal class permeates the literature and hence needs to be taught. The emphasis of the book is to teach concepts rather than rules, which necessarily require rote learning – I am uncertain as to whether I have done enough to minimize the use of conventions.

My interest in crystallography stems primarily from research on solid-state phase transformations. I have benefitted enormously from the writings of the late J. W. Christian and C. M. Wayman. And from incisive questions posed by undergraduates. I have been able to tap routinely into the knowledge and wisdom of John Leake and Kevin Knowles whenever I felt confused about the fine detail of crystallography. To all of these characters, I shall remain perpetually grateful for the education and camaraderie.

H. K. D. H. Bhadeshia
Cambridge

Author

Harshad K. D. H. Bhadeshia is the Tata Steel Professor of Metallurgy at the University of Cambridge. His main interest has been on the theory of solid-state phase transformations with emphasis on the prediction and verification of microstructural development in complex metallic alloys, particularly multicomponent steels. He has authored or co-authored some 700 publications including several books. He is a Fellow of the Royal Society, the Royal Academy of Engineering, and a Foreign Fellow of the National Academy of Engineering (India). Professor Bhadeshia was in 2015 appointed a Knights Bachelor in the Queen's Birthday Honours for services to Science and Technology.

Acronyms

bcc Body-centered cubic

bct Body-centered tetragonal

fcc Face-centered cubic

hcp Hexagonal close-packed

IPS Invariant-plane strain

ILS Invariant-line strain

P Primitive

I Body-centered

F Face-centered

R Trigonal

Part I

Basic Crystallography

1

Introduction and Point Groups

Abstract

The term "order" is a fundamental feature of the definition of a crystal, a feature that distinguishes it from random aggregates of atoms where the description of the structure requires a specification of the coordinates of every single atom that exists in the material. In contrast, the crystal can be described completely in terms of a repeat unit that usually consists of a "handful" of atoms. Another consequence of the order is that crystals have properties that vary with direction, unlike amorphous materials that tend to be isotropic. Certain tools are required to deal with the regularity of crystals, some of which are introduced in this chapter.

1.1 Introduction

Amorphous solids are homogeneous and isotropic because there is no long range order or periodicity in their internal atomic arrangement. In contrast, the crystalline state is characterized by a regular arrangement of atoms over large distances. Crystals are therefore anisotropic — their properties vary with direction. For example, the interatomic spacing varies with orientation within the crystal, as does the elastic response to an applied stress.

Crystals can be two or three dimensional; graphene is a two-dimensional crystal if we neglect the fact that it is one atom thick (Figure 1.1a), whereas polonium (Figure 1.1b) is obviously three dimensional. There is talk about one-dimensional crystals where a string of equally spaced atoms is in a straight line, made by confining the atoms into a carbon nanotube; however, there are interactions between these atoms and the nanotube which violate the definition.[1] Four-dimensional crystals which are periodic in both time and space have been postulated [2]. For example, a cylindrical crystal with a periodic arrangement of atoms that repeats as time is stepped. Such a structure would require perpetual motion, i.e., must not radiate rotational energy so there are difficulties in explaining how such a crystal could exist in a non-stationary ground state, a subject of discussion in the associated literature. Some experimen-

[1] To quote Ziman [1], "the hypothetical ordered linear chain is physically unrealizable as a genuine one-dimensional system".

tal evidence now exists for discrete time crystals [3, 4]; the question remains as to whether the "time crystals" are everlasting [5].

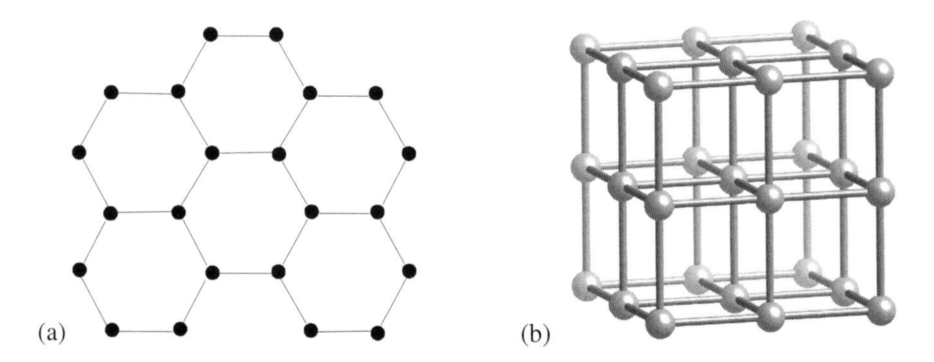

FIGURE 1.1
(a) The arrangement of carbon atoms in the two-dimensional structure of graphene. (b) The arrangement of atoms in the three-dimensional structure of polonium.

The crystalline form usually conjures visions of beautiful facetted forms, but it is worth noting at the outset that they can be solid or liquid,[2] and can have arbitrary shapes. The familiar elegant shapes are a consequence of the minimization of surface energy but in engineering applications the crystals are produced in functional shapes (Figure 1.2).

The beautiful shapes of many crystals that are found in nature represent an attempt at the minimization of surface energy per unit volume of material when at equilibrium, or because the rate of growth varies with direction. For isotropic materials, a sphere would represent the equilibrium shape but crystals are not isotropic; some planes of atoms have a lower surface energy than others so there will be a tendency to maximize the area of those planes, in which case the equilibrium shape is no longer spherical [7].

Engineering materials usually are space-filling aggregates of many crystals of varying sizes and shapes; these *polycrystalline* materials have properties which depend on the nature of the individual crystals, but also on aggregate properties such as the size and shape distributions of the crystals, and the orientation relationships between the individual crystals. The degree of randomness in the orientations of the crystals relative to a fixed frame is a measure of *texture*, which has to be controlled in the manufacture of transformer steels, uranium fuel rods, and beverage cans.

Connecting a pair of crystals together to form a bi-crystal requires a process involving at least five degrees of freedom; two for the choice of the plane along which the pair are connected and three for the relative orientations of the two crystals. The connecting plane is referred to as a "boundary" or an "interface" in general. The crystallography of interfaces connecting adjacent crystals can determine the deformation

[2]Liquid crystals contain order that is associated with anisotropic molecules that can flow past each other and yet maintain orientational order [6]. Normal liquids are isotropic.

(a)

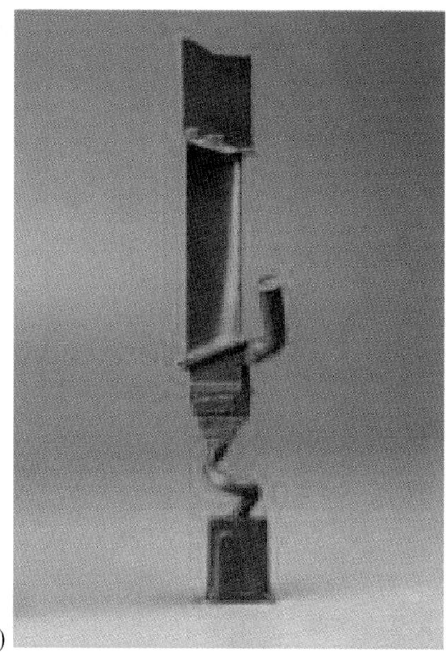
(b)

FIGURE 1.2
(a) Clusters of naturally occurring pyrite crystals (FeS_2) from Peru. The structure belongs to the cubic crystal system. We shall see later that the octahedral shapes with triangular facets are consistent with the defining symmetry of a cube. (b) A nickel alloy in the shape of a turbine blade; the part above the spiral is the single crystal. After removal of the appendages associated with the manufacturing process, the shape is consistent with the aerodynamic form required for a turbine engine. Blades of this kind serve in harsh environments and are subjected to large stresses. This can lead to slow, stress-driven lengthening of the blade driven by the movement of atoms, eventually making it unfit for service. In polycrystalline blades, the boundaries between the crystals are easy paths for atom movement and hence have shorter service lives.

behavior of the polycrystalline aggregate; it can influence mechanical properties such as toughness through its effect on the degree of segregation of impurities to such interfaces [8].

1.2 The lattice

Imagine an infinite, two-dimensional pattern of points such that an observer located at any single point would see the same pattern all around when translating to any other point in the pattern. Such a pattern would be periodic, i.e., a basic unit of the pattern repeats indefinitely to generate the whole pattern. Each of the points in the pattern therefore has the same environment and could be selected as the origin of the pattern. Points in this pattern that have identical environments are known as the *lattice points*.

The set of these lattice points constitutes a three-dimensional lattice. A *unit cell* may be defined within this lattice as a space-filling parallelepiped with origin at a lattice point, and with its edges given by three non-coplanar *basis vectors* $\mathbf{a}_1$, $\mathbf{a}_2$, and $\mathbf{a}_3$, each of which represents translations between two adjacent lattice points. The entire lattice can then be generated by stacking unit cells in three dimensions. Any vector representing a translation between lattice points is called a *lattice vector*, Figure 1.3.

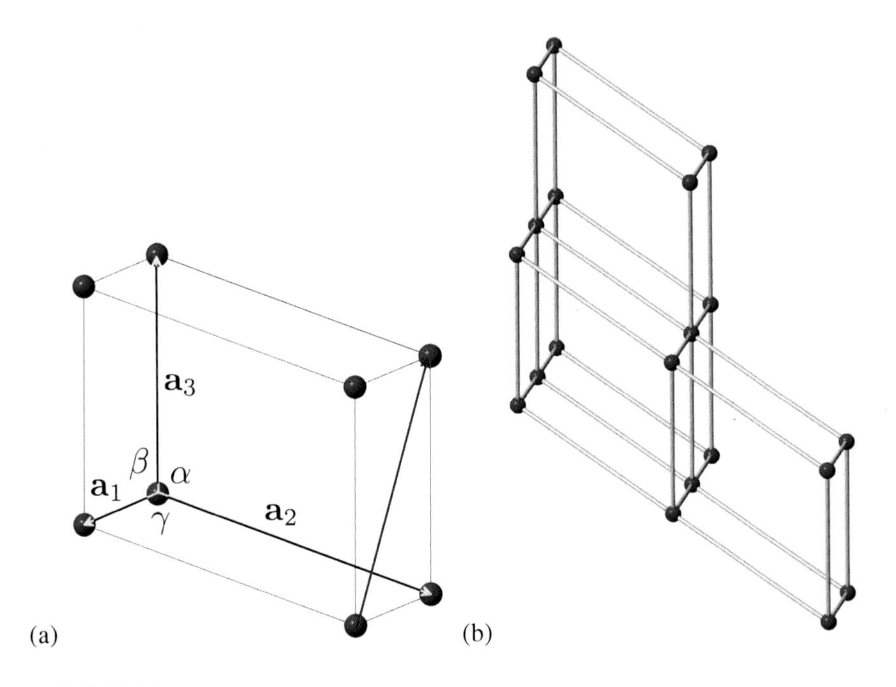

FIGURE 1.3
(a) Definition of a unit cell; the lattice points are drawn as blue spheres for clarity and the red vector is a lattice vector joining two lattice points. (b) The stacking of unit cells to produce a periodic pattern; the stacking can be continued indefinitely to fill the space available without leaving unfilled regions.

The unit cell defined above has lattice points located at its corners. Since these are shared with seven other such cells, and since each cell has eight corners, there is only one lattice point per unit cell. Such a unit cell is *primitive* and has the lattice symbol “P”.

Non-primitive unit cells can have two or more lattice points, in which case, the additional ones will be located at positions other than the corners of the cell. A cell with additional lattice points located at the centers of all its faces has the lattice symbol “F”; such a cell would contain four lattice points. Not all the faces of the cell need to have face-centering lattice points; when a cell containing two lattice points has the additional point located at the center of the face defined by $\mathbf{a}_2$ and $\mathbf{a}_3$, the lattice symbol is "A" and the cell is said to be A-centered. B-centered and C-centered cells have the additional lattice point located on the face defined by $\mathbf{a}_3$ and $\mathbf{a}_1$ or $\mathbf{a}_1$ and $\mathbf{a}_2$ respectively. A unit cell with two lattice points can alternatively have the additional lattice point at the body-center of the cell, in which case the lattice symbol is *I*. The lattice symbol "R" is for a trigonal cell; the cell usually is defined such that it contains three lattice points. The detailed shapes of unit cells are considered in Section 1.3.

The basis vectors $\mathbf{a}_1$, $\mathbf{a}_2$, and $\mathbf{a}_3$ define the unit cell; their magnitudes a_1, a_2, and a_3, respectively, are the *lattice parameters* of the unit cell. The angles $\widehat{\mathbf{a}_1\ \mathbf{a}_2}$, $\widehat{\mathbf{a}_2\ \mathbf{a}_3}$, and $\widehat{\mathbf{a}_3\ \mathbf{a}_1}$ are conventionally labelled γ, α, and β, respectively. Vectors within this framework can be represented by their components with respect to the basis vectors. For a particular vector, this set of components defines the *Miller indices* of that vector.[3]

Our initial choice of the basis vectors was arbitrary since there is an infinite number of lattice vectors which could have been used in defining the unit cell. The preferred choice includes small basis vectors which are as equal as possible, provided the shape of the cell reflects the essential symmetry of the lattice.

Example 1.1: Primitive representation of Cubic-F

The face-centered cubic (Cubic-F) cell has four lattice points per cell and hence is not primitive. Construct its primitive version bearing in mind that the definition of the unit cell requires it to be space filling. Identify its lattice type and calculate its relative volume.

In what follows, directions are represented by their components (Miller indices) along the basis vectors, enclosed within square brackets. The primitive cell is illustrated in Figure 1.4 with:

$$[100]_\mathrm{P} \equiv \frac{1}{2}[\bar{1}10]_\mathrm{F}$$
$$[010]_\mathrm{P} \equiv \frac{1}{2}[\bar{1}01]_\mathrm{F}$$
$$[001]_\mathrm{P} \equiv \frac{1}{2}[011]_\mathrm{F}.$$

The primitive cell is trigonal, $a_1 = a_2 = a_3$, the angle $\alpha = \beta = \gamma = 60°$, containing a single triad along $[111]_\mathrm{P}$. This is in contrast to the four triads along the body diagonals of the cubic cell. As discussed later (Table 1.1), the defining symmetry of a cube is four triads whereas that of a trigonal cell is just one triad.

Since the primitive cell has a quarter of the lattice points of the face-centered cubic cell, its volume is also smaller by a factor of four.

[3] See [9] for a discussion of the history of indexing.

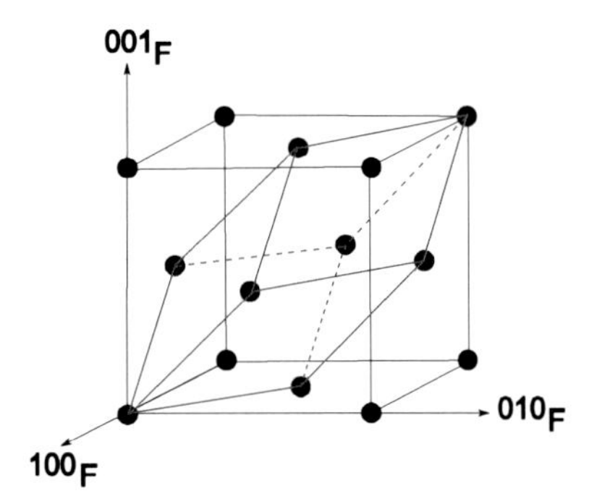

FIGURE 1.4
The conventional face-centered cubic unit cell is in black and its primitive version in blue.

1.3 Bravais lattices

The number of ways in which points can be arranged regularly in three dimensions, such that the stacking of unit cells fills space, is not limitless; Bravais showed in 1848 that all possible arrangements can be represented by just 14 lattices.

The 14 lattices can be categorized into seven *crystal systems* (cubic, tetragonal, orthorhombic, trigonal, hexagonal, monoclinic, and triclinic, Table 1.1); the cubic system contains, for example, the cubic-P, cubic-F, and cubic-I lattices. Each crystal system can be characterized uniquely by a set of defining symmetry elements, which any crystal within that system must possess as a minimum requirement (Table 1.1). The Bravais lattices are illustrated in Figure 1.5.

1.3.1 Two-dimensional crystals and surfaces

The loss of the third dimension reduces the number of periodic patterns of points that can be used to define unit cells to just the five possibilities illustrated in Figure 1.7. All of these can be derived from the projections shown in Figure 1.6 in the x-y plane – all the cubic lattices, for example, reduce to square lattices.

Two-dimensional atomic crystals are rare and frequently do not survive for long in uncontrolled environments. Graphene is a single layer of carbon atoms with a hexagonal two-dimensional lattice. It can be extracted from graphite, which consists of parallel layers of graphene with weak bonding between the layers. There are other examples of layered structures which are amenable to the preparation of two-dimensional crystals that can reach lateral dimensions that are in the micrometer range: BN, MoS_2 (often used as a lubricant because of its layered structure), $NbSe_2$, $Bi_2Sr_2CaCu_2O_x$ [10]. Complex molecules can precipitate on substrates, from fluid, as two-dimensional arrays that are arranged periodically, for example the two-dimensional chiral crystals of phospholipid [11]. There are numerous ex-

TABLE 1.1
The crystal systems.

System	Conventional unit cell		Defining symmetry
Triclinic	$a_1 \neq a_2 \neq a_3$	$\alpha \neq \beta \neq \gamma$	Monad
Monoclinic	$a_1 \neq a_2 \neq a_3$	$\alpha = \gamma = 90°,\ \beta \geq 90°$	1 diad
Orthorhombic	$a_1 \neq a_2 \neq a_3$	$\alpha = \beta = \gamma = 90°$	3 diads
Tetragonal	$a_1 = a_2 \neq a_3$	$\alpha = \beta = \gamma = 90°$	1 tetrad
Trigonal	$a_1 = a_2 = a_3$	$\alpha = \beta = \gamma \neq 90°$	1 triad
Hexagonal	$a_1 = a_2 \neq a_3$	$\alpha = \beta = 90°, \gamma = 120°$	1 hexad
Cubic	$a_1 = a_2 = a_3$	$\alpha = \beta = \gamma = 90°$	4 triads

Note: The symmetry axes monad, diad, triad, tetrad, and hexad involve rotations of 360°, 180°, 120°, 90°, and 60°, respectively. These particular rotations leave the lattice in a state that cannot be distinguished from its starting configuration. a_i are the magnitudes of the basis vectors $\mathbf{a}_i$ (which form a right-handed set). The defining symmetry cannot be achieved unless the conditions highlighted in red are satisfied.

amples of periodic patterns developing when molecules (e.g., water) or particles (polystyrene spheres) are deposited on substrates – whether these can be regarded as two-dimensional crystals, as they often are, remains doubtful since such patterned objects cannot be isolated in their own right. Nevertheless, from the point of view of effects such as diffraction and electrical properties, the periodicity of the particles deposited on substrates is influential.

The structure of a three-dimensional crystal is in general assumed to persist to the very surfaces of the crystals, but the forces there are different from those in the bulk of the material. As a consequence, surfaces are known to reconstruct into one of the five lattice types illustrated in Figure 1.7. This of course is important in processes that rely on surface structure, such as catalysis [12].

1.4 Directions

A vector $\mathbf{u}$ can be represented as a linear combination of the basis vectors $\mathbf{a}_i$ of the unit cell ($i = 1, 2, 3$):

$$\mathbf{u} = u_1\mathbf{a}_1 + u_2\mathbf{a}_2 + u_3\mathbf{a}_3 \tag{1.1}$$

and the scalar quantities u_1, u_2, and u_3 are the components of the vector $\mathbf{u}$ with respect to the basis vectors $\mathbf{a}_1$, $\mathbf{a}_2$, and $\mathbf{a}_3$. Once the unit cell is defined, any direction $\mathbf{u}$ within the lattice can be identified uniquely by its components $[u_1\ u_2\ u_3]$, and the components are called the Miller indices of that direction. They are by convention enclosed in square brackets (Figure 1.8).

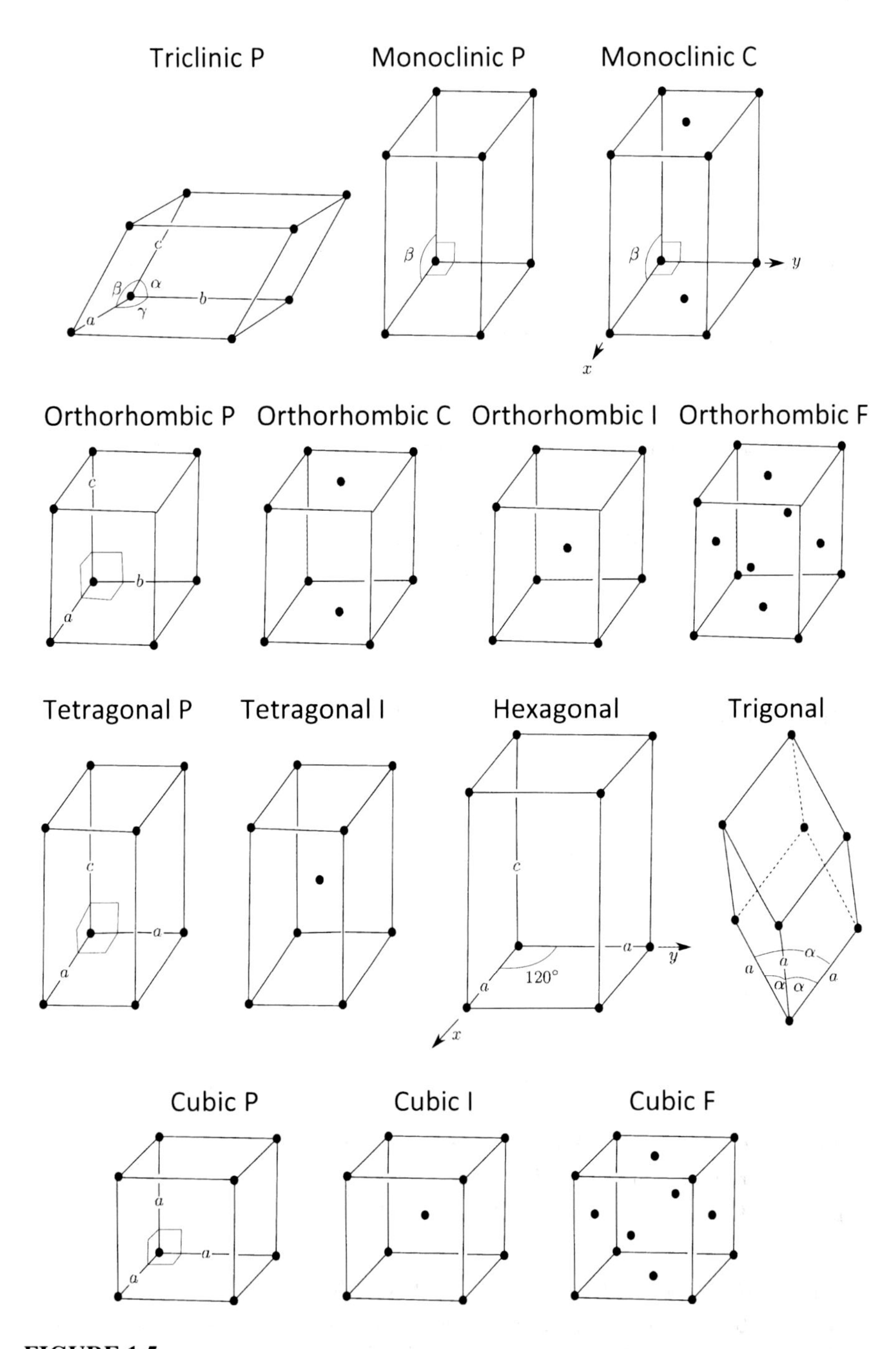

FIGURE 1.5
The 14 three-dimensional Bravais lattices.

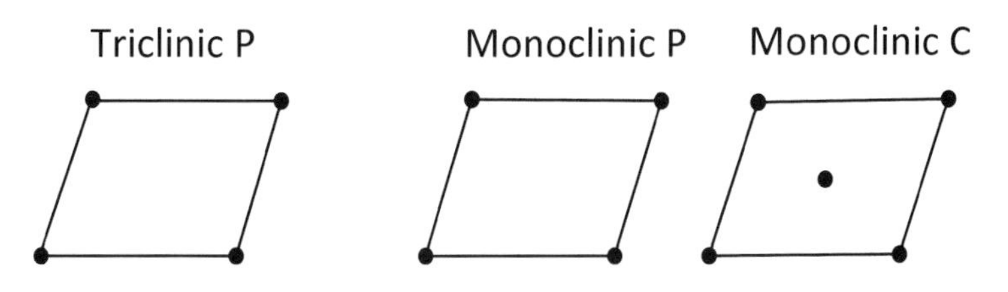

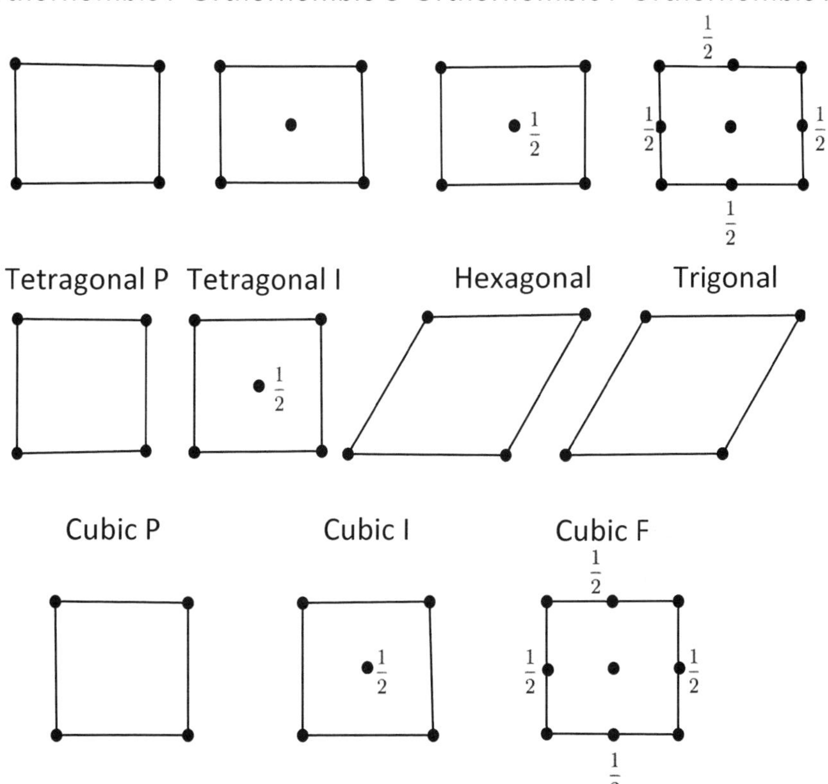

FIGURE 1.6
Projection of the three-dimensional shapes of the 14 Bravais lattices on to the $\mathbf{a}_1$-$\mathbf{a}_2$ plane. The numbers indicate the coordinates of lattice points relative to the $\mathbf{a}_3$ axis; the unlabelled lattice points are by implication located at coordinates 0 and 1 with respect to the $\mathbf{a}_3$ axis.

It is sometimes the case that the properties along two or more different directions are identical. These directions are said to be *equivalent* and the crystal is said to possess *symmetry*. For example, the $[1\ 0\ 0]$ direction for a cubic lattice is equivalent to the $[0\ 1\ 0]$, $[0\ 0\ 1]$, $[0\ \bar{1}\ 0]$, $[0\ 0\ \bar{1}]$, and $[\bar{1}\ 0\ 0]$ directions; the bar on top of the number implies that the index is negative.

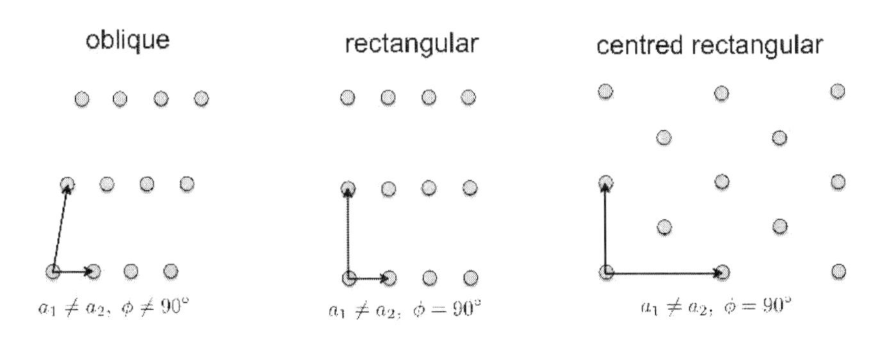

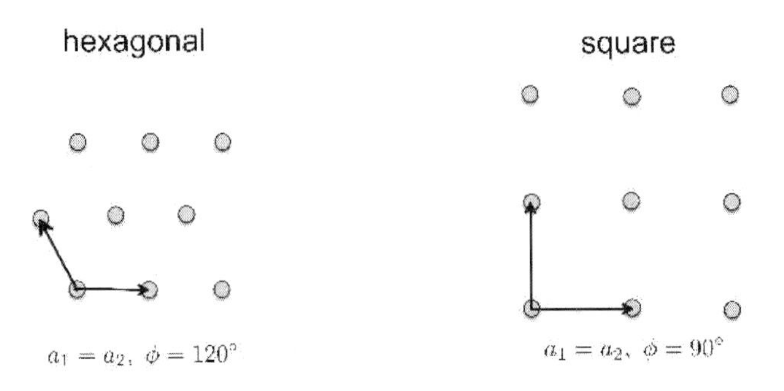

FIGURE 1.7
The five two-dimensional lattices.

The indices of directions of the same form are conventionally enclosed in special brackets, e.g., $\langle 1\ 0\ 0 \rangle$. The number of equivalent directions within the form is called the multiplicity of that direction, which in this case is 6.

Example 1.2: Number of equivalent indices

Given a direction with indices $[u\ v\ w]$ such that $u \neq v \neq w \neq 0$, calculate the number of equivalent directions possible, for the following lattice types: (a) cubic, (b) tetragonal.

(a) In the cubic system where $a_1 = a_2 = a_3$, there are six choices for the first index, i.e., u, $-u$, v, $-v$, w, $-w$. Suppose that v is selected as the first index, then the choice for the second index is limited to one of four options u, $-u$, w, $-w$. If the second index is $-u$, then the choice for the final index becomes w or $-w$. The number of equivalent directions of the form $\langle u\ v\ w \rangle$, $6 \times 4 \times 2 = 48$. However, if antiparallel options such as $[u\ v\ w]$ and $[-u\ -v\ -w]$ are not counted then the number of equivalent directions reduces to 24.

Note that if $u = v$ then the number of equivalent directions reduces to $6 \times$

$1 \times 2 = 12$ because the choice of the first index leaves only one choice for the second. An index that is zero also reduces the options by a factor of two since there is then no negative or positive aspect.Therefore, in the cubic system, there are 6 variations of $\langle 1\ 1\ 0 \rangle$: $[1\ 1\ 0]$, $[1\ \bar{1}\ 0]$, $[0\ 1\ 1]$, $[0\ 1\ \bar{1}]$, $[1\ 0\ 1]$, and $[1\ 0\ \bar{1}]$ if antiparallel directions are not counted.

(b) For the tetragonal system, $a_1 = a_2 \neq a_3$. Therefore, the choice of the first index is limited to four options, u, $-u$, v, $-v$,; if u is selected as the first index then the choice of the second is just two (v, $-v$) and of the third w or $-w$. Therefore, the number of equivalent directions of the form $\langle u\ v\ w \rangle$ becomes $6 \times 2 \times 2 = 24$, reducing to 12 if antiparallel options are removed.

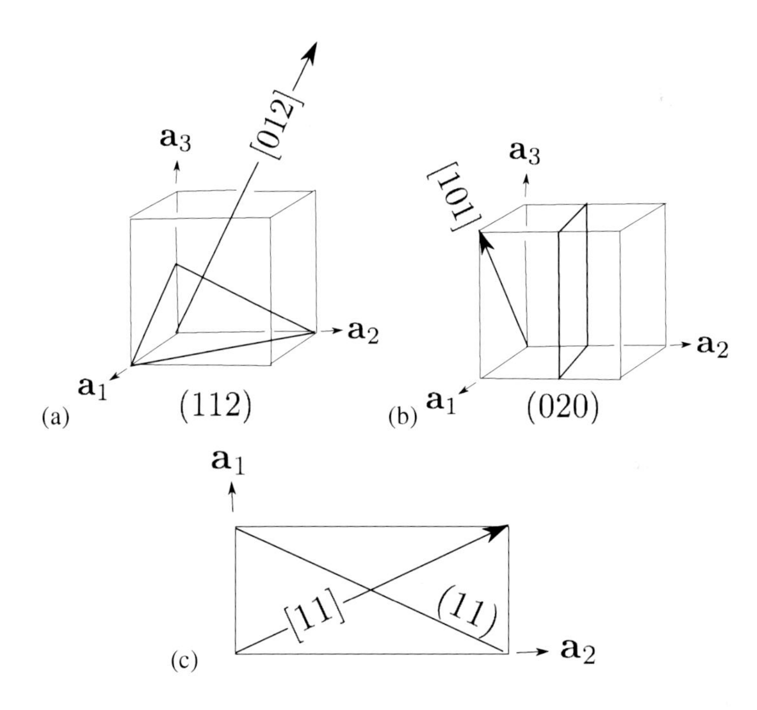

FIGURE 1.8
(a,b) Miller indices for directions and planes. (c) Notice that in non-cubic systems, a direction with indices identical to those of a plane is not necessarily normal to that plane.

1.5 Planes

If a plane intersects the $\mathbf{a}_1$, $\mathbf{a}_2$, and $\mathbf{a}_3$ axes at distances x_1, x_2, and x_3, respectively, relative to the origin, then the Miller indices of that plane are given by $(h_1\ h_2\ h_3)$

where:

$$h_1 = \phi a_1/x_1, \quad h_2 = \phi a_2/x_2, \quad h_3 = \phi a_3/x_3 \tag{1.2}$$

ϕ is a scalar which clears the numbers h_i off fractions or common factors. Note that x_i are negative when measured in the $-\mathbf{a}_i$ directions. The intercept of the plane with an axis may occur at infinity (∞), in which case the plane is parallel to that axis and the corresponding Miller index will be zero (Figure 1.8).

Miller indices for planes are by convention written using round brackets: $(h_1\ h_2\ h_3)$ with braces being used to indicate planes of the same form: $\{h_1\ h_2\ h_3\}$.

1.6 Weiss zone law

This law states that if a direction $[u_1\ u_2\ u_3]$ lies in a plane $(h_1\ h_2\ h_3)$ then

$$u_1h_1 + u_2h_2 + u_3h_3 = 0 \tag{1.3}$$

and the law applies to any crystal system. This will be proven when we deal with the reciprocal lattice (Chapter 5).

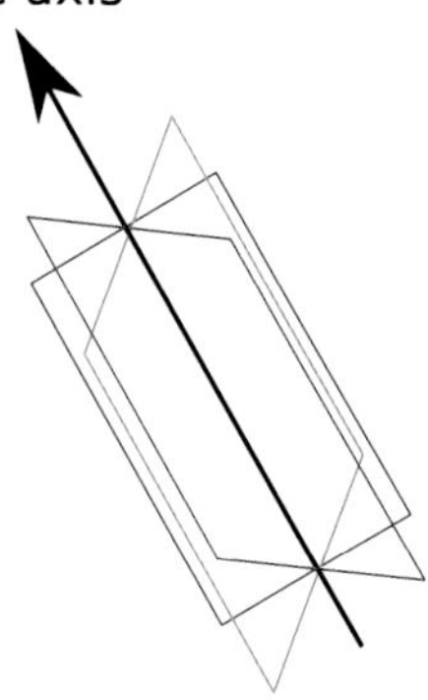

FIGURE 1.9
A *zone* refers to a set of planes which share a common direction, which in turn is known as a zone axis. The direction illustrated would satisfy the Weiss law for all the planes shown. For example, the planes $(1\ 1\ 1)$, $(1\ 1\ 2)$, and $(1\ 1\ 0)$ all belong to the zone $[1\ \bar{1}\ 0]$.

1.7 Symmetry

Although the properties of a crystal can be anisotropic, there may be different directions along which they are identical. These directions are said to be *equivalent* and the crystal is said to possess *symmetry*.

That a particular edge of a cube cannot be distinguished from any other is a measure of its symmetry; an orthorhombic parallelepiped has lower symmetry, since its edges can be distinguished by length.

Some symmetry operations are illustrated in Figure 1.10; in essence, they transform a spatial arrangement into another that is indistinguishable from the original. The rotation of a cubic lattice through 90° about an axis along the edge of the unit cell is an example of a symmetry operation, since the lattice points of the final and original lattice coincide in space and cannot consequently be distinguished.

We have encountered translational symmetry when defining the lattice; since the environment of each lattice point is identical, translation between lattice points has the effect of shifting the origin. Dislocations whose Burgers vectors are lattice vectors therefore accomplish slip without changing the crystal structure. Deformations like these are known as *lattice-invariant deformations* because they accomplish a shape change without altering the nature of the lattice. In contrast, *partial* dislocations have Burgers vectors that are not lattice vectors – they therefore change the local stacking fault sequence of the planes on which they glide.

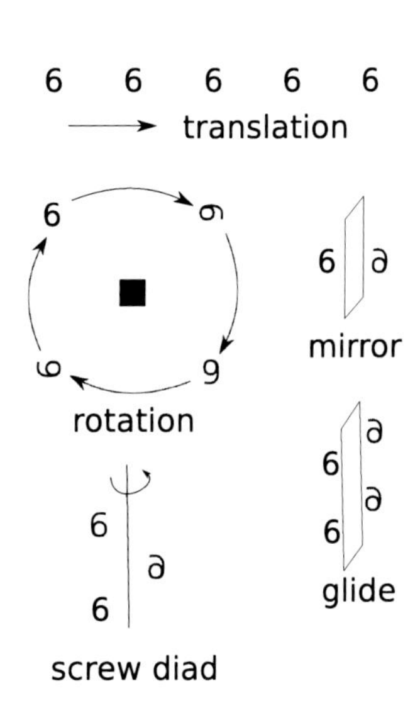

FIGURE 1.10
Some symmetry operations. The mirror plane involves reflection symmetry, whereas the glide plane is a combination of reflection and translation parallel to the mirror. Translational symmetry is intrinsic to all lattices whereas the other elements may or may not exist for all lattices. For example, the triclinic lattice has no mirror plane.

Note that the translations involved in the screw axis and glide plane are rational fractions of the repeat distance along the translation direction. The screw axis illustrated involves a rotation of 180° (≡ diad) combined with a translation that is half of the repeat distance parallel to the axis.

1.8 Symmetry operations

An object possesses an n-fold axis of rotational symmetry if it coincides with itself upon rotation about the axis through an angle $360°/n$. The possible angles of rotation, which are consistent with the translational symmetry of the lattice, are 360°, 180°, 120°, 90°, and 60° for values of n equal to 1, 2, 3, 4, and 6, respectively. A five-fold axis of rotation does not preserve the translational symmetry of the lattice and hence is forbidden. A one-fold axis of rotation is called a monad and the terms diad, triad, tetrad, and hexad correspond to $n = 2, 3, 4$, and 6, respectively.

All of the Bravais lattices have a *center of symmetry*, Figure 1.11. An observer at the center of symmetry sees no difference in arrangement between the directions $[u_1\, u_2\, u_3]$ and $[\overline{u}_1\, \overline{u}_2\, \overline{u}_3]$. The center of symmetry is such that inversion through that point produces an identical arrangement but in the opposite sense. A *rotoinversion* axis of symmetry rotates a point through a specified angle and then inverts it through the center of symmetry such that the arrangements before and after this combined operation are in coincidence. For example, a three-fold inversion axis involves a rotation through 120° combined with an inversion, the axis being labelled $\overline{3}$. Some of the symbols representing rotation and rotoinversion axis are as follows, where $\overline{2}$ is omitted because it is, as will be shown in Chapter 2, equivalent to a mirror operation:

$2 \quad 3 \quad 4 \quad 6 \quad \overline{1} \quad \overline{3} \quad \overline{4} \quad \overline{6}$

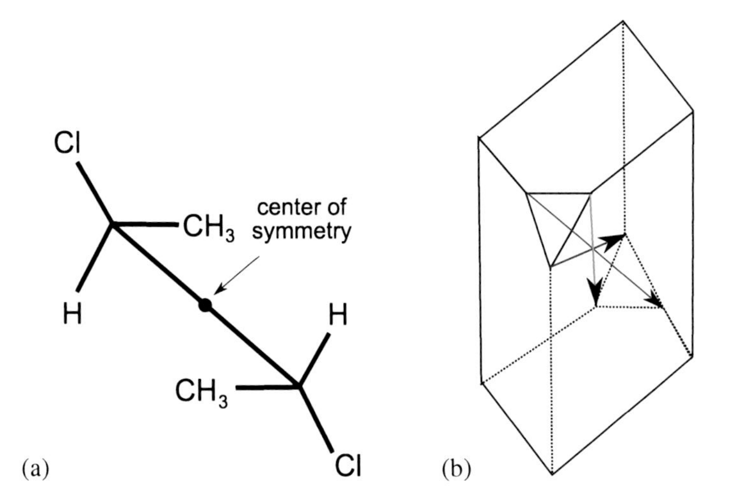

FIGURE 1.11
Illustration of center of symmetry. (a) A molecule with the center of symmetry identified by the dot. (b) The common point where the arrows intersect is the center of symmetry.

A rotation operation can be combined with a translation parallel to that axis to generate a *screw axis* of symmetry. The magnitude of the translation is a fraction of the lattice repeat distance along the axis concerned. A 3_1 screw axis would rotate a point through 120° and translate it through a distance $t/3$, where t is the magnitude of the shortest lattice vector along the axis. A 3_2 operation involves a rotation through 120° followed by a translation through $2t/3$ along the axis. For a right-handed screw axis, the sense of rotation is anticlockwise when the translation is along the positive direction of the axis.

A plane of *mirror symmetry* implies arrangements which are mirror images. Our left and right hands are (approximately) mirror images. The operation of a $\overline{2}$ axis produces a result which is equivalent to a reflection through a mirror plane normal to that axis.

The operation of a *glide plane* combines a reflection with a translation parallel to the plane, through a distance which is a fraction of the lattice repeat in the direction concerned. The translation may be parallel to a unit cell edge, in which case the glide is *axial*; the term *diagonal* glide refers to translation along a face or body diagonal of the unit cell. In the latter case, the translation is through a distance which is half the length of the diagonal concerned, except for *diamond* glide, where it is a quarter of the diagonal length.

Example 1.3: Five-fold rotation

Show that a five-fold rotation axis is inconsistent with the translational symmetry of a lattice.

In Figure 1.12, the a is the repeat distance between adjacent lattice points. The operation of a five-fold rotation axis by $\theta = 72°$ would lead to lattice points separated by a distance x. In order for this rotation to be a symmetry operation, x must equal na, where n is an integer. Since $x = a - 2a\cos\theta$, the governing equation that defines possible values of θ that are consistent with symmetry operations becomes:

$$\begin{aligned} x = na &= a - 2a\cos\theta \\ \cos\theta &= \frac{1-n}{2} \quad \text{with} \quad -1 \leq \cos\theta \leq 1 \end{aligned} \tag{1.4}$$

It follows that for $\cos\theta = -1, -\frac{1}{2}, 0, \frac{1}{2}$ and 1, θ takes the values 180°, 120°, 90°, 60° corresponding to twofold, three-fold, four-fold, and six-fold rotation axes. Equation 1.4 shows, therefore, that a five-, or indeed seven-fold rotation are not symmetry axes. Another way of approaching this problem is on page 100.

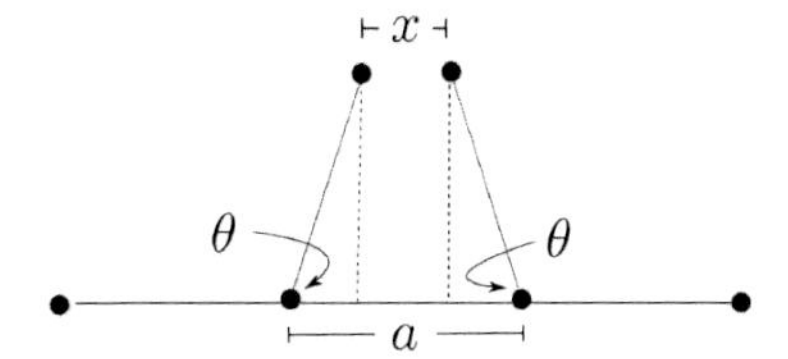

FIGURE 1.12
For θ to represent a rotation consistent with the translational symmetry of the lattice, x must equal an integer multiplied by the translation a. The proposed rotation axis is normal to the plane of the figure.

We note in passing that ordered structures exist that are not periodic but can yield diffraction patterns that have five-fold or ten-fold symmetries. These structures are known as quasicrystals; they lack translational symmetry [13].

1.9 Crystal structure

Lattices are regular arrays of imaginary points in space. A real crystal has atoms associated with these points. Consider the projection of the primitive cubic cell illustrated in Figure 1.13a. In the sequence Figure 1.13b-e, a pair of atoms (Cu at 0,0,0 and Zn at $\frac{1}{2}, \frac{1}{2}, \frac{1}{2}$), which also is known as the *motif*, is placed at each lattice point of Figure 1.13a in order to build up the crystal structure of β-brass (Figure 1.13f).

$$\text{lattice} + \text{motif} = \text{crystal structure} \tag{1.5}$$

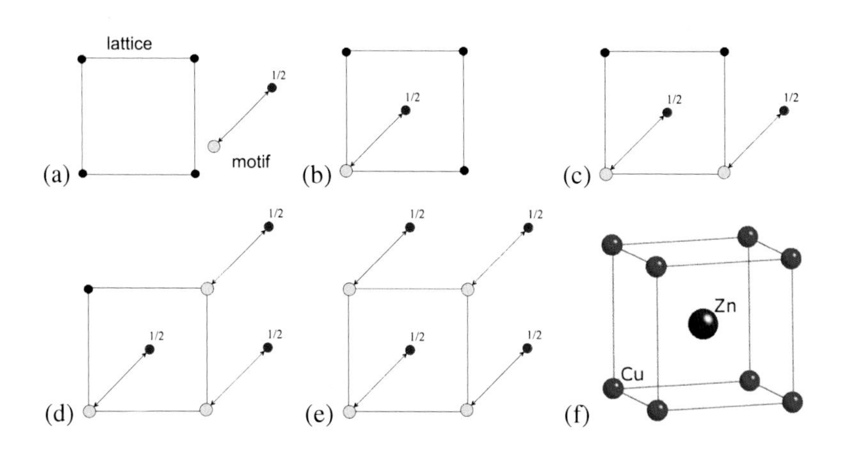

FIGURE 1.13
Building up the crystal structure of β-brass by placing a motif consisting of an appropriate pair of Cu and Zn atoms at each lattice point of a primitive cubic lattice. (a) Represents the primitive lattice with the motif consisting of a pair of distinct atoms that have yet to be placed. (b-e) The motif is placed at each lattice point. (f) The three-dimensional representation of the final structure.

The location of an atom of copper at each lattice point of a cubic-F lattice generates the crystal structure of copper, with four copper atoms per unit cell (one per lattice point), representing the actual arrangement of copper atoms in space.

Consider now a motif consisting of a pair of carbon atoms, with coordinates $[0\ 0\ 0]$ and $[\frac{1}{4}\ \frac{1}{4}\ \frac{1}{4}]$ relative to a lattice point. Placing this motif at each lattice point of the cubic-F lattice generates the diamond crystal structure (Figure 1.14), with each unit cell containing 8 carbon atoms (2 carbon atoms per lattice point). The figure also shows the tetrahedral bonding which makes diamond a giant molecule with strong covalent bonds in three dimensions. Diamonds do contain dislocations [14] but they have complex cores, are dissociated and incredibly difficult to move, making it very hard. The structure of silicon is obtained simply by replacing the carbon atoms by silicon (cubic-F, motif of a pair of Si atoms at 0,0,0 and $\frac{1}{4}, \frac{1}{4}, \frac{1}{4}$); the strong directional bonding also results in a large Peierls barrier to dislocation mobility [15], making silicon single-crystals rather brittle.

Figure 1.15 shows how a cubic-F cell changes into a cubic-P cell when the nickel and aluminium atoms order in the classical γ/γ' superalloy system.

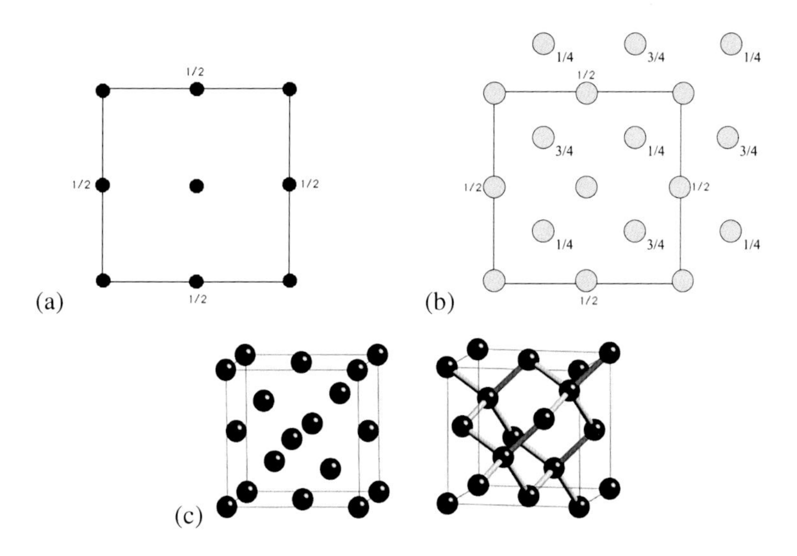

FIGURE 1.14
(a) Projection of the cubic-F lattice. (b) Projection of cubic-F lattice with a motif of a pair of carbon atoms at $0, 0, 0$ and $\frac{1}{4}, \frac{1}{4}, \frac{1}{4}$ placed at each lattice point. (c) Perspective of the same structure illustrating the tetrahedral bonding of the carbon atoms.

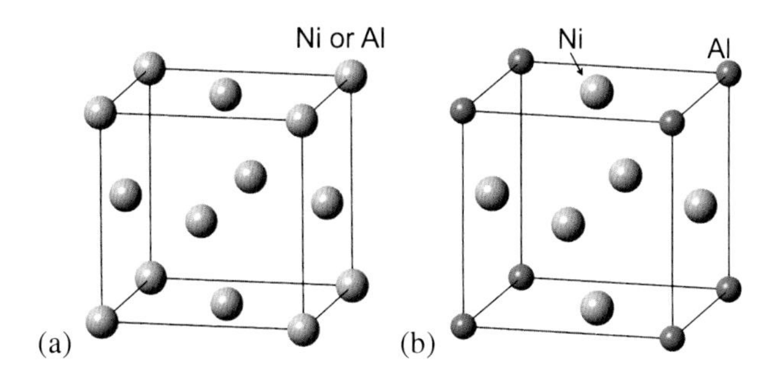

FIGURE 1.15
Nickel-based superalloys. (a) The face-centered cubic crystal structure of disordered γ. (b) The primitive cubic crystal structure of γ'.

Example 1.4: Structure of graphene

Figure 1.16 shows the atomic arrangement of carbon atoms in a two-dimensional layer of graphene. Identify the unit cell and lattice type, and calculate the lattice parameter given that the C-C distance is 0.142 nm (at room temperature).

A single layer of copper atoms is created by depositing copper atoms on an appropriate substrate. The layer corresponds to the $\{111\}$ plane of the three-dimensional structure of copper. Identify the unit cell and lattice type describing the two-dimensional arrangement of copper atoms and calculate its lattice parameter given that the three-dimensional form of copper has the lattice parameter 0.361 nm at room temperature.

Explain why the $\{111\}_{\mathrm{Cu}}$ plane would make a good substrate for the chemical vap deposition of single-layer graphene. Giving reasons, explain the orientation relationship you might expect between the copper and graphene layers.

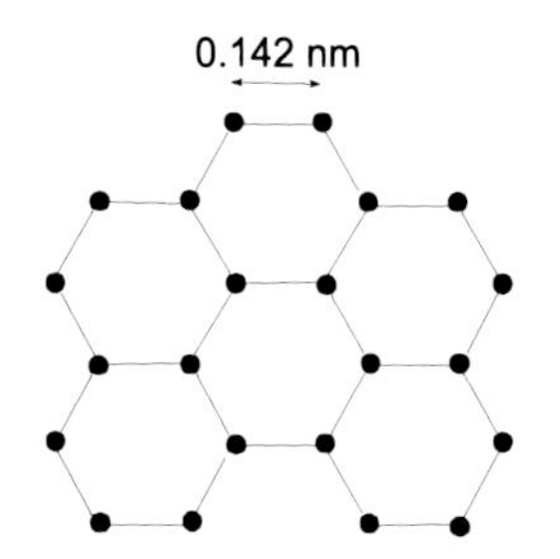

FIGURE 1.16
Atomic structure of graphene sheet without defects. Entropy effects mean that optically visible samples cannot ever be made pristine [16].

The unit cell, identified blue in Figure 1.17 is primitive hexagonal with a motif of two carbon atoms per lattice point. The environment about the black atom is not identical to that around the red atom. The lattice parameter is $0.142 \times 2 \times \cos 30° = 0.245$ nm.

Copper is cubic close-packed. The arrangement of atoms on the $\{111\}$ close-packed plane is hexagonal, primitive. The two-dimensional cell has the edges equal to $\frac{a}{2}\langle 110\rangle$, so the lattice parameter is 0.255 nm. This closely matches that of graphene.

Given that the unit cell edges of the two-dimensional structures of graphene and copper match to 4% $(100[0.255 - 0.245]/0.255)$, it is logical that their unit cell edges are exactly aligned, $[10]_{\mathrm{graphene}} \parallel [10]_{\mathrm{Cu}}$ and $[01]_{\mathrm{graphene}} \parallel [01]_{\mathrm{Cu}}$.

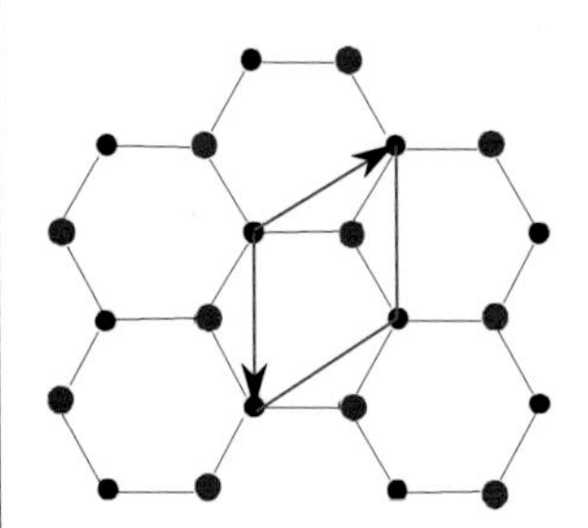

FIGURE 1.17
The blue lines represent the unit cell, with a motif of a pair of carbon atoms colored black and red, per lattice point.

1.10 Point group symmetry

Consider a molecule such as that illustrated in Figure 1.18. The symmetry operations on the molecule constitute a collection known as a *point group* because there always is one point in space that is left unchanged by every symmetry operation in that group. The point group symmetry of the molecule is illustrated in Figure 1.18a. When exposed to infrared light the molecules absorb energy and vibrate as illustrated in Figure 1.18b. These vibrations result in a characteristic spectrum showing peaks at certain energies which depend on the modes illustrated. Other molecules with the same point group symmetry, such as sulfur tetrafluoride, will exhibit similar spectroscopic properties. The point group of a crystal is also the symmetry common to all of its *macroscopic* properties.

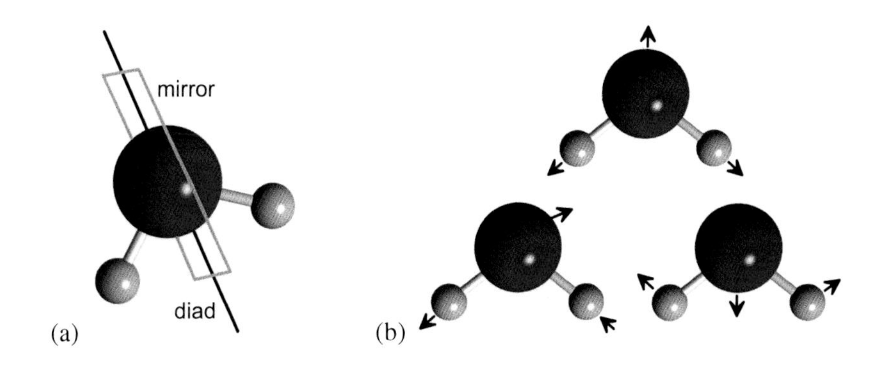

FIGURE 1.18
(a) A molecule with point group symmetry $2m$ (diad and a mirror plane parallel to the diad). (b) Vibration modes (symmetrical stretch, asymmetrical stretch, and bending) of the molecule when appropriately stimulated. There is an additional mirror plane passing through the centers of all three atoms if account is taken of the electron clouds associated with the molecule.

Translations therefore are necessarily excluded in a *point* group. Rotations, mirror planes, center of symmetry, and inversion axes are permitted. There are 32 point groups in three dimensions, classified within the seven crystal classes (Table 1.2). The triclinic, monoclinic, and orthorhombic groups do not contain triads, tetrads, or hexads. For those systems, when the point group symbol contains three elements (e.g., $2mm$), then the symbols are presented in the order of the symmetry elements parallel to the x, y, and z axes, respectively, as illustrated in Figure 1.19a for an object in the orthorhombic class.

For crystal systems with higher order axes, the z direction is assigned to that higher order axis, with the second symbol corresponding to equivalent *secondary* axes that are normal to z, and the third also normal to z to equivalent *tertiary* di-

TABLE 1.2
Point group symmetries associated with the seven crystal classes.

Class	Non-centrosymmetric	Centrosymmetric	
Cubic	$23, 432, \bar{4}3m$	$m3, m3m$	With high order axes
Hexagonal	$6, \bar{6}, 622, 6mm, \bar{6}m2$	$6/m, 6/mmm$	
Trigonal	$3, 32, 3m$	$\bar{3}, \bar{3}m$	
Tetragonal	$4, \bar{4}, 422, 4mm, \bar{4}2m$	$4/m, 4/mmm$	
Orthorhombic	$222, 2mm$	mmm	Without high order axes
Monoclinic	$2, m$	$2/m$	
Triclinic	1	$\bar{1}$	

rections passing between the secondary ones. Figure 1.19b shows an example for an object in the tetragonal class, with the tetrad placed along the z axis, the two mirror planes with normals along the x and y axes, and the additional mirror planes generated by this symmetry also illustrated.

In the case of the cubic system the triad is always the second symbol given that the defining symmetry is four triads.

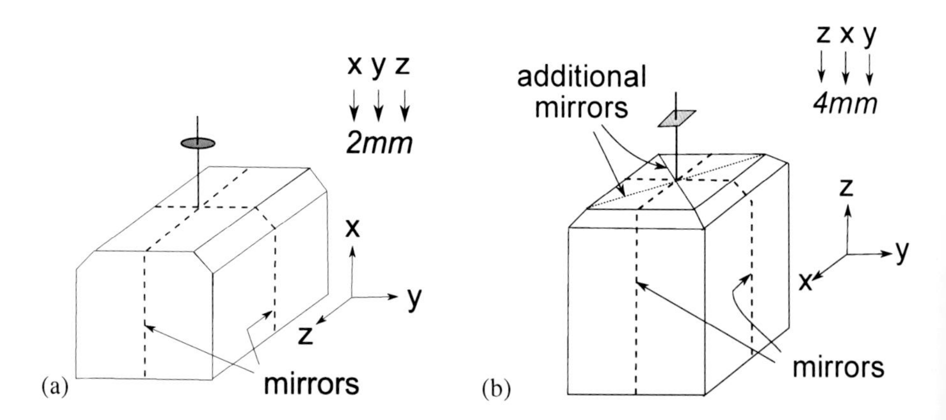

FIGURE 1.19
Convention for point group notation. (a) Groups without high order axes. (b) Groups with high order axes.

Some further details on the notation are as follows:

- $m \equiv \bar{2}$ represents a mirror plane.
- 2, 3, 4, 6 are 2-fold, 3-fold, 4-fold, and 6-fold rotation axes, respectively.

- $\overline{1}$, $\overline{2}$, $\overline{3}$ etc. are inversion axes; the one-fold inversion axis is equivalent to a center of symmetry; $\overline{2}$ signifies a rotation of $360°/2$ combined with an inversion through the center point.
- $\frac{4}{m}mm$ is a 4-fold axis with a mirror normal to it and four mirror planes containing it. The symbol is usually written without the space as $4/mmm$.
- 432 refers to a point group with 3-fold axes which are not parallel to the z axis, where a unit cell has x, y, and z axes. This is limited to the cubic system.

Example 1.5: Point symmetry of chess-pieces
Determine the point group symmetries of the chess pieces illustrated in Figure 1.20 and identify the crystal class to which each belongs. Both shapes have been machined from solid cubes.

The bishop has a three-fold axis passing through the cut corner of the cube and a mirror parallel to that axis, so the point group symmetry is $3m$ which fits the trigonal crystal class. The castle clearly has a single tetrad with two mirrors parallel to the tetrad so in this case, the point group is $4mm$ which is tetragonal class (Table 1.2).

(a) 

(b) 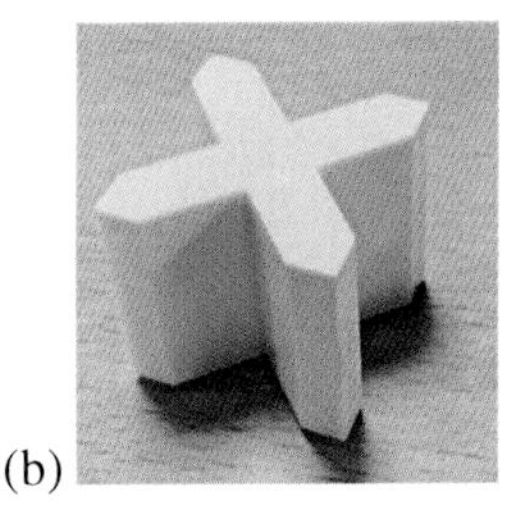

FIGURE 1.20
Avant garde chess pieces. (a) Bishop. (b) Castle.

Some of the point group notation comes from the classification of the macroscopic shapes of crystals. If a crystal exhibits well-developed faces, its point group symmetry can be derived from its external form. Figure 1.21 illustrates this for two cases. In the case of the gypsum there is a diad normal to (010) and a mirror plane normal to that diad so that the shape has the point group $2/m$ belonging to monoclinic class. With the epsomite there is a diad along the vertical axis of the diagram and two diads perpendicular to the vertical axis passing through the vertical edges between the planes (110) and $(1\overline{1}0)$, i.e., it belongs to the orthorhombic crystal class.

With appropriate analysis of facetted crystals it therefore becomes possible to deduce the crystal class simply from symmetry. Figure 1.22 shows two such examples. The quartz crystal clearly has a hexad which is the defining symmetry of the hexagonal crystal class. The celestite has three mutually perpendicular diads, placing it in the orthorhombic system.

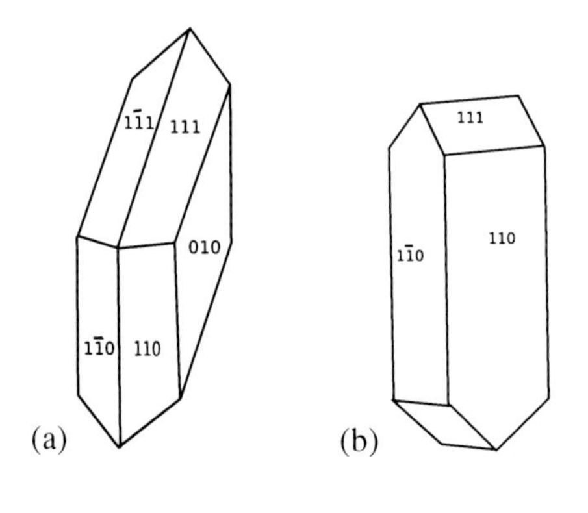

FIGURE 1.21
(a) Shape of gypsum ($CaSO_4.2H_2O$) with point group symmetry $2/m$. (b) Shape of epsomite ($MgSO_4.7H_2O$) with point group symmetry 222. Note that the angle between the (110) and $(1\bar{1}0)$ faces is 89.37°C, because the lattice parameters a, b, and c are 1.1866, 1.1998, and 0.6855 nm, respectively.

FIGURE 1.22
Each figure shows an idealized model on the left and the corresponding crystal on the right. (a) Quartz. (b) Celestite. Photographs taken at the Colorado School of Mines.

Example 1.6: Octahedral interstices in iron
Draw structure projections of the unit cells of austenite (cubic-F) and ferrite (cubic-I) in iron and mark on these the locations of the octahedral interstices. Identify the point group symmetry of an octahedral interstice in austenitic iron. Repeat this for the octahedral interstice in ferritic iron. Identify using Table 1.2, the symmetry class to which each interstice belongs.

Why does carbon strengthen ferrite much more than it does austenite?

The octahedral interstices in the two structures are illustrated in Figure 1.23 and the following points are revealing:

- As can be seen from the structure projection in Figure 1.23a, there is one octahedral interstice per iron atom in austenite. In contrast, there are three octahedral interstices per iron atom in the ferrite. This has consequences when the lattice of austenite is homogeneously deformed into that of ferrite because carbon atoms distributed at random in the austenite end up in just one subset of octahedral sites in the ferrite, in effect leading to an ordering of the carbon along one of the cube axes, causing the ferrite to become tetragonal rather than cubic [17]. This reduction in symmetry has consequences on the thermodynamics of the lattice, leading to an increase in the solubility of carbon in tetragonal ferrite that is in equilibrium with the austenite [18, 19].

- The octahedral interstice in austenite is regular (all sides of equal length) with a point group symmetry $m3m$. In contrast, the three axes of the irregular octahedral interstice in ferrite are a (magnitude of the lattice parameter), $\sqrt{2}a$, and $\sqrt{2}a$ with the tetragonal point group $4/mmm$.

- For both austenite and ferrite, the octahedral interstice is smaller than a carbon atom, resulting in a straining of the lattice. In the austenite, the cubic symmetry of the interstice means that the strain is isotropic whereas in the ferrite it is anisotropic (tetragonal). An isotropic strain interacts weakly with dislocation strain fields so the hardening caused by carbon in austenite is also weak. This is because the hydrostatic components of the strain fields of dislocations are much smaller than the deviatoric terms. Given this, there is a powerful interaction of the tetragonal strain with dislocation strain fields, leading to intense hardening.

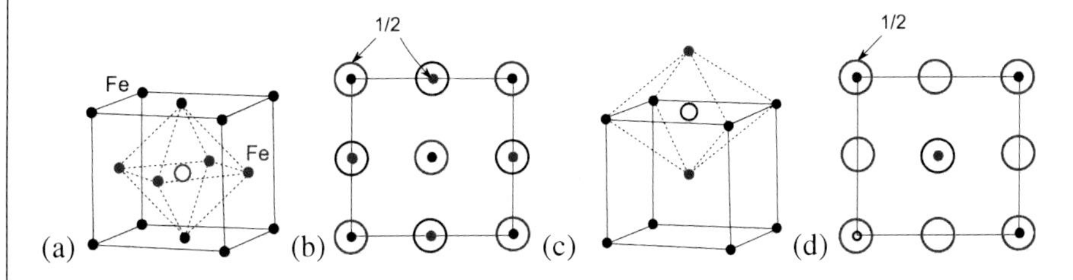

FIGURE 1.23
Octahedral interstices in austenite (cubic-F) and ferrite (cubic-I). The red objects are located at a height half, with filled circles representing iron atoms and open circles the positions of the octahedral interstices. (a) A regular-octahedral interstice in austenite; (b) projection of the austenite unit cell to show the positions of all such interstices. (a) An irregular-octahedral interstice in ferrite; (b) projection of the ferrite unit cell to show the positions of all such interstices.

1.11 Summary

It is extraordinary that all crystals without exception can be described in terms of just 14 imaginary lattices, the points of which can then be loaded identically with motifs that may contain one or more atoms that are not necessarily identical. On the same logic, the apparently great variety of wallpapers available for decorative purposes have just five basic lattices, with the variations generated by changing the motif. A sixth is of course possible if random two-dimensional patterns can be printed. One advantage of a random-printed wallpaper is that there would be no effort required in matching designs at the edges.

The idea that there is an identical configuration of atoms around each lattice point in a crystal does not in fact sit well with the vast majority of real crystals. Materials are not pure; random mixtures of atoms in solid solutions cannot result in an identical environment at each lattice point. Translational periodicity does not then exist, rather, there is a probability of finding a particular species at a particular location. The consequence is that the crystal must be considered in terms of its averaged properties and this suffices for most purposes, for example in X-ray diffraction, where an averaged atomic scattering factor is used to calculate intensities. The strain fields around solute atoms in a random solid solution may then lead to X-ray peak broadening.

The anisotropy of crystals, moderated by symmetry, has major consequences on how they can be exploited for engineering purposes. The crystallographic direction along which single-crystal turbine blades are grown is selected in order to minimize the possibility of vibrations during the operation of gas turbines.

References

1. J. M. Ziman: *Models of disorder*: Cambridge, U. K.: Cambridge University Press, 1979.

2. F. Wilczek: Quantum time crystals, *Physical Review Letters*, 2012, **109**, 160401.

3. J. Zhang, P. W. Hess, A. Kyprianidis, P. Becker, A. Lee, J. Smith, G. Pagano, I.-D. Potirniche, A. C. Potter, A. Vishwanath, N. Y. Yao, and C. Monroe: Observation of a discrete time crystal, *Nature*, 2017, **543**, 217–220.

4. S. Choi, J. Choi, R. Landig, G. Kucsko, H. Zhou, J. Isoya, F. Jelezko, S. Onoda, H. Sumiya, V. Khemani, C. von Keyserlingk, N. Y. Yao, E. Demler, and M. D. Lukin: Observation of discrete time-crystalline order in a disordered dipolar many-body system, *Nature*, 2017, **543**, 221–225.

5. C. Nayak: Time crystals: new form of matter through to break laws of physics created by scientists: Quoted in *The Independent* newspaper, 2017.

6. P. G. de Gennes, and J. Prost: *The physics of liquid crystals*: 2nd ed., Oxford, U.K.: Oxford University Press, 2003.

7. C. Herring: The use of classical macroscopic concepts in surface energy problems, In: R. Gomer, and C. S. Smith, eds. *Structure and properties of solid interfaces*. Chicago, IL: University of Chicago Press, 1953:5–81.

8. T. Watanabe: Grain boundary engineering: historical perspective and future prospects, *Journal of Materials Science*, 2011, **46**, 4095–4115.

9. R. J. Howarth: History of the stereographic projection and its early use in geology, *Tera Nova*, 1996, **8**, 499–513.

10. K. S. Novoselov, D. Jiang, F. Schedin, T. J. Booth, V. V. Khotkevich, S. V. Morozov, and A. K. Geim: Two-dimensional atomic crystals, *Proceedings of the National Academy of Sciences*, 2005, **120**, 10451–10453.

11. R. M. Weis, and H. . McConnell: Two-dimensional chiral crystals of phospholipid, *Nature*, 1983, **310**, 47–49.

12. G. A. Somorjai: The surface structure of and catalysis by platinum single crystal surfaces, *Catalysis Reviews*, 1972, **7**, 87–120.

13. D. Shechtman, I. Blech, D. Gratias, and J. W. Cahn: Metallic phase with long-range orientational order and no translational symmetry, *Physical Review Letters*, 1984, **53**, 1951–1953.

14. F. C. Frank: Observation by X-ray diffraction of dislocations in a diamond, *Philosophical Magazine*, 1959, **39**, 383–384.

15. V. V. Bulatov, S. Yip, and A. S. Argon: Atomic modes of dislocation mobility in silicon, *Philosophical Magazine*, 1995, **72**, 453–496.

16. H. K. D. H. Bhadeshia: Large chunks of very strong steel, *Materials Science and Technology*, 2005, **21**, 1293–1302.

17. E. Honda, and Z. Nishiyama: On the nature of the tetragonal and cubic martensites, *Science Reports of Tohoku Imperial University*, 1932, **21**, 299–331.

18. J. H. Jang, H. K. D. H. Bhadeshia, and D. W. Suh: Solubility of carbon in tetragonal ferrite in equilibrium with austenite, *Scripta Materialia*, 2012, **68**, 195–198.

19. H. K. D. H. Bhadeshia: Carbon in cubic and tetragonal ferrite, *Philosophical Magazine*, 2013, **93**, 3714–3715.

2

Stereographic Projections

Abstract

A stereographic projection maps angular relationships on to a two-dimensional surface; it can illustrate the beauty of symmetry and its consequences, or when combined with an appropriate net, allow the quantitative determination of angles between plane normals or directions. The focus of this chapter is on the particular kind of projection that retains angular truth. Many more applications of stereographic projections will become apparent in the chapters that follow.

2.1 Introduction

We have seen already that the projection of a three-dimensional crystal structure into two dimensions, along the z-coordinate, can simplify the perception and representation of atomic arrangements. There is no loss of information given that the fractional z-coordinate is identified clearly on the projection. This can be seen in Figure 1.14 where the three-dimensional representation of the structure of diamond lacks clarity whereas the projected cell is readily visualized. The ease of interpretation using the projected structure is illustrated for the more complex structure of ε-carbide, which has a chemical formula between Fe_2C and Fe_3C, in Figure 2.1. There are six iron atoms in the cell (3 at $z = \frac{1}{2}$ and the six others that are shared at faces with other cells at $z = 0, 1$ and therefore contributing a further three to the cell). The illustrated cell contains a full complement of carbon atoms but to achieve the composition $Fe_{2.4}$, the sites colored blue would only be occupied partially [1].

The projection described in Figure 2.1 is a straightforward linear operation and there are no distortions of that linearity. In contrast, Figure 2.2 shows non-linear projections of circular arcs, in one case a simple extrapolation onto a horizontal line, and in the other case via a pole located below the horizontal line where the projections are recorded. The purpose here is to record angles rather than spatial coordinates.

Imagine now that instead of quadrants of a circle, the diagrams in Figure 2.2 represent the surfaces of spheres. A circle drawn on the surface of the sphere would in general project as an ellipse on the horizontal plane of Figure 2.2a, whereas it would project as a true circle in the case of the projection method used in Figure 2.2b.

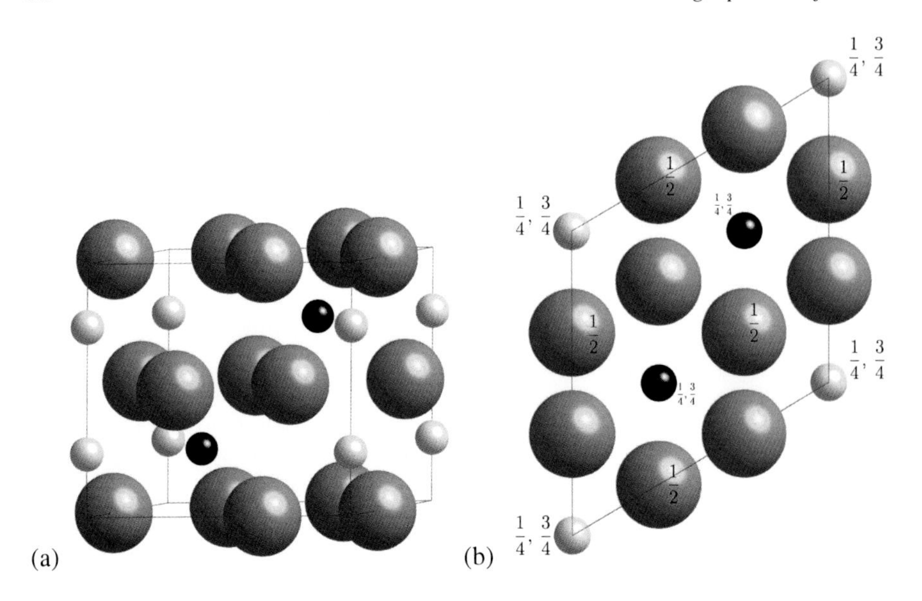

FIGURE 2.1
The structure of hexagonal ε-carbide with lattice parameters $a = 0.4767\,\text{nm}$ and $c = 0.4353\,\text{nm}$. The large atoms are iron. The small atoms are carbon but not all the sites designated blue are occupied. (a) The three-dimensional cell with the z-axis vertical. (b) Projection of the cell along the z-axis, with the fractional z-coordinates listed.

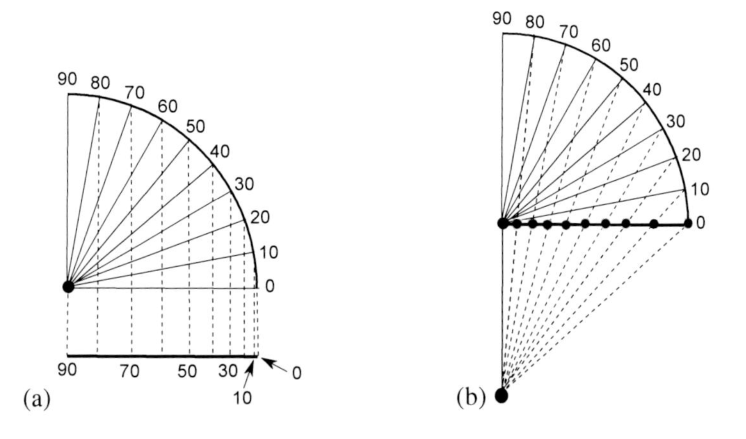

FIGURE 2.2
Projections of angles made by lines intersecting a circular quadrant onto a horizontal line.

Example 2.1: Projection of small circle
A sphere has north and south poles. Show that if a circle is constructed on the

northern hemisphere with each point on its locus projected on to the equatorial plane by lines originating from the south pole, then the projection itself will remain a circle. Figure 2.3a illustrates the circle ab on the surface of the sphere, together with its projection $a'b'$ on the equatorial plane. A cross-section of the same diagram through the center of the circle ab is shown in Figure 2.3b. The conjugate circular section cd is symmetrically inclined to the axis ST.

The line eb is constructed to be parallel to the equatorial plane. The arcs eS and Sb are equal in length so it follows that $\phi_1 = \phi_2$. Since cd is symmetrically inclined to ST, it follows that $\phi_3 = \phi_2$ and therefore $\phi_3 = \phi_1$. In other words, the circle cd is parallel to the equatorial plane. If the circle cd is now translated toward the equatorial plane, it remains a circle and ends up as a circle $a'b'$.

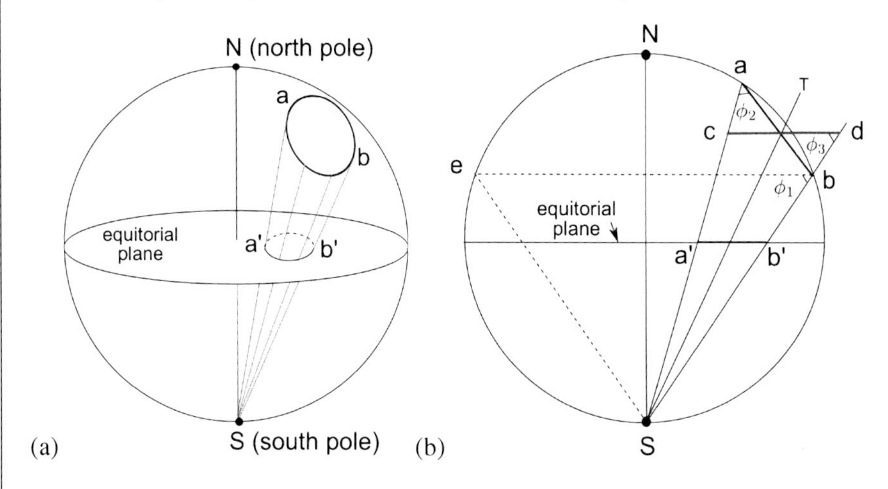

FIGURE 2.3
(a) Circle on the surface of a sphere projected onto the equatorial plane via the south pole. (b) Construction showing that the projection is also a circle.

This example shows that the *stereographic* projection method illustrated in Figure 2.3 preserves angular truth; if two circles (e.g., a longitude and latitude) inscribed on the sphere cross at 90° then the projections of those two circles will also cross at that angle. We shall now consider these stereographic projections in more detail, following a brief description of why they are of importance.

2.2 Utility of stereographic projections

There is an excellent review of the history of the stereographic projection by Howarth [2], tracing the origin to the second century BCE in the context of astronomy. Howarth describes how projections were then used in mineralogy and structural geology, where "the spatial orientations of a crystal face, betting plane, fault surface etc. can be considered by imagining a plane passing through the center of a sphere".

Stereographic projections are two-dimensional representations which now have found numerous applications in materials science, with the method often embedded in the software that controls experimental crystallography, for example,

- the deformation of single crystals whence crystal planes rotate in order to comply with the external stress; such rotations in polycrystalline materials lead to non-random aggregates of crystals, the properties of which are in between those of single and random polycrystals. In both of these cases, it becomes necessary to define the orientations of individual crystals relative to the deformation axes, which is where stereographic projections become useful [3].
- When phase transformations occur in the solid state, the probability is that the product phase will form such that the atomic arrangements match as much as possible, along the parent/product interface. This often leads to a reproducible orientation relationship between the two crystals, one which can be represented on a stereographic projection in order to understand the mechanism of transformation or the deformation behavior of the two-phase mixture. Stereographic projections can be used to decide whether the orientation relationship is reproducible or occurs by chance [4].
- Diffraction data can be presented on stereographic projections. It is now routine to to examine both structure and crystallographic information from polycrystalline samples on a single image [5], with the crystallographic data presented in the form of corresponding colors on the microstructural image and stereographic projection.

2.3 Stereographic projection: construction and characteristics

Imagine a sphere. Any plane that passes through its center will intersect the surface of the sphere at a circle whose diameter is that of the sphere; this is known as a *great circle*, Figure 2.4a. The primitive (great) circle is where the equatorial plane intersects the sphere. The diameter of a great circle is the same as that of the sphere. Lines of longitude are all great circles in the spherical earth approximation.

When an intersecting plane does not pass through the center of the sphere it results in a *small circle* at the intersection, as illustrated in Figure 2.4b. Lines of latitude are in general small circles, the exception being the equator.

The shortest distance between two points on a sphere is along the great circle which passes through both of the points.

Figure 2.5a shows a crystal with a cubic lattice, placed at the center of a sphere with the $[100] \parallel x$-axis, $[010] \parallel y$-axis, and $[001] \parallel z$-axis of the sphere; this is the conventional orientation of the crystal in the sphere. Similarly, the orientation of an

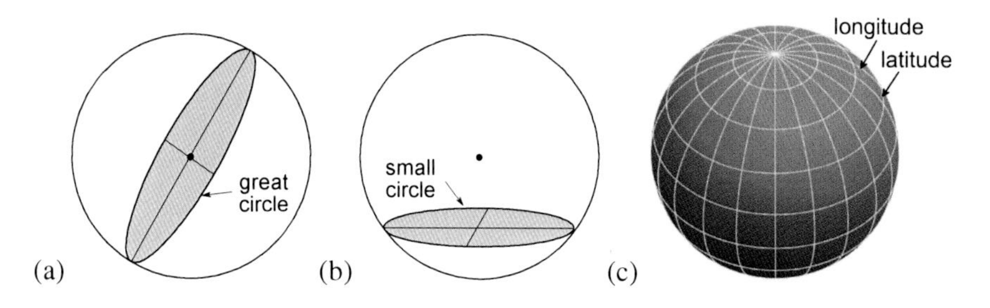

FIGURE 2.4
Intersections of planes with sphere. (a) Great circle inclined with respect to the north and south poles of the sphere, (b) small circle; (c) latitude and longitude lines (image courtesy of vectortemplates.com).

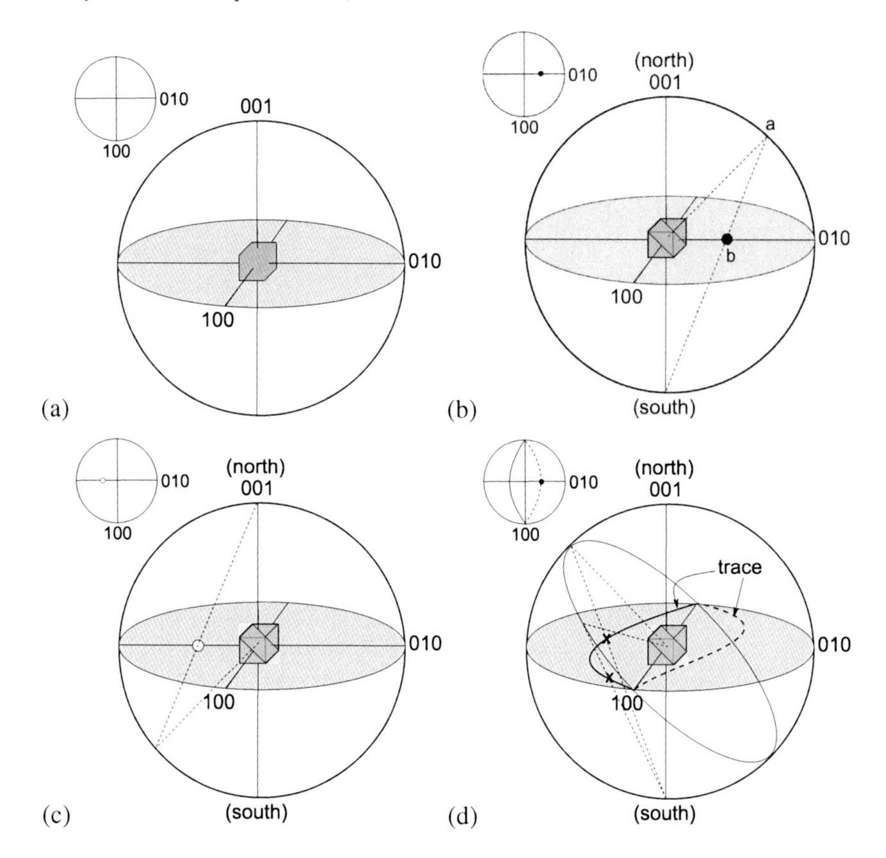

FIGURE 2.5
Plotting plane normals on a stereographic projection. The insets, top-left in each case, show a plan view of the equatorial plane. (a) Crystal placed at the center of the sphere. (b) Projection of the pole of (011). (c) Projection of $(0\overline{1}\overline{1})$. (d) Projection of the trace of the great circle which is (011) on to the equatorial plane. The trace of the plane in the southern hemisphere is marked as a dashed curve. The crosses identify the intersections of the projection lines with the equatorial plane.

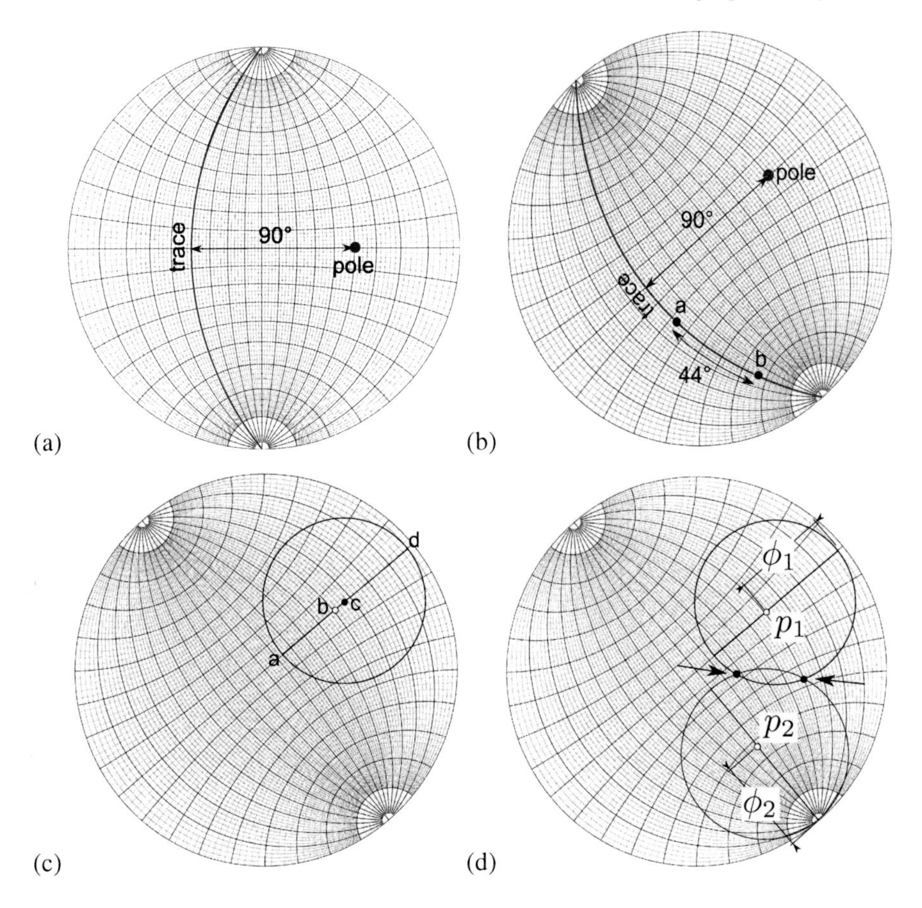

FIGURE 2.6
The Wulff net. (a) Relation between pole and trace of a plane. (b) The net is rotated until the poles a and b lie on the same great circle to measure the angle in between. (c) The geometrical center c of the small circle is different from its angular center b. The angles ab and bd as measured on the great circle are identical. The distances ac and cd are identical when measured on a ruler. (d) The poles identified with arrows are both located at angles ϕ_1 to p_1 and ϕ_2 to p_2.

arbitrary plane can be defined by its normal, which in general is known as the *pole* of the plane. Suppose now that the normal to the (011) plane is projected so that it intersects the sphere at the point a (Figure 2.5b), which then is projected to the south pole. The intersection on the equatorial plane being the point b which represents the pole of the plane on the stereographic projection.

The plane (011) when extended so that it intersects the sphere will result in a great circle on the surface of the sphere, the trace of which can be projected onto the equatorial plane as illustrated in Figure 2.5d. The part of the plane that is in the northern hemisphere projects as a line, and that from the southern hemisphere as a dashed-line. All points on the trace are at 90° to the pole of (011) since the pole represents the normal to the plane, another indication of how the stereographic projection maintains angular truth.

A projection of the longitudes and latitudes illustrated in Figure 2.4c on to the equatorial plane would lead to a net, known as the "Wulff net" in which the resulting great circles are spaced at equal angles as illustrated in Figure 2.6 where the angular intervals are 2°. The net can be used to measure angles between two poles on a stereographic projection by rotating the net such that both poles lie on the same great circle, followed by counting the degrees separating them. Note that the angle between two great circles will be the same as the angle between their poles.

Example 2.2: Radius of trace of great circle on Wulff net

The curve representing a great circle on a Wulff net is itself an arc of a circle, a result which follows from the fact that the projection method preserves angular truth. This arc is from a circle that in general has a radius r greater than that of the net (r_o). Derive the relationship between r, r_o and the offset x of the trace of the great circle from the origin.

Referring to the figure below, the distance ab is $2r_o$; the Pythagoras theorem then gives

$$r^2 - r_o^2 = (r - x)^2$$

so it follows that

$$r = (x^2 + r_o^2)/2x \quad \text{with} \quad r_o \leq r \leq \infty.$$

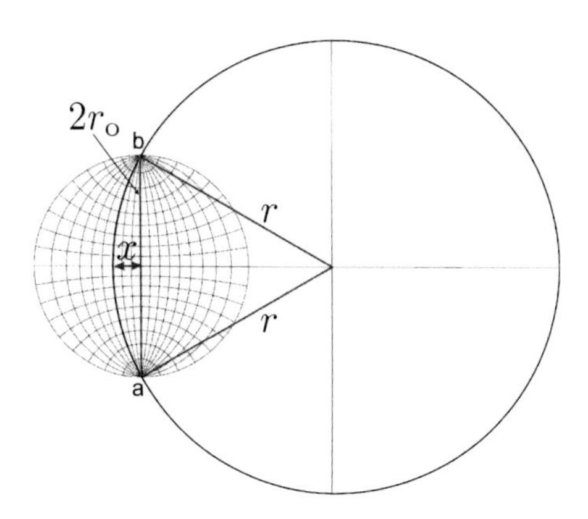

Figure 2.7a,b show the stereographic projection for the cubic system, with the rotational symmetry elements related to the crystallographic directions about which the rotations occur. Figure 2.7c shows the stereographic projection being used to illustrate how the elastic modulus of a single crystal of ferritic iron varies between 131 and 284 GPa as a function of the crystallographic orientation, in a manner consistent with its crystalline symmetry. Even aggregates of crystals often do not show collective macroscopic-properties that are isotropic because the crystals are not distributed at random. Polycrystalline ferritic steel sheet in which the crystals are aligned with the $\langle 001 \rangle$ direction and with the $\{110\}$ planes aligned to the rolling plane, exhibit

modulus variation in the range 140 to 210 GPa [6]. The level of anisotropy in aggregates can be controlled, as we shall see in later chapters, to suit specific purposes such as the formability of metal sheets or the magnetic properties of transformer steels.

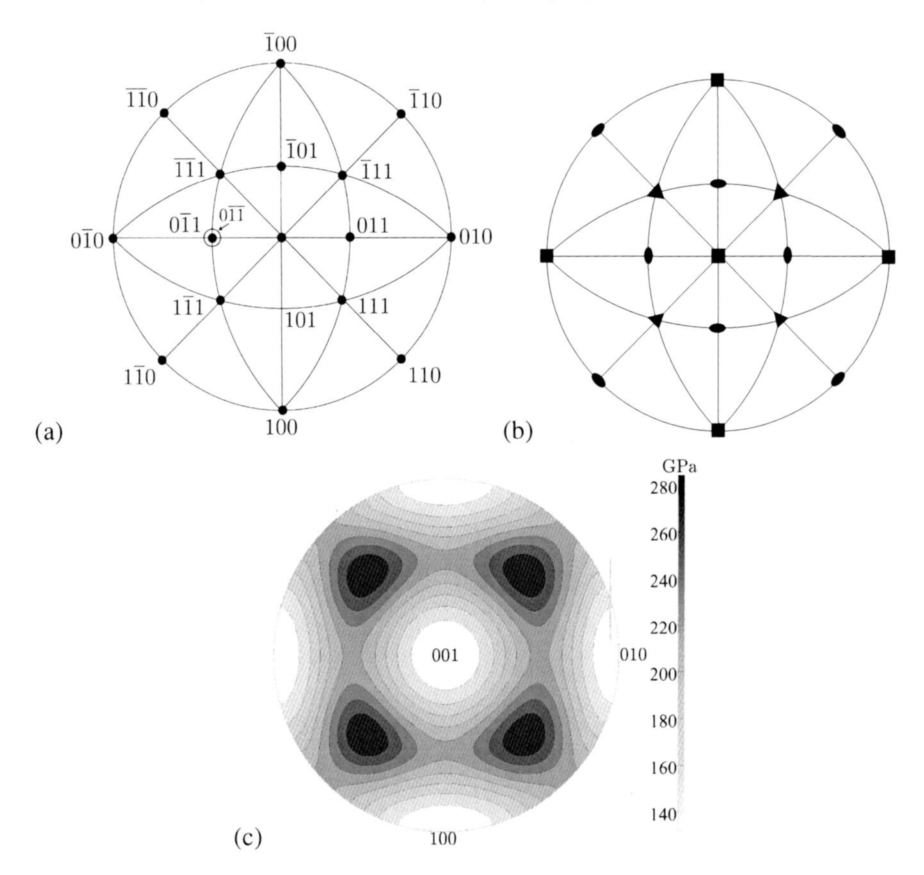

FIGURE 2.7
Stereographic projections for the cubic system. (a) Showing the standard way of drawing the projection with the locations of [100] and [010] on the perimeter and [001] at the center. (b) The symmetry elements. The filled squares are tetrads, the ellipses are diads, and triangles are triads. The tetrads are at $\langle 001 \rangle$, triads at $\langle 111 \rangle$, and diads at $\langle 011 \rangle$. Anything with cubic symmetry must have four triads. (c) Variation in the elastic modulus of a single crystal of bcc iron as a function of orientation. Diagram courtesy of Shaumik Lenka, constructed using data from Dieter [7].

Example 2.3: Traces of plates

Meteorites are essentially iron-nickel alloys that cool ridiculously slowly from red heat as they traverse through space. Their structure at high temperatures is austenitic (cubic-F) but on cooling, plates of ferrite (cubic-I) precipitate to form a Widmanstätten pattern as illustrated in Figure 2.8. Because of the slow cooling rate, the coarse structure that forms is visible to the naked eye when the meteorite is sectioned and etched. Assuming that these plates precipitate exactly on $\{111\}_F$, and that all the plates are of identical dimensions and shapes, sketch the macrostructures that would be observed when the meteorite is sectioned on the $(001)_F$, $(111)_F$, and $(112)_F$ planes.

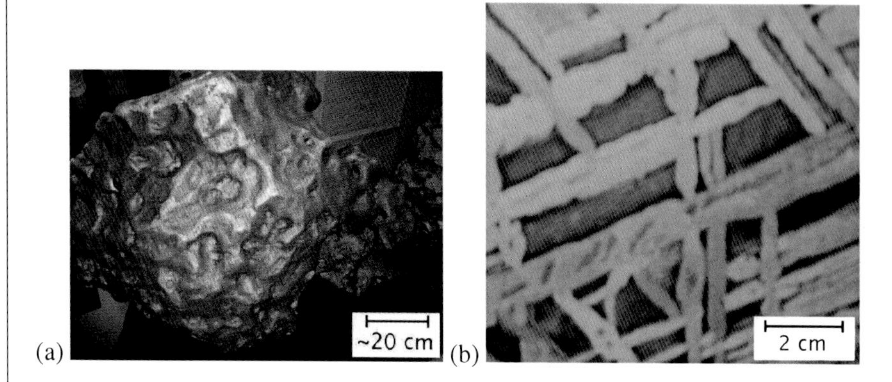

FIGURE 2.8
(a) A meteorite at the Smithsonian Museum in Washington. (b) The Widmanstätten pattern observed in the meteorite, of ferrite plates in an austenitic matrix.

Figure 2.9 shows the stereographic projection referred to the crystallographic axes of austenite, containing traces drawn in blue, of the $\{111\}_F$ planes on which the Widmanstätten ferrite precipitates. The intersections of these $\{111\}_F$ planes with the plane of observation defines the orientations of the plates that would be observed. Such intersections are identified by letters a,b,c,. . . on the stereographic projection.

The $(001)_F$ plane will reveal a square pattern of plates, because pairs of $\{111\}_F$ planes have common intersections at the points marked "a" and "d".

Plates precipitating on the $(111)_F$ appear as discs when observed on that plane. The $(\bar{1}11)_F$, $(\bar{1}\bar{1}1)_F$, and $(1\bar{1}1)_F$ intersect $(111)_F$ at "b", "a" and "c", respectively, with these three intersections making angles of 60° to each other, generating the equilateral arrangement of plates in addition to the disc.

On the $(112)_F$ plane of observation, there will be an arrangement of three plates corresponding to the intersections at "a", "e", and "f", making angles 67° (between "a" and "e"), 46° (between "e" and "f"), and 67° (between 'e' and 'a').

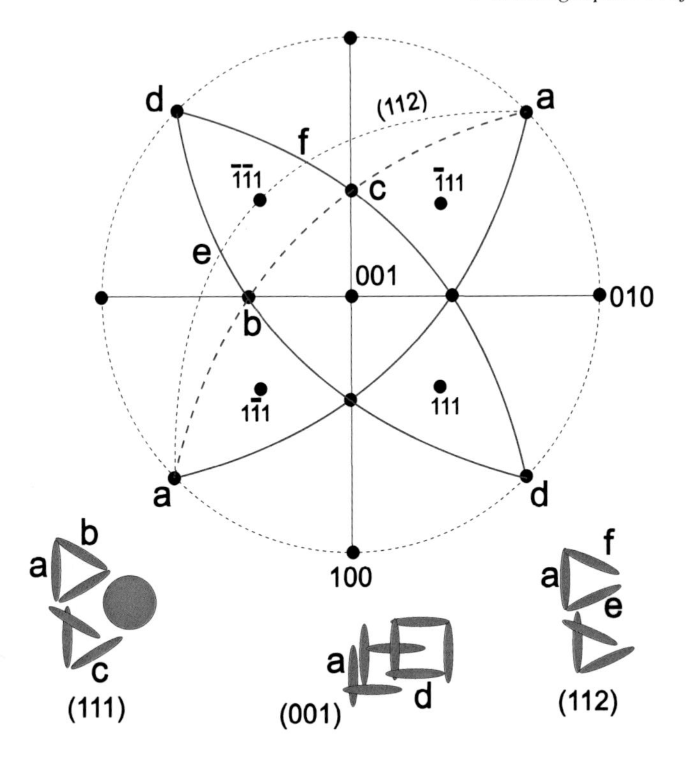

FIGURE 2.9
Stereographic projection referred to the crystallographic axes of the austenite. The blue lines are traces of $\{111\}_F$ planes, and the dashed lines the planes on which the macrostructure is observed. The insets at the bottom show what the macrostructures should look like if all the plates are of identical dimension and shape, as a function of the plane of observation.

2.4 Stereographic representation of point groups

Example 2.4: Mirror plane equivalent to $\overline{2}$
Show using sketch stereograms that the operation of a diad followed by an inversion (i.e., $\overline{2}$) is equivalent to a mirror plane. Do the operations $\overline{3}$ and $\overline{6}$ lead to centers of symmetry?

A general pole on a stereographic projection is one that is not located at particular symmetry elements such as the rotation axis (other than a monad) or on a mirror plane. The black dot on Figure 2.10a is a general pole which on the operation of a diad results in a second pole as shown in Figure 2.10b. When inverted through the center, this second pole intersects the stereographic sphere in the southern hemisphere and hence is shown as an open circle in Figure 2.10c, which shows that the net operation of a diad followed by an inversion is a mirror plane parallel to the plane of the diagram. It is seen that $\overline{2} \equiv m$.

The operation of a triad on a general pole leads to three such poles, which after inversion result in the situation illustrated in Figure 2.10c, which shows that the point group $\overline{3}$ belongs to a class that has centers of symmetry whereas $\overline{2}$ and $\overline{6}$ do not.

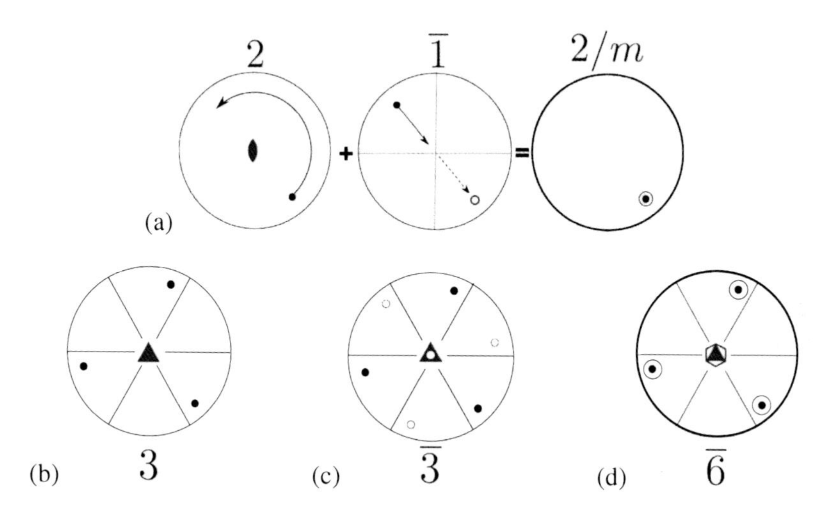

FIGURE 2.10
(a–c) An arbitrary pole is rotated by 180° about the diad normal to the plane of the diagram, to generate the second pole. This second pole is then inverted through the center into the southern hemisphere to result in inversion diad $\overline{2}$ that is equivalent to a mirror normal to that axis. (b) Triad, (c) inversion triad, and (d) inversion hexad. Mirror planes in (a) and (d) are identified using bold circles.

Example 2.5: The point groups $3m$ and $m3$

Referring to Table 1.2, explain using sketch stereograms, the difference between the point groups $3m$ and $m3$.

According to the conventions described on page 22, the point group $3m$ implies that there is a single triad with the secondary axis being the normal of the mirror plane. The triad ensures that there would be three such mirror planes. There is no mirror plane normal to the triad (Figure 2.11a), which means that the atomic arrangement along one end of the triad is different from the opposite end. Such a crystal may display spontaneous polarization and the point group $3m$ is said to be a *polar point group*, which includes 1 (triclinic); 2, m (monoclinic); $mm2$ (orthorhombic); 4, $4mm$ (tetragonal); 3, $3m$ (trigonal); 6, $6mm$ (hexagonal). There are no polar groups in cubic crystals. Gallium nitride as a polar material ($6mm$) is of huge interest in the manufacture of light emitting diodes – Figure 2.11b shows that the top surface normal to the hexad is different from the bottom surface. During deposition of GaN films on substrates, the layers deposit predominantly with uniform polarity except some regions which have opposite polarity and are regarded as defects that lead to poor control of electrical properties [8].

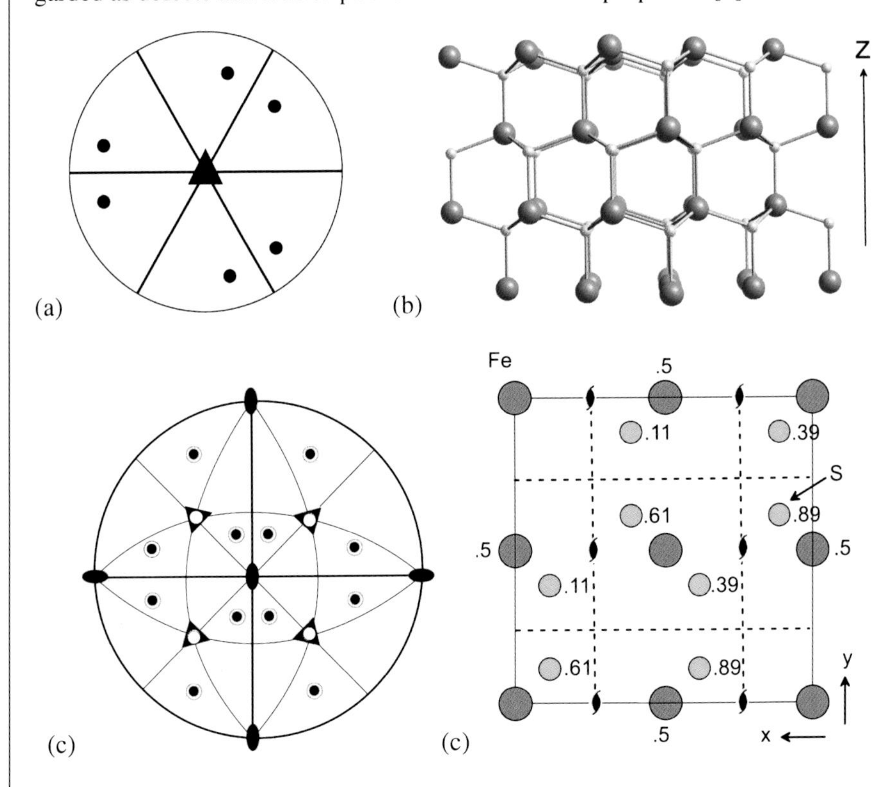

FIGURE 2.11
Sketch stereograms showing what happens to a general pole on the operation of the following point groups (bold lines represent mirror planes): (a) trigonal $3m$. (b) The structure of GaN with point group symmetry $6mm$, viewed with the 6-fold axis along z. The gallium atoms are labelled green. (c) Cubic $m3$. (d) Projection of the unit cell of pyrite (FeS_2) with the fractional z coordinates indicated, glide planes marked with dashed lines and screw diads. Not all the symmetry elements are illustrated for simplicity. For example, there will be glide planes parallel to the plane of the diagram at heights $\frac{1}{4}z$ and $\frac{3}{4}z$.

The point group $m3$ belongs to the cubic crystal class with the first symbol m parallel to the z axis (page 22) and the mandatory symbol "3" to follow since the defining symmetry of a cubic system is four triads, in this case inversion triads. It is evident from the plotting of a general pole and its symmetry related poles shown in Figure 2.11c that there is a center of symmetry. Pyrite belongs to the cubic class with point group $m3$; the actual crystals are illustrated in Figure 1.2 where the four triads associated with the regular octahedral shape of the crystals are obvious. Figure 2.11d shows a projection of the crystal structure of pyrite, which is primitive cubic. The structure can be compared with the stereogram showing the point group symmetry in Figure 2.11c. Bearing in mind that point groups do not include translations, the glide planes and screw axes indicated in Figure 2.11d become mirrors and diads respectively in Figure 2.11c.

Crystals which lack a center of symmetry can be piezoelectric, i.e., they develop a dipole on being deformed and will change shape on the application of an electrical field. Under normal conditions the crystal has positive and negative electrical charges which are symmetrically distributed, leaving it in a neutral state when homogeneously deformed. The application of a stress causes charge asymmetry, and the development of a voltage across the crystal. $PbTiO_3$ is such a substance (Figure 2.12) where, for example, the titanium atom is not located at exactly $\frac{1}{2}, \frac{1}{2}, \frac{1}{2}$. The structure is primitive tetragonal, with point group $4mm$.[1]

The stereogram illustrated in Figure 2.12b shows that the structure does not have a center of symmetry. Second, if an atom is placed at a general position (marked a) then there will be seven other identical atoms in the unit cell, inconsistent with the structure illustrated in Figure 2.12a. An atom placed on a tetrad (marked c) will generate just one atom per unit cell; thus, there is just one Ti at $\frac{1}{2}, \frac{1}{2}, 0.572$. The oxygens at $\frac{1}{2}, \frac{1}{2}, 0$ and $\frac{1}{2}, \frac{1}{2}, 1$ also lie on a tetrad, and since they are shared between two cells, there is only one such oxygen in the unit cell. An atom, such as the oxygen at $0, \frac{1}{2}, \frac{1}{2}$, placed on the mirror plane as illustrated by the poles marked b, would generate two such atoms in the cell (the four on the vertical faces are each shared by two cells).

This example shows how the placing of an atom within a unit cell must be consistent with the symmetry elements of the cell because there must be identical atoms at all symmetry related positions.

There are many other consequences of whether or not a crystal has a center of symmetry. The free energy of a small volume element containing a one-dimensional composition gradient varies as follows [9]:

$$g = g\{\overline{c}\} + \kappa_1 \frac{dc}{dz} + \kappa_2 \frac{d^2c}{dz^2} + \kappa_3 \left(\frac{dc}{dz}\right)^2 \tag{2.1}$$

where c is the concentration, $g\{\overline{c}\}$ is the free energy of a homogeneous solution with composition $\overline{c}$, z is the distance, and κ_i are specific functions of the dependence of free energy on the gradients of concentration as described elsewhere [9]. In this, κ_1 is zero for a centrosymmetric crystal since the free energy must then be invariant to a change in the sign of the coordinate z.

[1] That is, by the convention quoted on page 22 the z direction is assigned to the fourfold rotation axis followed by m for the mirrors with normals parallel to the x and y axes, and the final symbol for the tertiary mirrors with normals between the x and y axes.

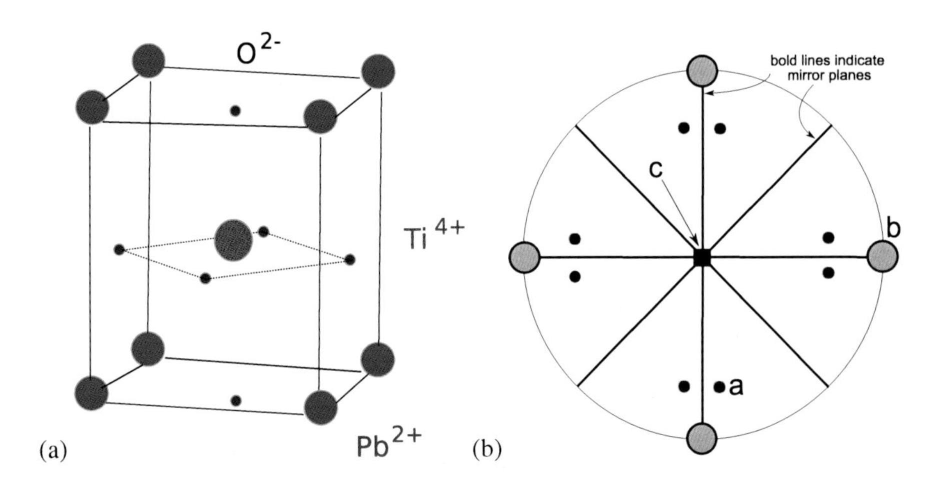

FIGURE 2.12
(a) Lead titanate ($PbTiO_3$) is tetragonal with $a = 0.3904\,\text{nm}$ and $c = 0.4152\,\text{nm}$. The Pb is located at coordinates (0,0,0.116), the Ti at $(\frac{1}{2}, \frac{1}{2}, 0.572)$, an oxygen atom at $(0, \frac{1}{2}, \frac{1}{2})$ and another oxygen at $(\frac{1}{2}, \frac{1}{2}, 0)$. The structure is primitive tetragonal with point group $4mm$. Note that each of the lead atoms is on a cell edge (and hence shared with three other adjacent cells), not a cell corner so there is only one lead atom per cell. (b) Stereogram showing the symmetry operations associated with the point group $4mm$.

2.5 Summary

A circle on a sphere projects as a circle on the stereographic projection; angular relationships are therefore preserved. The method is therefore well suited to look at angles between planes or directions in a crystal. The symmetry of the crystal in its three dimensions becomes easy to visualize by plotting the symmetry elements in their correct angular orientations on a stereographic projection. The placing of atoms inside the unit cell during attempts to solve structure must clearly be consistent with the symmetry determined, for example, from the external form of well-formed crystals.

The stereographic projection can also be used quantitatively, by superimposing a Wulff net to directly measure angles between a pair of poles. The net consists of projections of great and small circles, rather like the longitudes and latitudes on a globe representing the earth, The shortest distance between two points on the surface of a sphere is when the two points are located on an arc of a circle which has the same diameter as the sphere. When using a Wulff net to measure the angle, it therefore is necessary to locate the two poles on a great circle.

We have seen that some quite elementary considerations of symmetry can indicate the occurrence of phenomena such as piezoelectricity, or polar materials. There is much more to be explored in the chapters that follow.

References

1. S. Nagakura: Study of metallic carbides by electron diffraction part III. iron carbides, *Journal of the Physical Society of Japan*, 1959, **14**, 186–195.

2. R. J. Howarth: History of the stereographic projection and its early use in geology, *Tera Nova*, 1996, **8**, 499–513.

3. S. Suwas, and R. K. Ray: *Crystallographic texture of materials*: New York: Springer, 2014.

4. H. S. Fong: Chance occurrence of orientation relationships between phases in solid transformations and distribution of grain misorientations in a polycrystalline single phase material, *Metallurgical Transactions A*, 1981, **12**, 2057–2062.

5. A. F. Gourgues-Lorenzon: Application of electron backscatter diffraction to the study of phase transformations, *International Materials Reviews*, 2007, **52**, 65–128.

6. S. Pramanik, S. Suwas, and R. K. Ray: Influence of crystallographic texture and microstructure on elastic modulus of steels, *Canadian Metallurgical Quarterly*, 2014, **53**, 274–281.

7. G. E. Dieter: *Mechanical metallurgy*: McGraw-Hill, 1988.

8. K. Bertness: GaN nanowires: knowing which end is up: 2015: URL http://www.nist.gov/pml/div686/grp10/polarity-of-gan-nanowires.cfm.

9. J. E. Hilliard: Spinodal decomposition, In: V. F. Zackay, and H. I. Aaronson, eds. *Phase transformations*. Metals Park, OH: ASM International, 1970:497–560.

3

Stereograms for Low Symmetry Systems

Abstract

Most of the metals in the periodic table have a cubic structure under ambient conditions, but there are many that are in the hexagonal class. The stereographic projection for non-cubic systems depends on the lattice parameters and the angles between the basis vectors. And it can no longer be assumed that directions and planes with the same indices are parallel. We shall see that in the case of the hexagonal system, it can be useful to use four basis vectors to define directions and planes.

3.1 Introduction

In the cubic system, a plane normal (pole) with indices hkl is parallel to a direction uvw when $h = u,\ k = v,\ l = w$; thus the normal to (123) is parallel to [123]. This is generally not the case for less symmetrical systems. Figure 3.1 illustrates this for a two-dimensional unit cell in which $b > a$.

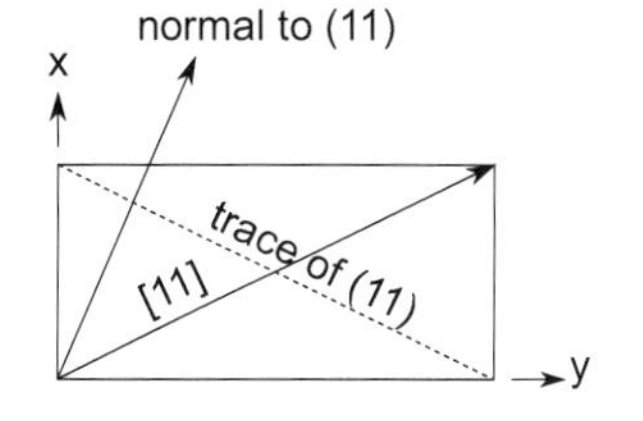

FIGURE 3.1
The direction [11] is not parallel to the normal to the plane (11). On the other hand, the directions parallel to the cell edges, i.e., [01] and [10] are parallel to the normals of (01) and (10), respectively.

The principles for the plotting of poles on stereographic projections are nevertheless identical for any lattice. However, the cubic system exhibits extraordinary symmetry so angular relationships on the stereographic projection are not dependent on the lattice parameter. This is not the case for other lattices with lesser symmetry, the stereograms of which will not only appear different from a distance, but for the same lattice type, the angular positions become dependent on the lattice parameters. Figures 3.2a,b show how the projections differ in the case of the cubic and orthorhombic lattices where the absence of tetrads in the latter is obvious. Figure 3.2c shows how the projection changes when one of the lattice parameters of the orthorhombic lattice is changed relative to that illustrated in Figure 3.2b.

The fact that directions are not in general parallel to plane normals with the same indices is apparent from a comparison of Figure 3.2c,d.

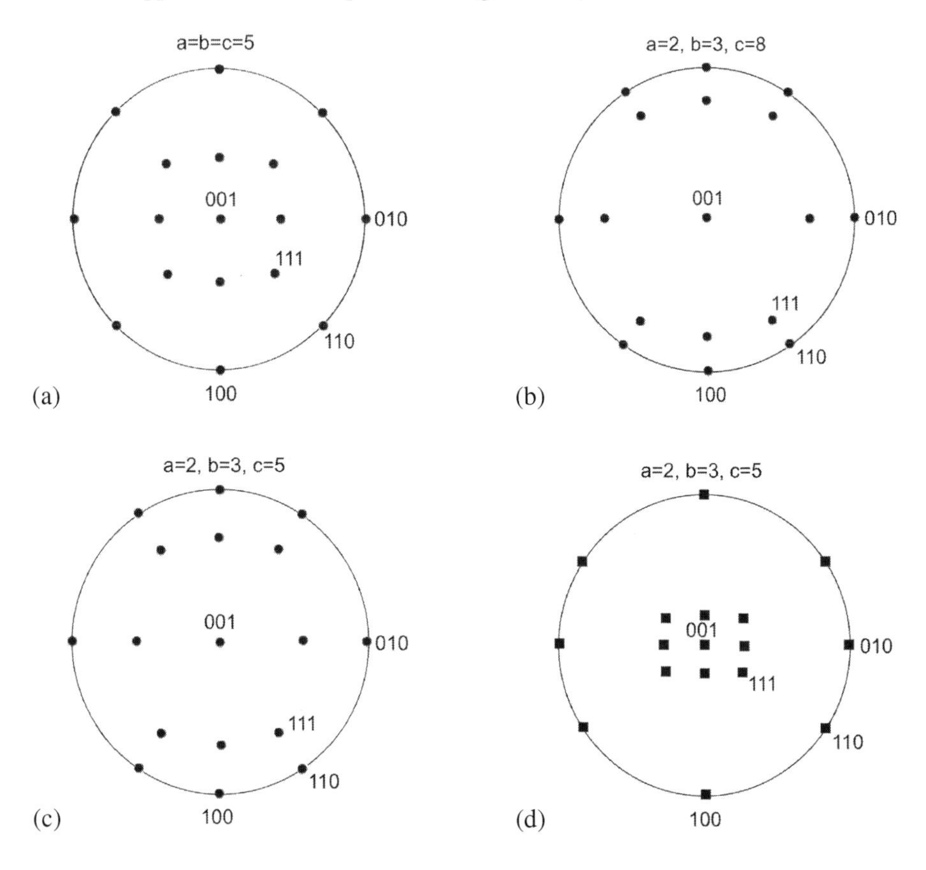

FIGURE 3.2
Stereographic projections centered on the 001 pole. (a) Cubic, lattice parameter 5 units. (b) Orthorhombic, lattice parameters 2, 3, 8 units. (c) Orthorhombic, lattice parameters 2, 3, 5 units. (d) The orthorhombic cell as in (c) but with directions plotted instead of poles.

3.2 Hexagonal system

The unit cell has $a = b \neq c$ and $\alpha = \beta = 90°$ with $\gamma = 120°$. The normals to the (100) and (010) planes therefore make an angle of 60° (Figure 3.3a). This is shown on the projection in Figure 3.3, which highlights a special difficulty with the hexagonal system.

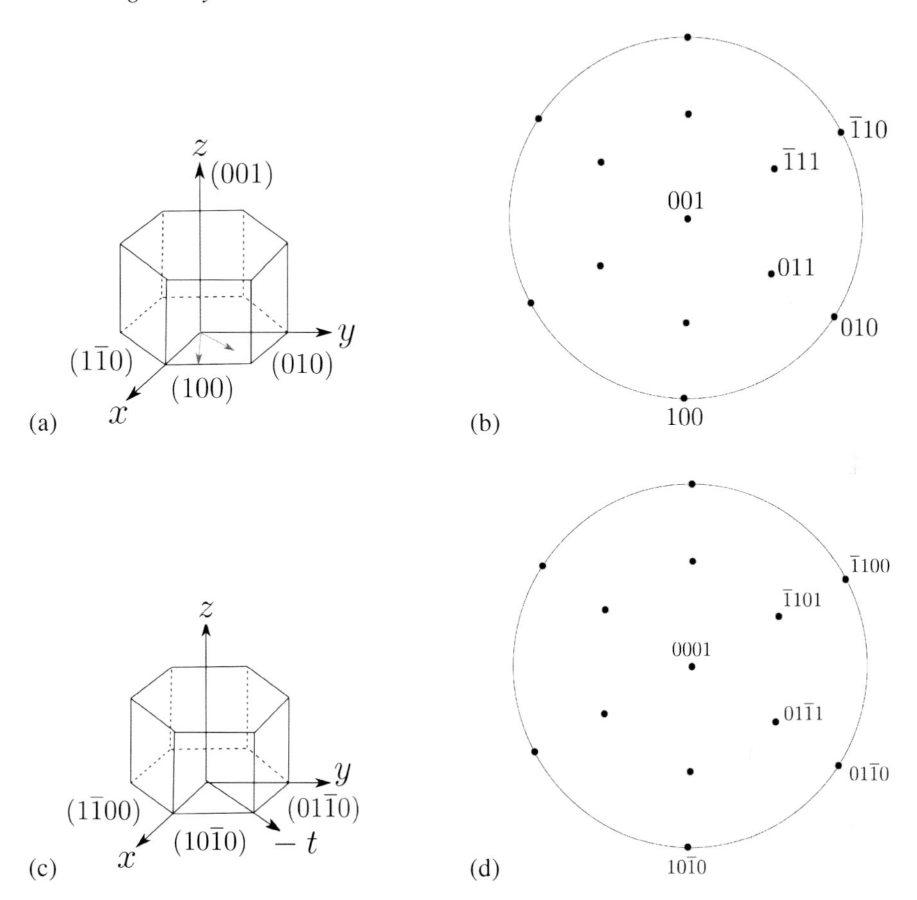

FIGURE 3.3
(a) Showing the hexagonal cell with the normals to the (100) and (010) planes identified with blue-colored arrows. (b) Stereographic projection of hexagonal structure (lattice parameters $a = 2, b = 2$, and $c = 3$ units with $\gamma = 120°$) centered on the 001 pole. (c) Showing the hexagonal cell, now in four index notation. (d) Corresponding stereogram in four index notation.

It is evident from Figure 3.3a that the (100), (010), and $(\bar{1}10)$ are planes belonging to the same form $\{100\}$, i.e., they are crystallographically equivalent. And yet, they have different indices. To eliminate this discrepancy, a four index system is frequently used for hexagonal lattices by introducing a fourth axis (labelled t in Figure 3.3c), which gives plane indices as $hkil$, where from geometry, $i = -(h+k)$. This has the effect of giving equivalent planes the same combination of digits in their indices. Thus, (100), (010), and $(\bar{1}10)$ become $(10\bar{1}0)$, $(01\bar{1}0)$, and $(\bar{1}100)$, respectively. The four components are referred to as the Miller–Bravais indices and the system is reserved for hexagonal lattices [1].

A similar conversion for directions from uvw to $UVJW$ is a little more difficult because it is designed in such a way that $J = -(U + V)$ and at the same time, the Weiss zone law in which a plane hkl containing a direction uvw is satisfied

$$hu + kv + lw = 0 \tag{3.1}$$

so that

$$hU + kV + iJ + lW \qquad \text{should equal zero.}$$

On substituting for $i = -(h + k)$ we get,

$$\begin{aligned} hU + kV + (-h - k)J + lW &= 0 \\ h(U - J) + k(V - J) + lW &= 0 \end{aligned} \tag{3.2}$$

On comparing the coefficients of h, k, and l, and noting that $J = -(U + V)$,

$$\begin{array}{lcll} U - J = u & & u = 2U + V & & U = \frac{1}{3}(2u - v) \\ V - J = v & \rightarrow & v = U + 2V & \text{and} & V = \frac{1}{3}(2v - u) \\ W = w & & w = W & & J = -\frac{1}{3}(u + v) \end{array} \tag{3.3}$$

TABLE 3.1
Directions of the same form now have similar combinations of digits in the four figure notation.

Three-index notation	Four-index notation
[100]	$\frac{1}{3}[2\bar{1}\bar{1}0]$
[010]	$\frac{1}{3}[\bar{1}2\bar{1}0]$
[110]	$\frac{1}{3}[11\bar{2}0]$
[uvw]	$\frac{1}{3}[2u - v,\ 2v - u,\ -(u + v),\ 3w]$

Example 3.1: Angles in the hexagonal system

Iron-manganese alloys can be hexagonal, class $6/mmm$ with lattice parameters $a = 0.2951$ nm and $c = 0.4679$ nm. The material is ductile because slip can occur not only on the basal plane (0001) but also the prism $\{10\bar{1}0\}$ and pyramidal $\{10\bar{1}1\}$ planes. Calculate the angles that these planes make with the [0001] direction.

Draw a stereogram representing the upper hemisphere and mark and index the expected slip directions. Show and index the poles of the slip planes.

The normal method of taking scalar products between vectors in order to determine angles is difficult in non-cubic systems without a deeper understanding of real and reciprocal vectors, as explained in Chapter 9. Noting that angles between planes are also the angles between their normals, the simple geometry illustrated in Figure 3.4a,b shows that the required angle is $\arctan(c/a\cos 30) = 61.36°$. Note that the [0001] direction is normal to (0001) and parallel to the $(10\bar{1}0)$ plane, so the normals to these planes make angles of 0° and 90°, respectively to [0001].

The slip directions and slip planes are plotted on Figure 3.4c. The slip direction lying in a particular slip plane can be identified by applying the Weiss zone rule. Thus, $[2\bar{1}\bar{1}0]$ lies in each of the planes (0001), $(01\bar{1}1)$, and $(01\bar{1}0)$.

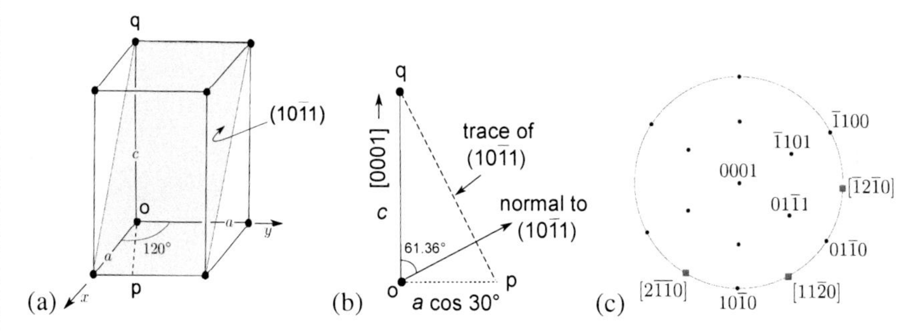

FIGURE 3.4
(a) Hexagonal unit cell showing the $(10\bar{1}1)$ plane. (b) The section *poq* which contains the normals to the (0001) and $(10\bar{1}1)$ planes. (c) Stereographic projection showing the poles of the slip planes (filled circles), and the slip directions (filled squares). The plot is based on the lattice parameters $a = 0.2951$ nm and $c = 0.4679$ nm.

Many polycrystalline metals with hexagonal close-packed crystal structures are brittle. This is because only three slip systems are active, the (0001) close-packed plane with the three close-packed $\langle 11\bar{2}0\rangle$ directions on that plane. To produce an arbitrary deformation requires five independent slip systems. Imagine a strain tensor representing strains with respect to a coordinate system:

$$\begin{pmatrix} \epsilon_{11} & \epsilon_{12} & \epsilon_{13} \\ \epsilon_{21} & \epsilon_{22} & \epsilon_{33} \\ \epsilon_{31} & \epsilon_{32} & \epsilon_{33} \end{pmatrix}$$

where ϵ_{ij} with $i = j$ are normal strain and the rest shear strains such that $\epsilon_{ij} = \epsilon_{ji}$. Assuming that there is no volume change, $\epsilon_{11} + \epsilon_{22} + \epsilon_{33} = 0$ so only two of the normal strains are independent and there are three shear strains. Therefore, there are five independent strains needed in order to produce an arbitrary deformation. Consider now a polycrystalline material. The application of stress will cause the individual crystals to respond differently because they have differing orientations relative to the stress. Their changes in shape are therefore arbitrary as far as neighboring crystals are concerned, so these differences must be accommodated, without which fracture occurs. Hence the need for five independent slip systems for a polycrystalline material to maintain structural integrity.

Iron, by far the most used metallic material in the world, under ambient conditions is cubic-I with one iron atom located at each lattice point. The common slip system is $\{0\bar{1}1\}\langle 111\rangle$, each containing two independent slip directions of the form $\langle 111\rangle$. There are therefore, twelve independent slip systems, making the iron ductile in polycrystalline form. The same can be demonstrated for the cubic-F form of iron. This makes iron very ductile.

The cubic-I form of iron is stabilized under ambient conditions by its ferromagnetic properties; ruthenium and osmium, which in terms of their outer-electron structure are iron analogues in the periodic table, do not exhibit ferromagnetism and hence have hexagonal crystal structures. In the absence of its ferromagnetic properties, the stable form of iron would also be hexagonal with a motif of a pair of identical Fe-atoms, one at 0,0,0 and another at $\frac{1}{3}, \frac{2}{3}, \frac{1}{2}$. The easy-slip system is then $\{0001\}\langle 11\bar{2}0\rangle$. Since there is only one (0001) plane in which there are three directions of the form $\langle 11\bar{2}0\rangle$, polycrystalline hexagonal iron is relatively brittle since the number of slip systems in each crystal is less than the required five. So in the absence of ferromagnetism, we would not have civilization in the form that we know today.

There are caveats to this story. Deformation can occur by mechanisms other than slip, for example, mechanical twinning, so that the five independent slip system requirement is not sufficient to determine the ductility of a polycrystalline material. And iron below its Curie temperature is not strictly cubic, but tetragonal because the magnetic spins in any given crystal tend to be aligned along a $\langle 100\rangle$ axis [2]. However, the tetragonality is quite small and can be neglected in most experiments, although it does manifest on a macroscopic scale via the magnetostrictive effect, which is the variation in the length of a sample subjected to a magnetic field [3].

Example 3.2: Growth direction of cementite laths

Cementite (Fe_3C, orthorhombic) can precipitate in certain iron alloys in the form of coarse laths at temperatures where the iron is in its austenitic (cubic-F) state. A lath is a shape where two of its dimensions are much smaller than its length, and therefore can be associated with a direction along which it exhibits rapid growth. An idealized shape of a lath of cementite is shown in Figure 3.5. The precipitates when observed on two-dimensional sections can give a misleading indication of the major growth direction. For example, when observed on the section "abcd", the longest axis of the section is not parallel to the actual growth direction.

To resolve this, a number of experiments revealed traces of the longest dimensions of cementite laths to be parallel to [011], [012], [013]. Plot these traces on a stereogram in order to see whether they intersect at a common direction, which corresponds to the true growth direction. Bearing in mind that cementite is orthorhombic with lattice parameters $a = 0.45165$ nm, $b = 0.50837$ nm, and $c = 0.67475$ nm. It is first necessary to calculate the angles that these traces make with respect to the basis vectors of the unit cell. This can be done by converting the components of each vector into an orthonormal basis:

$$\begin{aligned}
[100]_{\mathrm{Fe_3C}} &\equiv [a\,0\,0]_{\mathrm{orthonormal}} \\
[010]_{\mathrm{Fe_3C}} &\equiv [0\,b\,0]_{\mathrm{orthonormal}} \\
[001]_{\mathrm{Fe_3C}} &\equiv [0\,0\,c]_{\mathrm{orthonormal}} \\
[011]_{\mathrm{Fe_3C}} &\equiv [0\,b\,c]_{\mathrm{orthonormal}} \\
[012]_{\mathrm{Fe_3C}} &\equiv [0\,b\,2c]_{\mathrm{orthonormal}} \\
[013]_{\mathrm{Fe_3C}} &\equiv [0\,b\,3c]_{\mathrm{orthonormal}}
\end{aligned} \tag{3.4}$$

Vector scalar products can then be implemented using elementary methods as follows to calculate the required angles (ϕ in the case illustrated):

$$\cos\phi = \frac{[0\,0\,c]_{\mathrm{Fe_3C}} \cdot [0\,b\,c]_{\mathrm{orthonormal}}}{\sqrt{c^2 \times (b^2 + c^2)}} = \frac{c^2}{\sqrt{c^2 \times (b^2 + c^2)}} = 37^\circ$$

After calculating all three angles, a stereographic projection is constructed with the traces marked as in Figure 3.5; the traces intersect at $[100]_{\mathrm{Fe_3C}}$ The growth occurs most rapidly, therefore, along the $[100]_{\mathrm{Fe_3C}}$ direction (to avoid confusion in the order in which the axis are labelled, we note that this direction corresponds to the lattice parameter $a = 0.45165\,\mathrm{nm}$). This example is illustrative, but the growth direction deduced is as is observed experimentally for cementite laths precipitating in alloyed austenite at high temperature (650 °C) [4].

Finally, it is worth noting that the method used to calculate angles is cumbersome and limited to orthogonal lattices when compared with the use of metric tensors, as described in Chapter 9.

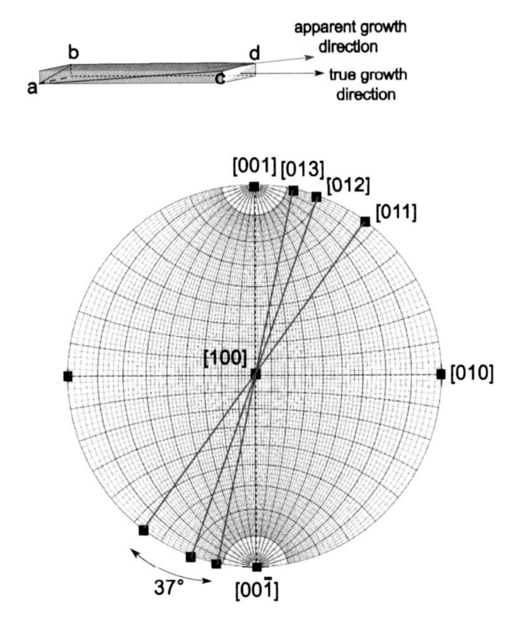

FIGURE 3.5
Discovery of the growth direction of cementite laths in austenitic iron using trace analysis. The stereogram shows directions in cementite that has the lattice parameters $a = 0.45165\,\text{nm}$, $b = 0.50837\,\text{nm}$, and $c = 0.67475\,\text{nm}$.

3.3 Summary

Complications arise once the defining symmetry of the cube, i.e., four triads, is lost. Directions and plane normals with the same indices are then not necessarily parallel and hence have to be distinguished on a stereographic projection. Furthermore, the extent of the discrepancy relative to the cubic system depends on the anisotropy of the lattice parameters and the angles between the basis vectors of the system concerned.

The hexagonal system poses a particular problem of communication because of the nature of the special symmetry of the hexad. For example, the crystallographically equivalent planes (100), (010), and $(1\bar{1}0)$ have indices that are not permutations of the same set of digits. The Miller–Bravais four-index system eliminates this ambiguity but requires additional work to ensure that the Weiss zone rule is satisfied. Whether the intellectual cost of the four-index system over the usual three-index notation is justified by its elegance is a matter of taste.

References

1. H. M. Otte, and A. G. Crocker: Crystallographic formulae for hexagonal lattices, *Physica Status Solidi B*, 1965, **9**, 441–450.

2. R. M. White: *Quantum theory of magnetism*: New York: Springer, 2007.

3. J. P. Joule: On the effects of magnetism, &c. upon the dimensions of iron and steel bars., *The London, Edinburgh, and Dublin Philosophical Magazine and Journal of Science*, 1847, **30**, 225–241.

4. M. V. Kral, and R. W. Fonda: The primary growth direction of Widmanstätten cementite laths, *Scripta Materialia*, 2000, **43**, 193–198.

4

Space Groups

Abstract

Some of the symmetry elements of a crystal involve translations that are simple fractions of a repeat distance. These translations have consequences, for example on the number of symmetry related positions in the unit cell, and hence on the number of atoms in the unit cell. This chapter deals with *space groups*, which, unlike point groups, include screw axes and glide planes.

4.1 Introduction

A point group is a collection of symmetry elements passing through a point, and therefore, necessarily does not include translations. *Space groups*, in contrast, include translations that are fractions of a repeat unit, for example, a 2_1 axis which involves a rotation of 180° followed by a translation of $\frac{1}{2}$ of the repeat distance along the axis. Glide planes involve reflections followed by fractional translations. These translations are small, and hence do not manifest when the point group is determined from the external shape of well-formed crystals. However, they have consequences in structure determination and on other properties of crystals. There are 230 space groups which are made from combinations of the 32 crystallographic point groups with the 14 Bravais lattices.

The relationships between the placing of atoms in the unit cell and the symmetry of the structure of lead titanate was described on page 41. The structure of $PbTiO_3$ is primitive tetragonal with point group $4mm$. Since there are no screw axes or glide planes, the space group symbol for the titanate is $P4mm$, where the first symbol is to identify the lattice type as primitive. In this chapter, we will consider a case where translational symmetry elements[1] are present.

4.2 Screw axes and glide planes

A proper rotation axis designated N brings a motif into coincidence after a rotation of $\frac{2\pi}{N}$. A screw axis, designated N_p, does the same but only when a translation of $\frac{p}{N} \times t$

[1] Other than the normal periodicity of a lattice.

follows the rotation by $\frac{2\pi}{N}$. Here t is the distance between lattice points parallel to the axis. A screw diad (2_1) is illustrated in Figure 4.1, involving a rotation of 180° followed by a translation of $\frac{1}{2}t$ parallel to the axis. Other rotation axes following the same logic in terms of notation are: 3_1, 3_2, 4_1, 4_2, 4_3, 6_1, 6_1, 6_2, 6_3, 6_4, and 6_5.

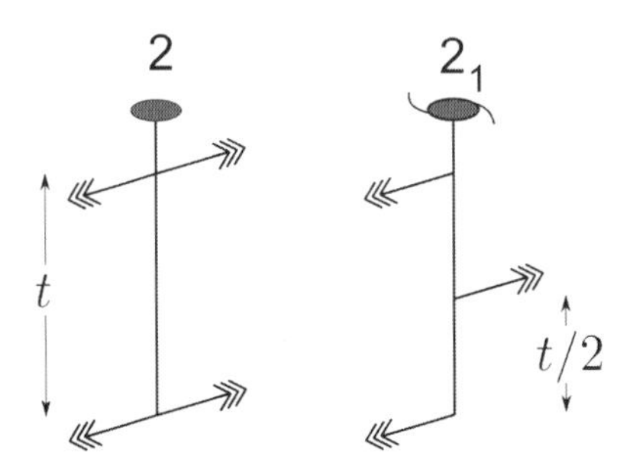

FIGURE 4.1
An illustration of the operation of a diad where the structure is restored by a rotation of 180°, and a screw diad which involves the same rotation but then a translation along the axis by 0.5 (hence the notation 2_1) of the repeat distance t.

Glide planes are like mirror planes but a reflection is followed by a translation in order to recover the symmetry. The terminology for glide planes is explained in Table 4.1; the translations following reflection are parallel to the glide plane, the exception begin the diamond glide in a body-centered cubic structure.

TABLE 4.1
Symbols for glide planes.

Glide character	Direction	Magnitude	Symbol
Axial glide	Parallel to a, b or c axis	$\frac{1}{2}(a,\ b,\ \text{or}\ c)$	a, b, or c
Diagonal glide	Parallel to face diagonal	$\frac{1}{2}(a+c)$, $\frac{1}{2}(b+c)$ or $\frac{1}{2}(a+b)$	n
Diamond glide (fcc)	Parallel to face diagonal	$\frac{1}{4}(a+c)$, $\frac{1}{4}(b+c)$ or $\frac{1}{4}(a+b)$	d
Diamond glide (bcc)	Parallel to body diagonal	$\frac{1}{4}(a+b+c)$	d

Note: “fcc” and “bcc” are abbreviations for face–centered cubic and body–centered cubic, respectively. The structure of pyrite, which is primitive cubic and has a point group symmetry $m3$ is illustrated on page 40, containing axial glide planes; its space group therefore is $Pa3$ where the symbol a replaces the m in the point group.

4.3 Cuprite

Consider now the structure of cuprite (Cu_2O) which is primitive cubic, with a motif of four copper atoms at 000, $\frac{1}{2}0\frac{1}{2}$, $0\frac{1}{2}\frac{1}{2}$ and $\frac{1}{2}\frac{1}{2}0$, and two oxygen atoms at $\frac{1}{4}\frac{1}{4}\frac{3}{4}$, and $\frac{3}{4}\frac{3}{4}\frac{1}{4}$ located at each lattice point.

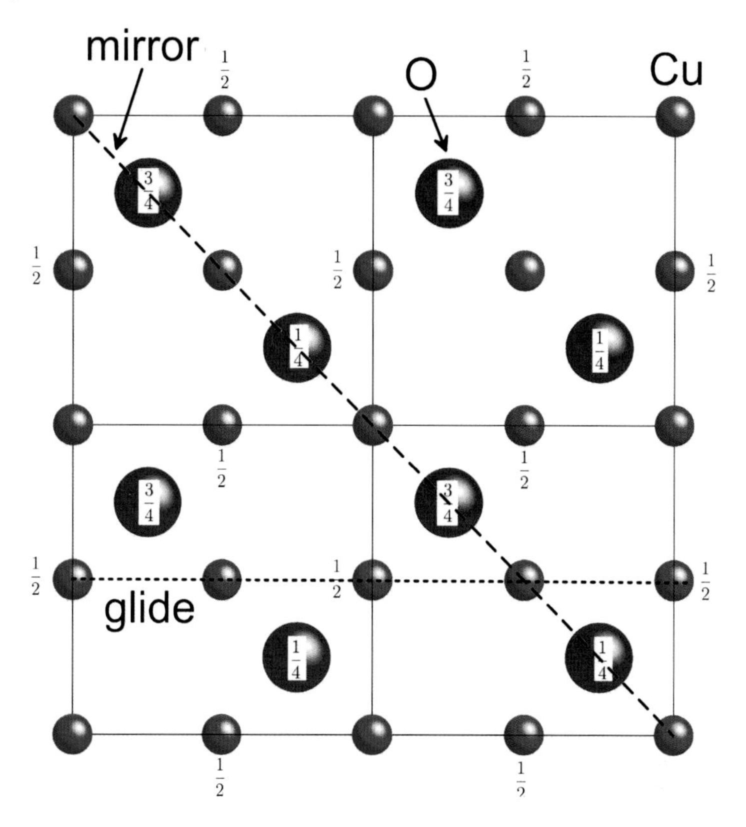

FIGURE 4.2
Projection along the [001] direction of four unit cells of cuprite (Cu_2O). The fractional height along [001] are indicated, with unlabelled atoms at 0 and 1.

Figure 4.2 shows a projection of four such unit cells along the [001] direction, with fractional coordinates along that direction indicated. The environment of each of the six atoms within the unit cell is clearly different, consistent with the cubic-P lattice type. There are mirror planes parallel to $\{110\}$ – the trace of only one of these is illustrated in Figure 4.2 for the sake of clarity. Note also that there are 4_2 axes located at $\frac{1}{4}\frac{1}{4}z$ and $\frac{3}{4}\frac{3}{4}z$ and their equivalent positions. A 4_2 axis involves a rotation of 90° followed by a translation by $\frac{2}{4}$ of the repeat distance along that axis. The diagram also shows that the structure has a center of symmetry at $\frac{1}{2}\,\frac{1}{2}\,\frac{1}{2}$.

The glide plane parallel to $\{100\}$ (Figure 4.2) involves a reflection followed by two orthogonal translations parallel to the plane by $\frac{1}{2}$ the lattice parameter. It is, therefore, an n-glide plane according to the notation in Table 4.1.

Since cuprite is cubic, the defining symmetry is four triads. However, Figure 4.3 shows that the triads are in fact inversion triads ($\overline{3}$), meaning that a rotation of 120° is followed by an inversion through the center to recover the structure. This is evident from the projection along $\langle 111 \rangle$ that there is no triad since the atoms marked with z coordinates are in two sets of three, with each set at a different height along z. A simple three-fold operation would not therefore work. On the other hand, the rotation combined with the inversion reproduces the structure.

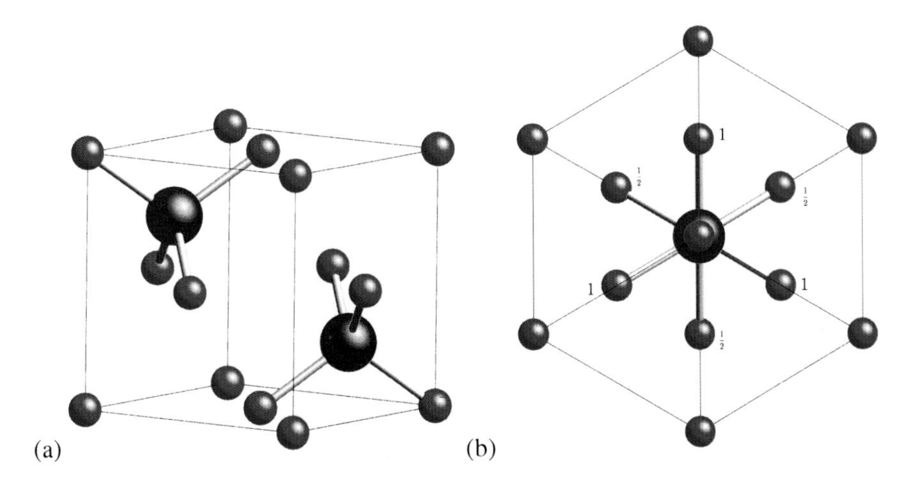

FIGURE 4.3
(a) The unit cell of cuprite. (b) A projection along $\langle 111 \rangle$, with fractional coordinates along [001] indicated for some of the atoms.

4.4 Location of atoms in cuprite cell

The number of atoms in the unit cell depends on where an atom is placed first because all other symmetry related equivalent positions must also contain that atom. The space group $Pn\overline{3}m$ reduces to the point group $m3m$ when translations are omitted. Figure 4.4a shows that the placing of an atom at a general location (i.e., not on a symmetry element) must lead to a total of 48 such atoms in the cell since there are that many crystallographically equivalent positions in the cell. On the other hand, by placing the atom on a mirror plane, the number of positions halves to 24 since an atom that is on a mirror plane does not have a reflection from that plane, Figure 4.4b.

Referring to Figure 4.4c,d, suppose that a copper atom is located at 0,0,0 which has the symmetry elements $\overline{3}m$ passing through it. The operation of the n-glide plane marked by the horizontal line generates the atom at the opposite face-center at a height $\frac{1}{2}$ and the other symmetry elements then generate atoms at all face-centers and corners of the cube. There are, therefore, four equivalent copper atoms in the unit cell.

Suppose that an oxygen atom is placed at the location $\frac{1}{4}, \frac{1}{4}, \frac{3}{4}$ which has the symmetry elements $\overline{4}3m$ passing through it. The operation of the same n-glide plane on

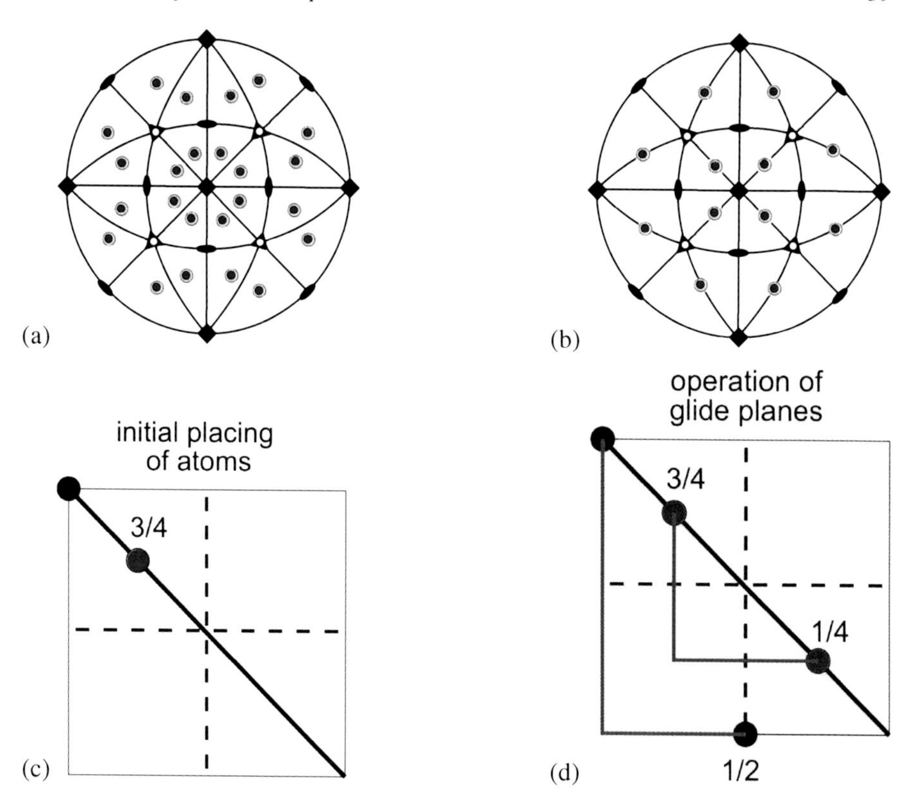

FIGURE 4.4
Stereograms consistent with the point group $m3m$, showing the number of equivalent positions when (a) an atom is placed at general location $x,\ y,\ z$ and all equivalent positions; (b) an atom is placed at general location $x,\ x,\ z$ and all equivalent positions. (c) Projection of the cuprite unit cell with the traces of glide planes marked as dashed lines and those of mirror planes as heavy lines. Initial placing of atoms copper in black and oxygen in red. (d) New set of atoms generated by the operation of the glide planes. The relation between the original and new atoms is marked by blue lines.

that oxygen atom leads to an oxygen atom at $\frac{3}{4}, \frac{3}{4}, \frac{1}{4}$. All other symmetry operations have the same effect so there are only two oxygen atoms per cell. The locations of the copper and oxygen atoms are therefore consistent with the chemical formula Cu_2O and the density of the compound can be measured to verify this. Placing a copper atom at the general position x, y, z or x, x, z would not lead to the correct density or stoichiometry. The results are summarized in Table 4.2. The International Union of Crystallography publishes much more comprehensive space group tables which list the multiplicity of coordinates as a function of symmetry at that coordinate, for all possible scenarios and for all space groups. Table 4.2 can be regarded as a much simplified version of this kind of information.

TABLE 4.2
Space group for cuprite, $Pn\overline{3}m$. The origin of the cell is set on a copper atom.

Point symmetry	Number	Coordinates			
1	48	x, y, z	$\overline{x}, \overline{y}, z$	$\overline{x}, y, \overline{z}$	...
m	24	x, x, z	$\overline{x}, \overline{x}, z$	$\overline{x}, x, \overline{z}$	...
⋮					
⋮					
$\overline{3}m$	4	0,0,0	$\frac{1}{2}, \frac{1}{2}, 0$	$\frac{1}{2}, 0, \frac{1}{2}$	$0, \frac{1}{2}, \frac{1}{2}$
$\overline{4}3m$	2	$\frac{1}{4}, \frac{1}{4}, \frac{3}{4}$	$\frac{3}{4}, \frac{3}{4}, \frac{1}{4}$		

Example 4.1: Space group of Fe-Si-U compound

- An intermetallic compound containing Fe, Si, and U has the lattice parameters $a = 0.7$ nm, $b = 0.6$ nm, $c = 0.5$ nm parallel to the x, y, and z axes, respectively, with the angle $\beta = 105°$ between the x and z axes. What crystal class does this belong to?

- The space group symmetry of the crystal is Pm with the mirror planes normal to b at $y = 0$ and $y = \frac{1}{2}$. Find the number of equivalent sites (multiplicity) for the following coordinates: (x, y, z), $(x, \frac{1}{2}, z)$, and $(x, 0, z)$.

- A silicon atom is placed at $0, 0, 0$, Fe at $(0.5, 0.1, 0.3)$, U at $(0.75, 0.5, 0.75)$. Accounting for multiplicity, draw a projection of the unit cell normal to the z axis. Hence write the chemical formula for the compound. What are the point group symmetries of the iron and uranium atoms?

- Describe a shear that could convert the monoclinic cell into one that is orthorhombic, stating the plane and direction of the shear. Calculate the magnitude of the shear.

The crystal class is monoclinic since none of the lattice parameters are equal, and one of the angles does not equal 90°. The cell is primitive with a mirror planes at $y = 0$ or $\frac{1}{2}$. Anything located at a general coordinate x, y, z therefore has a multiplicity of of 2, i.e., x, y, z and $x, -y, z$, due to reflection in mirror. Anything located on a mirror plane clearly does not have a reflection, and therefore, a multiplicity of just 1, which applies to $x, \frac{1}{2}, z$ and $x, 0, z$.

The resulting structure projection is illustrated below, giving the chemical formula Fe_2USi. Note that there are two iron atoms due to the mirror plane, but only one uranium atom that is located on the mirror plane. The point group symmetry of the uranium atom is therefore m. That of iron is just 1 (monad).

In order to convert the monoclinic cell into one that is orthorhombic, it is necessary to convert the angle β to 90°. The required shear is therefore on the (001) plane, in the [100] direction, with a magnitude $\tan(105 - 90)$ where the angles are expressed in degrees.

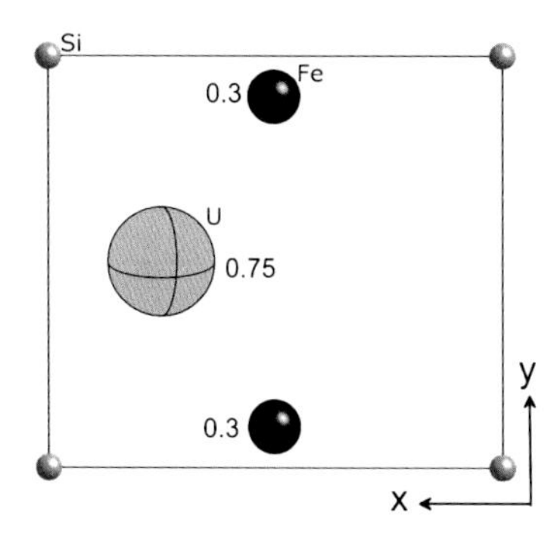

FIGURE 4.5
Projection of the unit cell of Fe_2USi normal to the z-axis.

Example 4.2: Cementite

Cementite is an iron carbide (Fe_3C) and the commonly used convention in metallurgical studies is to set the order of the lattice parameters as $a = 0.50837$ nm, $b = 0.67475$ nm, and $c = 0.45165$ nm. There are 12 atoms of iron in the unit cell and 4 of carbon, as illustrated in Figure 4.6. Deduce the space group of the structure.

The lattice type is primitive (P). There are n-glide planes normal to the x-axis, at $\frac{1}{4}x$ and $\frac{3}{4}x$ involving translations of $\frac{b}{2}+\frac{c}{2}$. There are mirror planes normal to the y-axis and a-glide planes normal to the z-axis, at heights $\frac{1}{4}z$ and $\frac{3}{4}z$ with fractional translations of $\frac{a}{2}$ parallel to the x-axis. The space group symbol is therefore $Pnma$ [1].

The structure of cementite clearly is very anisotropic so its properties, such as the elastic moduli, also vary strongly with the direction within the crystal [2]. The shear modulus c_{44} is exceptionally small, some two times smaller than the corresponding term for aluminium. Nevertheless, the cementite has an exceptionally large ideal shear strength because elastic deformation reduces its symmetry from orthorhombic to monoclinic (space group $P2_1/c$), with an accompanying increase in three-dimensional covalent bonding that stiffens the material [3].

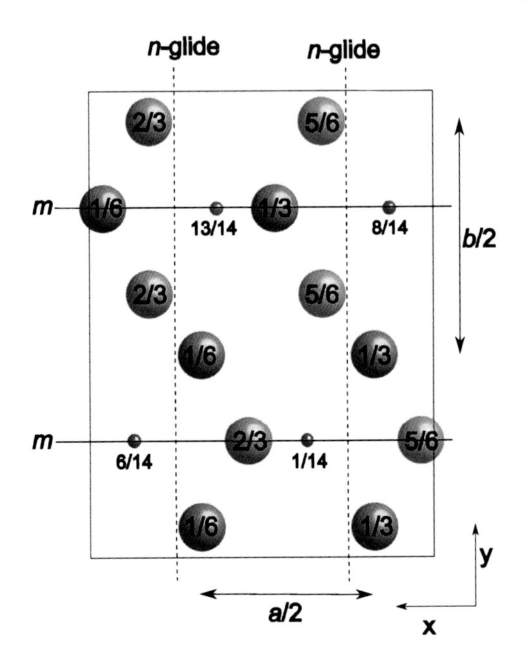

FIGURE 4.6
The crystal structure of cementite, consisting of 12 iron atoms (large) and 4 carbon atoms (small). The fractional z coordinates of the atoms are marked.

[a] In the context of plastic deformation, the slip systems on which dislocation glide has been observed include $(001)[100]$, $(100)[010]$, $(100)[001]$, $(010)[001]$, and $(010)[100]$ [4] but the complex arrangement of atoms means that slip is actually quite difficult to achieve, making the cementite very hard. There may be other slip systems that operate when the cementite is forced to deform in a phase mixture such as pearlite, but the Burgers vectors of slip dislocations would nevertheless be large when compared with dislocations in the allotropic forms of iron [5].

[a] It is important to note the labelling of the orthogonal axes of the unit cell, i.e., $a = 0.50837$ nm, $b = 0.67475$ nm, and $c = 0.45165$ nm. since different conventions are used in the published literature. The indices referred to below have been adjusted from the original sources, for the unit cell as defined here.

Example 4.3: Diamond and zinc sulfide

Both diamond and zinc sulfide have a cubic-F lattice; in diamond a motif of a pair of carbon atoms at 0,0,0 and $\frac{1}{4}, \frac{1}{4}, \frac{1}{4}$ is placed at each lattice point whereas in ZnS, the motif consists of a Zn atom and a S atom. The space group of diamond is $Fd\overline{3}m$ whereas that of ZnS is $F\overline{4}3m$. Draw structure projections and explain why the diamond glide is missing in the structure of ZnS.

Diamond consists only of carbon atoms, each of which is symmetrically related to all the others in the cell. For example, the operation of the diamond glide plane illustrated in Figure 4.7 on the carbon atom at 0,0,0 leads to the carbon atom at $\frac{1}{4}, \frac{1}{4}, \frac{1}{4}$. In ZnS, this operation would not lead to a symmetrically related Zn atom because a sulfur atom is located at $\frac{1}{4}, \frac{1}{4}, \frac{1}{4}$.

The space group of ZnS is in fact a sub-set of the space group of diamond, with all the symmetry elements of the diamond structure that lead a zinc atom into the position of a sulfur atom, missing. Therefore, the inversion triad in diamond is reduced to a triad and there is no center of symmetry. In diamond the center of symmetry is at $\frac{1}{8}, \frac{1}{8}, \frac{1}{8}$ and equivalent positions.

Referring to Figure 1.15, when the randomly arranged nickel and aluminium atoms in γ, which has the space group $Fm\overline{3}m$ on ordering to the γ' phase becomes the sub-group $Pm\overline{3}m$. In γ the point symmetry elements passing through each atom are $m\overline{3}m$, whereas in γ' the point group symmetries of the nickel atoms are $4/mmm$ and the aluminium atoms, $m\overline{3}m$. There is, therefore, no three-fold axis through the nickel atoms. The reduction in symmetry on ordering has huge consequences for the deformation behavior of γ' because larger displacements are necessary before the structure is reproduced. A typical Burgers vector of a dislocation doubles in length in the γ' relative to the γ.

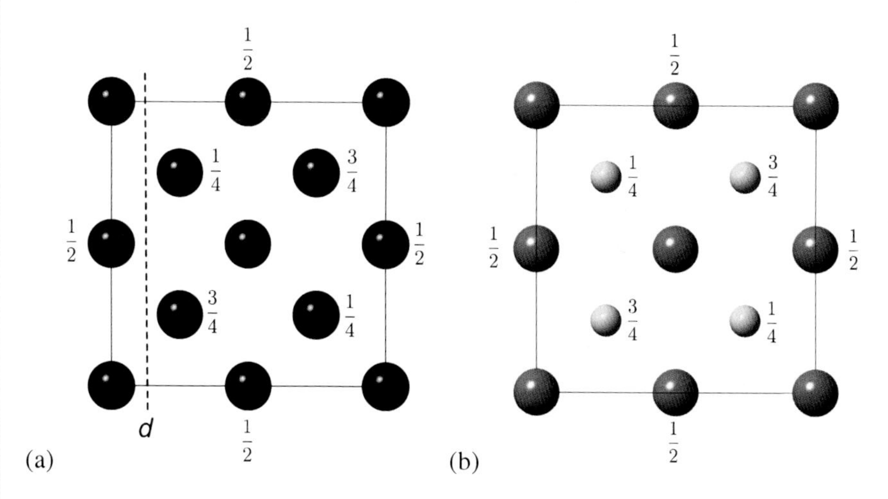

FIGURE 4.7
Structure projections. (a) Diamond (carbon). The dashed line shows a d-glide plane involving two orthogonal translations by a quarter of the repeat distance, parallel to the glide plane. (c) Cubic form of zinc sulfide.

4.5 Shape of precipitates

Crystallography is one of the characteristics that can influence the shape of a precipitate, but its influence is usually overwhelming when the precipitation occurs in the solid state [6–9]. In some circumstances, it is symmetry which determines the shape of precipitates which form inside a solid.

Suppose that G_m and G_p represent the symmetry groups of the matrix and precipitate, respectively. The common symmetry (intersection group) between these two phases in the observed orientation relationship is defined as H. Given a planar interface between them, symmetry operations consistent with H leads to other orientations of the interface until the particle is fully enclosed and its shape defined. Note that any operation of H is a symmetry operation of the parent and product and hence each crystal orientation is brought into congruence so the orientation of the two crystals is not affected by H.

Aluminium has the space group $Fm\overline{3}m$ and the silver–rich precipitate Ω has the space group $Fmmm$ [10]. The orientation relation has been measured to be:

$$[001]_\Omega \parallel [111]_{\mathrm{Al}}$$

$$[100]_\Omega \parallel [10\overline{1}]_{\mathrm{Al}}$$

Figure 4.8 shows the stereograms of the symmetry elements for the two crystals in the correct relative orientation. It is evident the common point group symmetry is $2/m$, which is consistent with the observed plate shape of the Ω both with respect to the diad and the mirror plane.

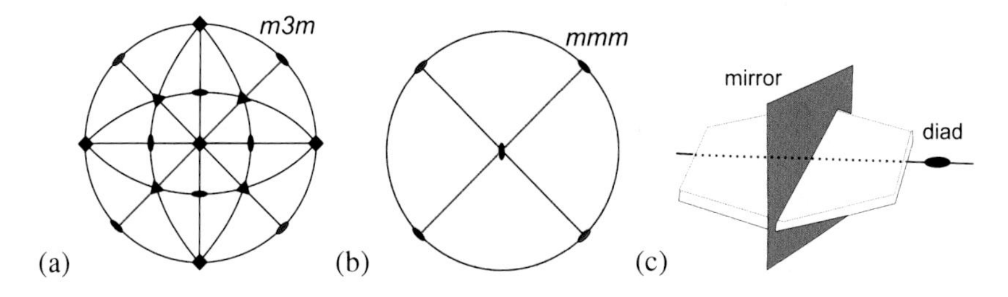

FIGURE 4.8
(a) Stereographic representation of the symmetry elements in the point group $m3m$ for aluminium. (b) Stereographic representation of the symmetry elements in the point group mmm for Ω. The symmetry elements shared by Al and Ω are highlighted in red. (c) The precipitate shape observed, corresponding to the shared symmetry group $2/m$.

4.6 Summary

We have touched on the notion of space groups which represent all the symmetry elements of the crystal. The appreciation of space groups is essential in the solution of structures, transformations which lead to changes in symmetry, and the properties of crystals. A great deal of the work in identifying the number of symmetry related positions as a function of the space group has already been documented, with the information routinely available either from the tables published by the International Union of Crystallography or via the World Wide Web.

References

1. E. J. Fasiska, and G. A. Jeffrey: On the cementite structure, *Acta Crystallographica*, 1965, **19**, 463–471.
2. C. Jiang, S. G. Srinivasan, A. Caro, and S. A. Maoly: Structural, elastic, and electronic properties of Fe_3C from first principles, *Journal of Applied Physics*, 2008, **103**, 043502.
3. C. Jiang, and S. G. Shrinivasan: Unexpected strain-stiffening in crystalline solids, *Nature*, 2013, **496**, 339–342.
4. A. Inoue, T. Ogura, and T. Muramatsu: Burgers vectors of dislocations in cementite crystal, *Scripta Metallurgica*, 1977, **11**, 1–5.
5. J. Gil-Sevillano: Room temperature plastic deformation of pearlitic cementite, *Materials Science & Engineering*, 1975, **21**, 221–225.
6. G. R. Hugo, and B. C. Muddle: Role of symmetry in determining precipitate morphology, *Materials Forum*, 1989, **13**, 147–152.
7. G. R. Hugo, and B. C. Muddle: The morphology of precipitates in an Al-Ge alloy-ii. analysis using symmetry, *Acta Materialia*, 1990, **38**, 365–374.
8. B. C. Muddle, and I. J. Polmear: The precipitate omega phase in Al Cu Mg Ag alloys, *Acta Metallurgica*, 1989, **37**, 777–789.
9. L. S. Chumbley, B. C. Muddle, and H. L. Fraser: Crystallography of precipitation of Ti_5Si_3 in Ti-Si alloys, *Acta Metallurgica*, 1988, **36**, 299–310.
10. K. M. Knowles, and W. M. Stobbs: The structure of $\{111\}$ age-hardening precipitates in Al-Cu-Mg-Ag alloys, *Acta Crystallographica B*, 1988, **B44**, 207–227.

5

The Reciprocal Lattice and Diffraction

Abstract

The Miller indices for directions and planes are defined differently. For directions the indices are simply the components of a vector corresponding to that direction with respect to the basis vectors, but those of a plane are defined rather strangely at first sight. The intercepts of the plane with the basis vectors are determined, and their reciprocals form the indices of the plane. The mystery behind this scheme is revealed in this chapter, whereby a plane is also defined by a vector that is normal to it and has a magnitude that is the reciprocal of the plane spacing. Diffraction too becomes easier to visualize when expressed in terms of reciprocal space.

5.1 The reciprocal basis

The basis vectors $\mathbf{a}_i$ were defined in Figure 1.3; a vector $\mathbf{u}$ in real space can then be written as a linear combination of the basis vectors:

$$\mathbf{u} = u_1\mathbf{a}_1 + u_2\mathbf{a}_2 + u_3\mathbf{a}_3,$$

where u_1, u_2, and u_3 are its components, when $\mathbf{u}$ is referred to the basis A.

The reciprocal lattice constitutes a special coordinate system, designed originally to simplify the study of diffraction phenomena. If we consider a lattice, represented by a basis symbol "A" and an arbitrary set of non-coplanar basis vectors $\mathbf{a}_1$, $\mathbf{a}_2$, and $\mathbf{a}_3$, then the corresponding reciprocal basis A^* has basis vectors $\mathbf{a}_1^*$, $\mathbf{a}_2^*$, and $\mathbf{a}_3^*$, defined by the following equations:

$$\begin{aligned} \mathbf{a}_1^* &= (\mathbf{a}_2 \wedge \mathbf{a}_3)/(\mathbf{a}_1.\mathbf{a}_2 \wedge \mathbf{a}_3) \\ \mathbf{a}_2^* &= (\mathbf{a}_3 \wedge \mathbf{a}_1)/(\mathbf{a}_1.\mathbf{a}_2 \wedge \mathbf{a}_3) \\ \mathbf{a}_3^* &= (\mathbf{a}_1 \wedge \mathbf{a}_2)/(\mathbf{a}_1.\mathbf{a}_2 \wedge \mathbf{a}_3) \end{aligned} \tag{5.1}$$

The term $(\mathbf{a}_1.\mathbf{a}_2 \wedge \mathbf{a}_3)$ represents the volume of the unit cell formed by $\mathbf{a}_i$, while the magnitude of the vector $(\mathbf{a}_2 \wedge \mathbf{a}_3)$ represents the area of the $(1\ 0\ 0)_A$ plane (Figure 5.1). Since $(\mathbf{a}_2 \wedge \mathbf{a}_3)$ points along the normal to the $(1\ 0\ 0)_A$ plane, it follows that $\mathbf{a}_1^*$ also points along the normal to $(1\ 0\ 0)_A$ and that its magnitude $|\mathbf{a}_1^*|$ is the reciprocal of the spacing of the $(1\ 0\ 0)_A$ planes (Figure 5.1).

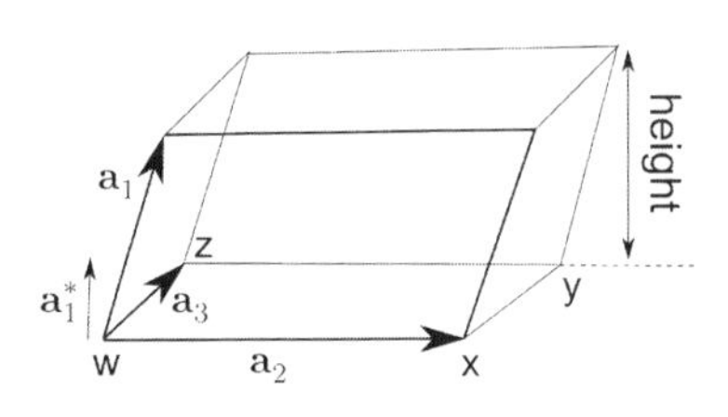

FIGURE 5.1
The relationship between $\mathbf{a}_1^*$ and $\mathbf{a}_i$. The vector $\mathbf{a}_1^*$ lies normal to the basal plane wxyz and the volume of the parallelepiped formed by the basis vectors $\mathbf{a}_i$ is given by $\mathbf{a}_1.\mathbf{a}_2\wedge\ \mathbf{a}_3$. The area of the basal plane is $|\mathbf{a}_2\wedge\ \mathbf{a}_3|$. The height is the spacing between planes parallel to the basal plane, given by $1/|\mathbf{a}_1^*|$.

The components of any vector referred to the reciprocal basis represent the Miller indices of a plane whose normal is along that vector, with the spacing of the plane given by the inverse of the magnitude of that vector. For example, the vector $(\mathbf{u}; A^*) = (1\ 2\ 3)$ is normal to planes with Miller indices $(1\ 2\ 3)$ and interplanar spacing $1/|\mathbf{u}|$. We see that

$$\mathbf{a}_i.\mathbf{a}_j^* = 1 \quad \text{when} \quad i = j, \quad \text{and} \quad \mathbf{a}_i.\mathbf{a}_j^* = 0 \quad \text{when} \quad i \neq j \tag{5.2}$$

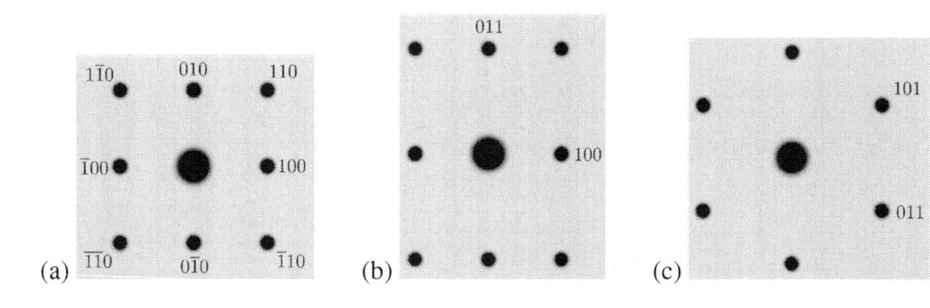

FIGURE 5.2
Sections of the reciprocal lattice of a cubic-P lattice. (a) Section normal to [001]. (b) Section normal to $[01\bar{1}]$. (c) Section normal to $[\bar{1}1\bar{1}]$. Notice how the distance varies inversely with the spacing of the planes. The origin of the reciprocal lattice is not labelled, but if necessary, can be designated 000. Each of the spots in these patterns should strictly be points, but they are weighted here with intensity to show the relationship with electron diffraction patterns discussed later.

Example 5.1: Weiss zone law
Using real and reciprocal lattice vectors, prove the Weiss zone rule that when a direction with real space indices $[u\ v\ w]$ lies in a plane with reciprocal space indices $(h\ k\ l)$,

$$uh + vk + wl = 0 \tag{5.3}$$

irrespective of the nature of the unit cell.

The area of the polygon defined by a pair of vectors is given by the magnitude of the cross-product, i.e., $|\mathbf{a} \wedge \mathbf{b}|$, $|\mathbf{b} \wedge \mathbf{c}|$, and $|\mathbf{c} \wedge \mathbf{a}|$. The cross-product leads to a vector which is normal to the plane enclosed by the pair of vectors.

The volume is given by the area of the base multiplied by the height. When the base is considered to be defined by **a** and **b**, the area is $|\mathbf{a} \wedge \mathbf{b}|$ and the height of the cell normal to the **a**-**b** plane is

$$\mathbf{c}.\frac{\mathbf{a} \wedge \mathbf{b}}{|\mathbf{a} \wedge \mathbf{b}|} \tag{5.4}$$

If follows that the volume of the cell is

$$\mathbf{c}.\frac{\mathbf{a} \wedge \mathbf{b}}{|\mathbf{a} \wedge \mathbf{b}|} \times |\mathbf{a} \wedge \mathbf{b}| = \mathbf{c}.\mathbf{a} \wedge \mathbf{b} \tag{5.5}$$

Every vector in a reciprocal lattice defines a normal to a plane of the same indices as the components of the reciprocal lattice vector, and the vector has a magnitude which is the reciprocal of the spacing of those planes. Therefore, for each of the basis vectors,

$$\mathbf{c}^* = \frac{\mathbf{a} \wedge \mathbf{b}}{\mathbf{c}.\mathbf{a} \wedge \mathbf{b}} \qquad \text{etc.} \tag{5.6}$$

With this equation it is evident that $\mathbf{a}_i.\mathbf{a}_j^* = 0$ when $i \neq j$ and $\mathbf{a}_i.\mathbf{a}^*{}_j = 1$ when $i = j$. The angle between **a** and $\mathbf{b}^*$ is clearly 90°. Therefore, if a direction $[u\ v\ w]$ lies in a plane $(h\ k\ l)$,

$$[u\mathbf{a} + v\mathbf{b} + w\mathbf{c}].(h\mathbf{a}^* + k\mathbf{b}^* + l\mathbf{c}^*) = 0 = uh + vk + wl \tag{5.7}$$

Note that this is a generic proof for the Weiss zone rule since there are no restrictions placed on the basis vectors.

A corollary to this example is a general method for taking a scalar product between two vectors. The magnitude of a vector **u** is given by

$$\begin{aligned} |\mathbf{u}|^2 &= (u_1\mathbf{a_1} + u_2\mathbf{a_2} + u_3\mathbf{a_3}).(u_1^*\mathbf{a_1^*} + u_2^*\mathbf{a_2^*} + u_3^*\mathbf{a_3^*}) \\ &= u_1u_1 * + u_2u_2^* + u_3u_3^* \equiv (\mathbf{u}; \mathrm{A})[\mathrm{A}^*; \mathbf{u}] \end{aligned} \tag{5.8}$$

where $(\mathbf{u}; \mathrm{A})$ is a row matrix of the components u_i of **u** referred to the basis vectors $\mathbf{a}_i$ represented by the basis symbol "A", and $[\mathrm{A}^*; \mathbf{u}]$ is a column matrix containing the components u_i^* of **u** referred to the reciprocal basis vectors $\mathbf{a}_i^*$ represented by the basis symbol "A*". This is an important result that gives a different interpretation to the scalar (or "dot") product between any two vectors referred to any basis. To evaluate the dot product, one of the vectors must be referred to the real basis and the other to the corresponding reciprocal basis.

5.2 Crystallography of diffraction

Consider waves of length λ incident on planes of atoms. The beams reflected from *different* planes in the parallel set illustrated in Figure 5.3 must be in phase to avoid destructive interference. The path difference between beams a and b, i.e., the distance xyz, must then be an integral number of wavelengths. Since $xyz = 2d \sin\theta$, the diffraction condition is

$$n\lambda = 2d \sin\theta \tag{5.9}$$

where n is an integer; this equation is the Bragg law [1], with θ designated the Bragg angle.

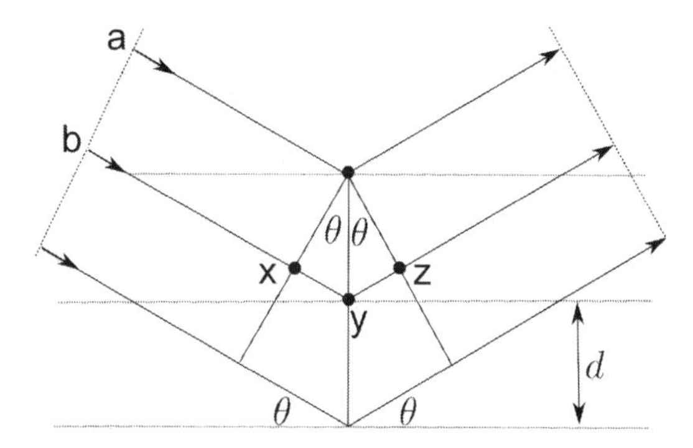

FIGURE 5.3
Electromagnetic waves incident on a set of parallel crystal planes with an interplanar spacing d. The angle of emergence of the scattered waves is the same as that of incidence.

The Bragg law can be expressed conveniently in terms of vectors in reciprocal space. In Figure 5.4, the Bragg law is satisfied if

$$\mathbf{k}' - \mathbf{k} = \mathbf{g} \tag{5.10}$$

Here **g** is a reciprocal lattice vector beginning at the origin (0,0,0) and representing the crystal which is being illuminated by the radiation **k**. From the geometry of the triangle, it is evident that $\sin\theta = \frac{1}{2}g/k$ so that $\sin\theta = (0.5/d)/(1/\lambda)$ which on rearrangement gives the Bragg law.

5.3 Intensities

The electrons from an individual atom can coherently scatter an X-ray beam; the atomic scattering factor f is the ratio of the amplitude scattered by an atom to that scattered by one electron. We are interested in coherent scattering by all the atoms in a unit cell, in which case the resultant amplitude for reflections from hkl planes is

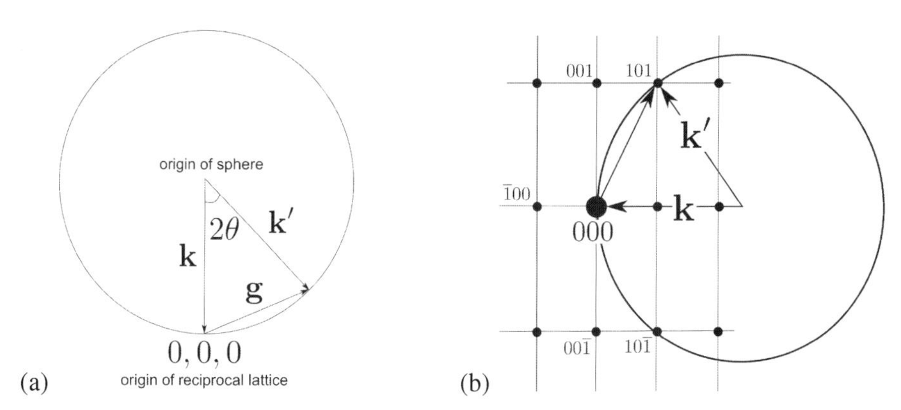

FIGURE 5.4
(a) The Ewald sphere construction in reciprocal space. **k** represents the direction of the incident beam, **k**′ that of the diffracted beam, and **g** the normal to the diffracting planes. Both **k** and **k**′ have a magnitude λ^{-1} and **g** has a magnitude d^{-1}. The beam is incident on a crystal which is represented by its reciprocal lattice with origin at 0,0,0. (b) Another illustration of the Ewald sphere superimposed on the reciprocal lattice. In this case only the 101 and $10\overline{1}$ planes are in Bragg condition. The reciprocal lattice vectors which do not touch the sphere are not in Bragg orientation.

given by a summation known as the *structure factor*:

$$
\begin{aligned}
F_{hkl} &= \sum_{1}^{n\ atoms} f_n \exp\{2\pi i(hu_n + kv_n + lw_n)\} \qquad (5.11)\\
&\equiv \sum_{1}^{n\ atoms} f_n[\cos 2\pi(hu_n + kv_n + lw_n) + i \sin 2\pi(h_{u_n} + kv_n + lw_n)]
\end{aligned}
$$

where f_n is the scattering factor of atom n and the sum is over all atoms in the unit cell.[1] The magnitude $|F|$ is now the ratio of the amplitude scattered by a unit cell to that scattered by one electron. $|F|^2$ is proportional to the scattered intensity. Notice that for a centrosymmetric system, the sine term can be neglected since that function is not symmetric about zero. As an example, for the structure of Cu (cubic close-packed, with four atoms in the cell at 000, $\frac{1}{2}\frac{1}{2}0$, $\frac{1}{2}0\frac{1}{2}$, and $0\frac{1}{2}\frac{1}{2}$),

$$F = f[1 + e^{\pi i(h+k)} + e^{\pi i(h+l)} + e^{\pi i(k+l)}] \qquad (5.12)$$

so that the $\{100\}$ and $\{110\}$ reflections would have zero structure factor.

Figure 5.5 shows how the electron diffraction patterns of austenite, which has a cubic-F lattice, have zero intensities for $\{100\}$ and $\{110\}$ planes, as expected by substitution of hkl into Equation 5.12.

[1] $e^{\pi i} = e^{3\pi i} = -1, \quad e^{2\pi i} = e^{4\pi i} = +1, \quad e^{\pi i/2} = i, \quad e^{3\pi i/2} = -i, \quad (1+i)(1-i) = 2$

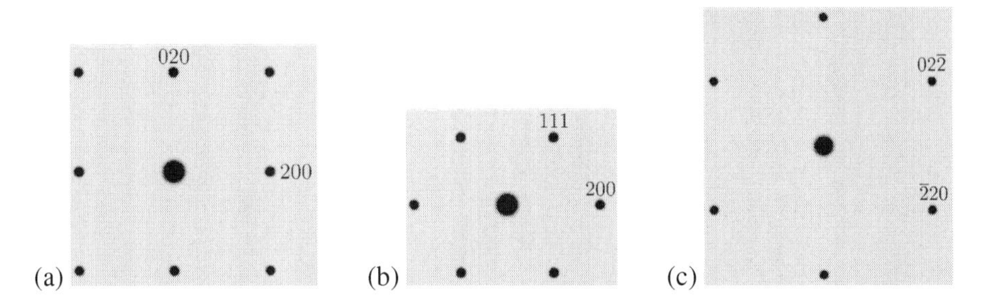

FIGURE 5.5
Electron diffraction patterns from austenite (cubic-F). (a) Zone axis parallel to [001]. (b) Zone axis parallel to $[0\bar{1}1]$. (c) Zone axis parallel to [111]. Notice that in any pattern, it is necessary to index just two reciprocal lattice vectors in order for all the others to be determined. For the cases illustrated, the symmetry of the patterns is consistent with that of the zone axes. Contrast these patterns with the reciprocal lattice sections presented in Figure 5.2.

Example 5.2: Solution of electron diffraction pattern
Figure 5.6a shows an electron diffraction pattern from a body-centered cubic lattice with a motif of just one atom per lattice point. Index the labelled g-vectors.

The pattern is rectangular so one possibility is that the axis normal to the pattern corresponds to a $\langle 110 \rangle$, a diad. The g-vectors normal to this direction could include the normals to $\{001\}$, $\{1\bar{1}0\}$, $\{1\bar{1}2\}$ and $\{1\bar{1}n\}$ where n is an integer. The d-spacing of planes in a cubic system is given by

$$\frac{1}{d^2} = \frac{h^2 + k^2 + l^2}{a^2}$$

where a is the lattice parameter. So the d-spacing decreases in the order $\{001\}$, $\{1\bar{1}0\}$, $\{1\bar{1}2\}$, and $\{1\bar{1}n\}$. Figure 5.6b shows that the intensity diffracted from the $\{001\}$ planes would be zero given that there is an atom at the body-center which would scatter out of phase with the atoms on the cube faces. However, $\{002\}$ planes would scatter in phase. If it is assumed that the g-vector labelled "a" is a $\{1\bar{1}0\}$ reflection (largest d-spacing and hence closest to the incident beam in reciprocal space) and that at the reflection at "b" is due to the $\{002\}$ planes then ratios of the lengths of the "a" and "b" and the 90° angle between gives consistent indexing. Any other choice would not.

Once two of the reflections are labelled, all others are linear combinations of those two. For example "c" is given by $002 + 1\bar{1}0 = 1\bar{1}2$.

In this particular example, it was easy by examining the structure projection to show that $\{001\}$ reflections would be systematically absent. However, if the indices are of higher order, e.g., $\{532\}$, then it is easier to use the structure factor equation to decide whether these planes would result in diffracted intensity.

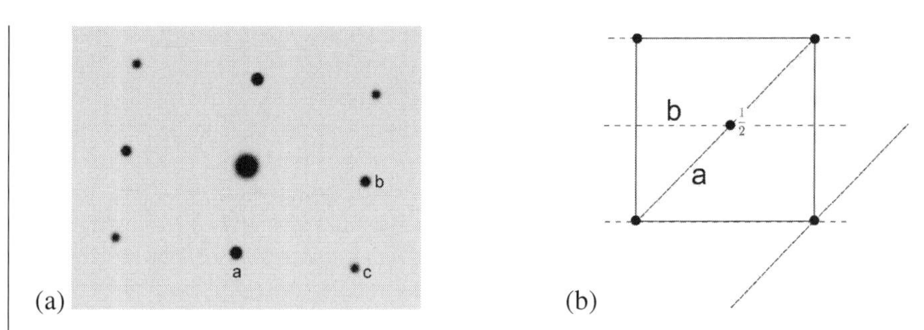

FIGURE 5.6
(a) An electron diffraction pattern from a body-centered cubic crystal. (b) Structure projection of bcc, with traces of $\{110\}$and $\{200\}$ planes indicated.

Example 5.3: Another diffraction pattern solution
The electron diffraction pattern in Figure 5.7 is from a cubic crystal. Index the pattern for all three types of cubic lattices.

The pattern has three-fold symmetry because there are three different g-vectors if their opposites are not counted. This indicates that there is a triad $\langle 111 \rangle$ normal to the pattern. All the g-vectors will be normal to the $\langle 111 \rangle$ axis and of equal length, so the following solutions can be deduced:

- For cubic-P, "a"$=\{01\bar{1}\}$ and "b"$=\{\bar{1}10\}$. The choice of the first solution as $\{\bar{1}10\}$ is arbitrarily made from the three possible, but the second will then be determined as being at 60° from the first. Their right-handed cross product gives $[111]$ so the beam direction is $[\overline{111}]$.
- For cubic-I, the solution would be the same as for cubic-P since $\{01\bar{1}\}$ type reflections are permitted.
- In the case of cubic-F, $\{01\bar{1}\}$ is a systematic absence but $\{02\bar{2}\}$ is not so all the indices should be doubled.

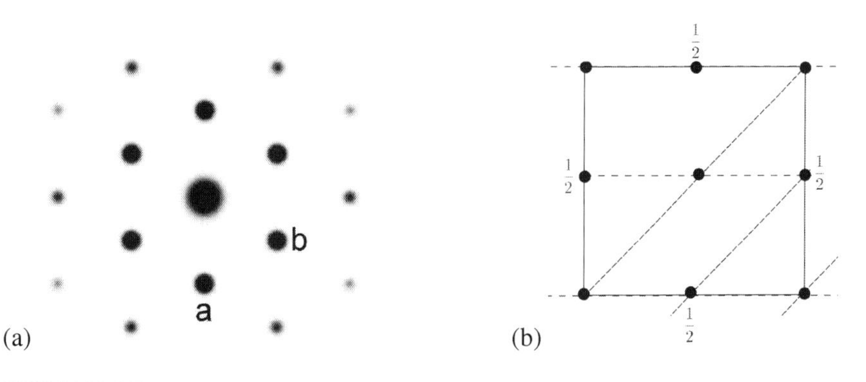

FIGURE 5.7

(a) An electron diffraction pattern from a crystal with a cubic structure. (b) Structure projection for cubic-F showing that $\{01\bar{1}\}$ would result in zero diffracted intensity because of the atoms in-between.

Notice that it is not possible, from the shape of the diffraction pattern alone, to determine the lattice type. However, if the lattice parameters of the different forms are known, then the measured d-spacings on the pattern can be compared with interplanar spacings calculated from the parameters to fix the lattice type. The relationship between the distance measured on the diffraction pattern and the camera length of the electron microscope is illustrated in Figure 5.8. It is evident that $\sin\theta = R/2L$ so with the Bragg Equation 5.9, it is easy to show that $Rd = L\lambda$, with the term $L\lambda$ referred to as the "camera constant".

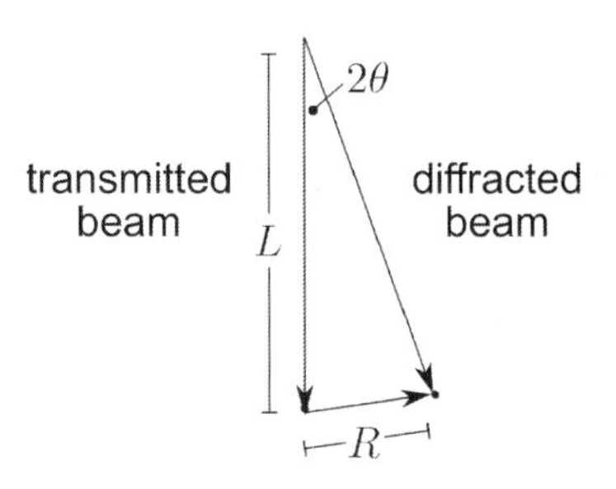

FIGURE 5.8
Geometry describing the relationship between the camera length on the transmission electron microscope and distance measured on an electron diffraction pattern.

Example 5.4: Disordered and ordered crystals

Austenitic Fe-Al-C alloys containing about 25 at% Al are heat-treated to precipitate the δ-phase within the aluminium-rich solid solution (γ). The δ-phase is an ordered compound with the chemical composition Al_3Fe. Both phases contain a total of four atoms per unit cell.

The γ has a cubic-F lattice; index its electron diffraction pattern. What is the shortest lattice vector? Judging from the symmetry of its diffraction pattern, indicate the possible lattice types for the δ phase.

Now assume that the correct lattice type for δ is that which contains four triads and its lattice parameter is almost identical to that of the solid solution γ. Bearing in mind its chemical composition, draw a projection of the crystal structure of δ along a convenient axis, identify the lattice type and the shortest lattice vector along the $\langle 110\rangle$ direction. Index the diffraction pattern.

Their electron diffraction patterns are presented in Figure 5.9 in the correct relative orientation. Comment on the orientation relationship between the two phases and why dislocations in γ would need to move in pairs in order to propagate through the δ-phase. Obtain expressions for the structure factors for the 100 and 200 reflections for both phases.

Similar ordering of atoms is observed in a wide range of alloys, often involving atoms with similar atomic numbers. Discuss the relative merits of different diffraction techniques that could be used to characterize order in such systems.

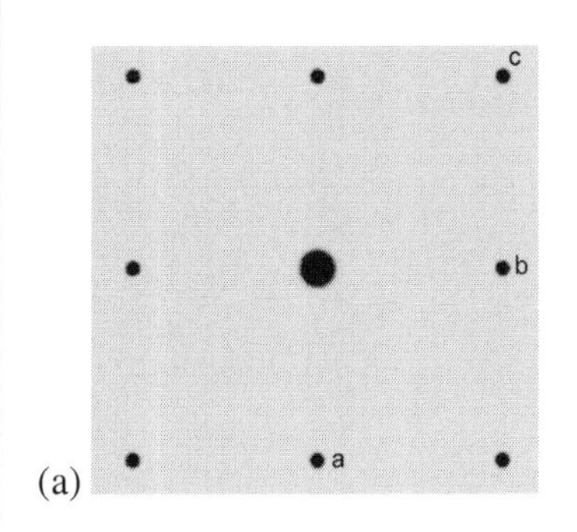

(a)

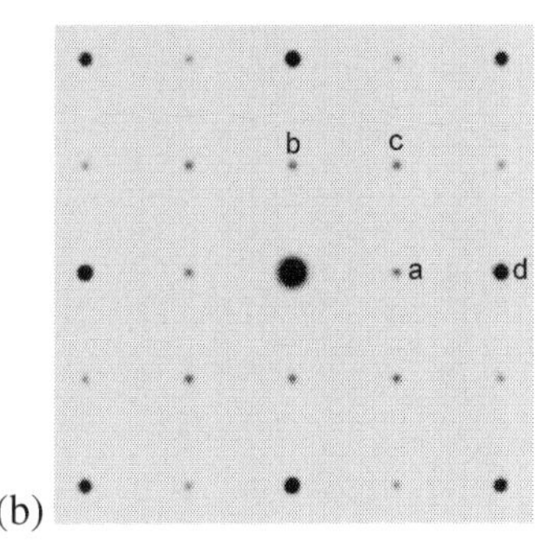

(b)

FIGURE 5.9
(a) Al-Fe solid solution (γ). (b) δ-phase. The relative orientations of the two patterns are preserved.

The γ is said to be cubic-F; since the pattern has four-fold symmetry, the direction normal to the pattern must be parallel to $\langle 001\rangle$ so the spots marked "a" and "b" can be labelled $\{020\}$ and $\{200\}$, respectively, with all other reflections obtained by linear combinations of these two. Therefore, "c" becomes $\{220\}$. The shortest lattice vector in cubic-F is $\frac{a}{2}\langle 110\rangle$. Note that in the cubic-F lattice, reflections of the type $\{100\}$ are systematic absences (Equation 5.12).

The pattern for the δ-phase also has four-fold symmetry and hence could come from a cubic or tetragonal lattice. However, it is stated that the lattice has four triads so it must be cubic-P, cubic-F, or cubic-I. The composition and number of atoms in the unit cell is consistent with cubic-P with a motif of four atoms per lattice point, iron at 000 and three aluminium atoms at $\frac{1}{2}\frac{1}{2}0$, $0\frac{1}{2}\frac{1}{2}$, and $\frac{1}{2}0\frac{1}{2}$. As a result, the spots labelled "a", "b", "c", and 'd' in Figure 5.9b are identified as $\{010\}$, $\{100\}$,$\{110\}$,$\{020\}$, respectively.

The structure projections of the two phases are shown in Figure 5.10, together with the shortest lattice vectors along the $\langle 110\rangle$ direction. The significance of this is that a dislocation with Burgers vector $\frac{a}{2}\langle 110\rangle$ in γ would create faults in the δ because it would not be a lattice vector in the ordered phase. However, a pair of γ dislocations moving together as a *superdislocation* in the δ would achieve slip without leaving a fault in their wake. The γ and δ are in a "cube-cube" orientation, i.e.,

$$[100]_\gamma \parallel [100]_\delta \quad \text{and} \quad [010]_\gamma \parallel [010]_\delta$$

and have similar lattice parameters, which would facilitate slip that initiates in the γ to traverse into δ with the caveat that lattice dislocations of γ would need to pair up when moving into δ.

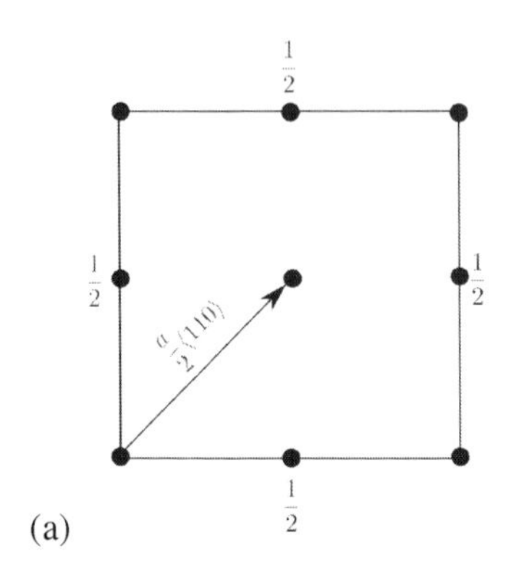

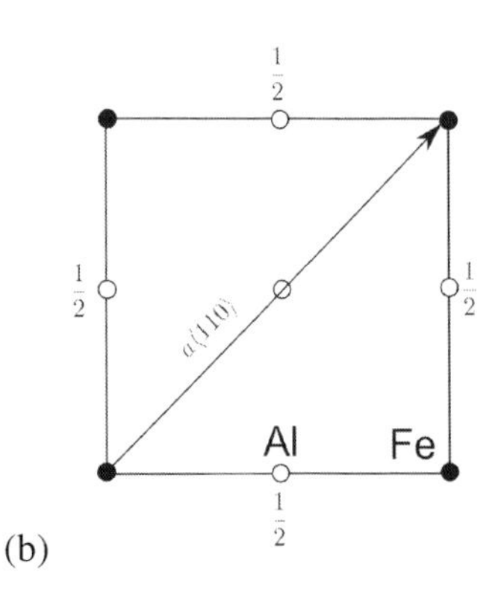

(a) (b)

FIGURE 5.10
(a) Structure projection of γ. (b) Structure projection of δ.

Using Equation 5.12 the required structure factors are as follows:

$$\begin{aligned}
F^{\gamma}_{100} &= \overline{f}[e^0 + e^0 + e^{\pi i} + e^{\pi i}] = 0 \\
F^{\gamma}_{200} &= \overline{f}[e^0 + e^0 + e^{2\pi i} + e^{2\pi i}] = 4\overline{f} \\
F^{\delta}_{100} &= f_{\mathrm{Fe}}e^0 + f_{\mathrm{Al}}e^{\pi i} + f_{\mathrm{Al}}e^0 + f_{\mathrm{Al}}e^{\pi i} = f_{\mathrm{Fe}} - f_{\mathrm{Al}} \\
F^{\delta}_{200} &= f_{\mathrm{Fe}}e^0 + f_{\mathrm{Al}}e^{2\pi i} + f_{\mathrm{Al}}e^0 + f_{\mathrm{Al}}e^{2\pi i} = f_{\mathrm{Fe}} + 3f_{\mathrm{Al}}
\end{aligned} \tag{5.13}$$

where $\overline{f}$ is an average atomic scattering factor for the random solid solution γ.

Since intensity is proportional to F^2, the $\{100\}_\delta$ *superlattice* reflection should be much weaker than $\{200\}_\delta$ as is evident in Figure 5.9b.

Techniques in which the atomic scattering factor scales with the atomic number include electron and X-ray diffraction. Reflections such as $\{100\}_\delta$ that are dependent on the differences in the atomic scattering factors of unlike species will therefore be weak. In neutron diffraction, the scattering factors can be quite different for atoms with similar atomic numbers so superlattice reflections can be detected more readily.

5.4 Diffraction from thin crystals

When an incident beam deviates by $\Delta\theta$ from the ideal Bragg angle θ consistent with Equation 5.9, it is possible to find another beam from within the depth of the crystal which has a path difference of $\frac{1}{2}\lambda$ and which will destructively interfere with it. This depth, needed to find the wave which is out of phase, increases as $\Delta\theta$ becomes smaller; hence for thin crystals, it may not be possible to destroy deviant rays. For thin crystals it becomes possible to obtain diffracted intensity even when the incident beam is not at the exact Bragg orientation.

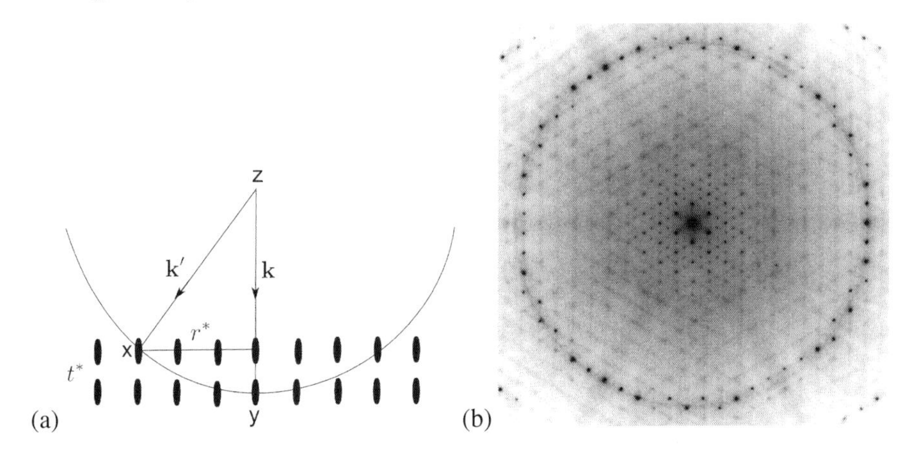

FIGURE 5.11
(a) Ewald sphere construction for diffraction from a thin foil. The origin of the reciprocal lattice is at "O" and the layer designated "P" is the next layer normal to the incident beam, a distance $1/t^*$ from the first layer. The arrows mark reciprocal lattice points which contribute to diffraction even though they are not at the exact Bragg orientation. (b) $\langle 111 \rangle$ axis ReO_3 showing the higher order Laue zone (the second layer of the type sketched in (a).) Pattern courtesy of A. Eggeman, T. Chang, and P. Midgley.

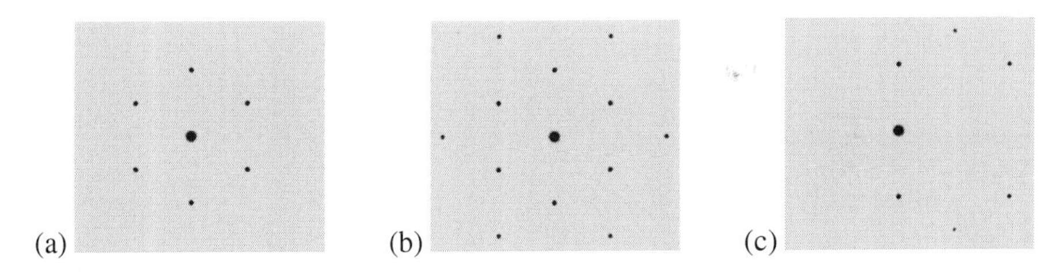

FIGURE 5.12
Fe with cubic-F structure. (a) Electron diffraction from a foil which is 200 nm thick, beam direction [111]. (b) Diffraction from the same foil after thinning it to just 50 nm thickness, beam direction [111]. Because of this minute thickness, it becomes possible to pick up reflections from planes of the form $\{224\}$. (c) as (a) but with the beam tilted toward [1 1 0.9]. The Ewald sphere is now tilted away from some of the reciprocal lattice points on the left-hand side.

The samples in transmission electron microscopy are constrained to be thin so that electrons can penetrate the typically < 100 nm thick foils. It follows that diffraction becomes possible even when the incident electron beam is not at the exact Bragg orientation. This can be expressed on the Ewald sphere construction by extending the reciprocal lattice points in the direction normal to the thin foil (Figure 5.11, 5.12, 5.13). The fact that electrons have a very small wavelength compared with X-rays

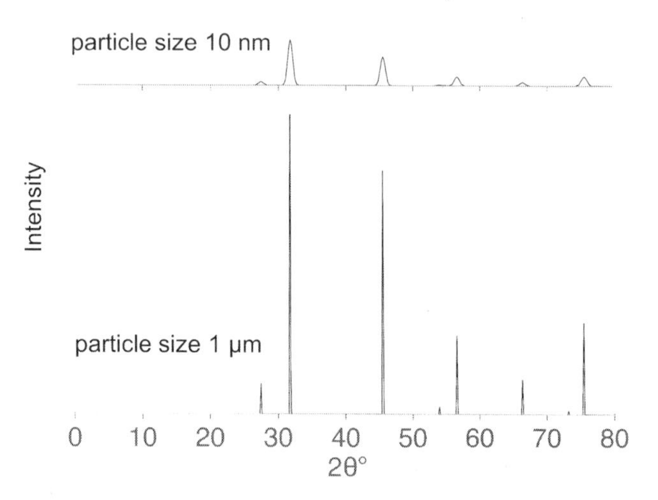

FIGURE 5.13
The effect of crystal size (in all directions) on an X-ray diffraction pattern. The material concerned is halite. The peaks broaden as the crystal gets smaller.

also helps achieve a greater chance of diffraction because the Ewald sphere is then relatively flat.

5.5 Neutron diffraction

Neutrons are scattered by the nuclei of atoms so the scattering factor is not sensitive to scattering angle; unlike X-rays, the scattering power does not diminish with the Bragg angle θ, so good results are obtained even at high θ. Also unlike X-rays, there is little correlation between the scattering factor and atomic number. Because the interactions are with the nuclei, it becomes possible to distinguish between isotopes because they now have different neutron scattering factors (for example, hydrogen and deuterium).

X-rays typically penetrate a few micrometers of the surface of a sample, whereas neutrons can penetrate several centimeters. Neutrons can therefore be used to determine the state of residual stress that exists in a macroscopic sample that is otherwise at equilibrium [2, 3]. Figure 5.14 shows residual stress contours that are determined by measuring lattice spacings, comparing them with a stress-free sample to calculate strains, and then converting strains into stresses using the elastic properties of the material.

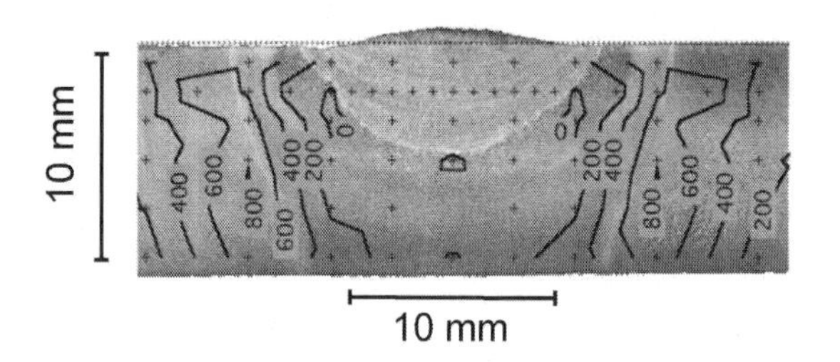

FIGURE 5.14
Stress distribution in a steel weld determined using neutron diffraction [4]. This would not be possible using X-rays. The numbers on the contours represent MPa.

5.6 Summary

It takes a leap of imagination to understand why it is necessary to deal with a space in which dimensions are inverted. But the fact is that in crystallography, it can be regarded simply as a lattice in which all reciprocal lattice vectors represent normals to planes in real space, with the magnitudes of these vectors representing the inverse of the spacing of those planes. Whereas a plane in real space is defined by a pair of vectors that lie in that plane, in reciprocal space the plane is represented by a vector. A plane in reciprocal space represents all real space planes that share a common direction (a zone axis).

Imagine now that we wish to find a direction that is parallel to a plane normal. This is trivial for a cubic system where plane normals and directions with the same indices are parallel. So a [123] direction is parallel to the normal of the (123) plane. This is not generally true for non-cubic systems, but there exists a metric tensor which makes it easy to find vectors in reciprocal space that are parallel to those in real space (and vice versa), a subject reserved for Chapter 9.

References

1. W. H. Bragg, and W. L. Bragg: The reflection of X-rays by crystals, *Proceedings fo the Royal Society A*, 1913, **88**, 428–438.

2. P. J. Withers, and H. K. D. H. Bhadeshia: Residual stress part 1 – measurement techniques, *Materials Science and Technology*, 2001, **17**, 355–365.

3. P. J. Withers, and H. K. D. H. Bhadeshia: Residual stress part 2 – nature and origins, *Materials Science and Technology*, 2001, **17**, 366–375.

4. J. A. Francis, H. J. Stone, S. Kundu, R. B. Rogge, H. K. D. H. Bhadeshia, P. J. Withers, and L. Karlsson: Transformation temperatures and welding residual stresses in ferritic steels, In: *Proceedings of PVP2007, ASME Pressure Vessels*

and Piping Division Conference. American Society of Mechanical Engineers, San Antonio, Texas: ASME, 2007:1–8.

6

Deformation and Texture

Abstract

Single crystals are now used routinely in engineering turbine blades for service in the hottest part of an aeroengine; it is astonishing that the design of the blade permits it to operate in an environment where the temperature is greater than its melting point. The deformation of a single crystal on the application of a stress or a system of stresses is interesting in its own right, but the understanding that emerges from single crystal studies can useful in explaining the rotation of crystals in a polycrystalline material during plastic deformation. Real polycrystalline materials rarely consist of crystals that are randomly oriented relative to the sample frame of reference. Instead, the long-range order in a polycrystalline material is somewhere between that of a hypothetical aggregate consisting of randomly oriented crystals, and a single crystal. This is because the crystallographic axes of the different grains in the polycrystal tend to align in specific ways. We shall learn in this chapter the methods for representing such *texture*.

6.1 Slip in a single-crystal

In a tensile test on the sample of cross-sectional area A as illustrated in Figure 6.1a, the normal to the slip plane lies at an angle ϕ to the force F, and the slip direction is at an angle λ to the tensile axis. The area of the slip plane is $A/\cos\phi$ and the force in the slip direction is $F\cos\lambda$ so that the shear stress on the slip plane in the slip direction is

$$\tau = \frac{F}{A} \underbrace{\cos\phi\cos\lambda}_{\text{Schmid factor}} \quad . \tag{6.1}$$

The Schmid factor [1–3] has a maximum value of 0.5 when $\phi = \lambda = 45°$. Suppose that a tensile axis is along [312] in a cubic system, as plotted on the shaded stereographic triangle in Figure 6.1b. There is a simple way of determining which slip system will have the largest resolved shear stress. The slip system will be defined by a slip plane and a slip direction with the latter lying in the former. This means that the two components of the slip system must come from different stereographic triangles. The two stereographic triangles that are adjacent to the shaded triangle satisfy this condition. Any other combination gives angles that are too large, giving Schmid factors that are smaller. The slip system with the highest resolved shear stress when

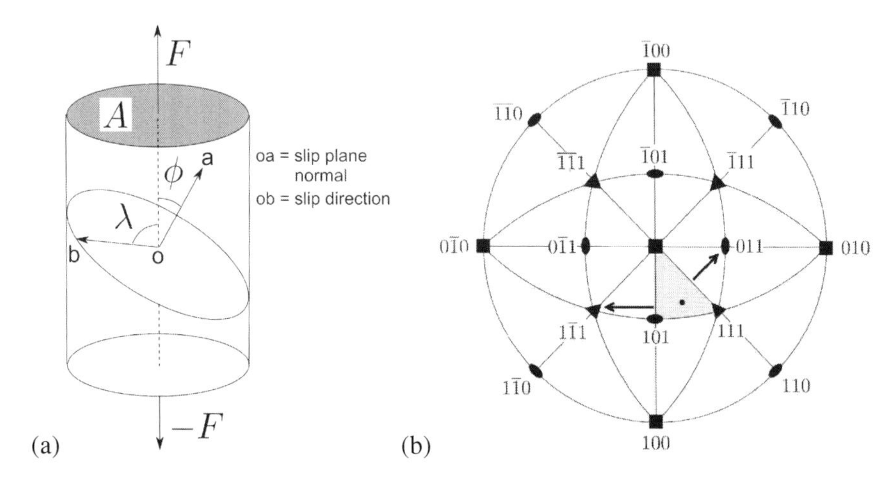

FIGURE 6.1
(a) Deformation of a single crystal. The force is resolved onto the slip plane and toward the slip direction. Dividing by the area of the slip plane gives the shear stress along the slip direction. (b) There are 24 stereographic triangles corresponding to the 24 slip systems of the type $\{111\}\langle 1\bar{1}0\rangle$. The orientation of a tensile axis is plotted in the shaded triangle.

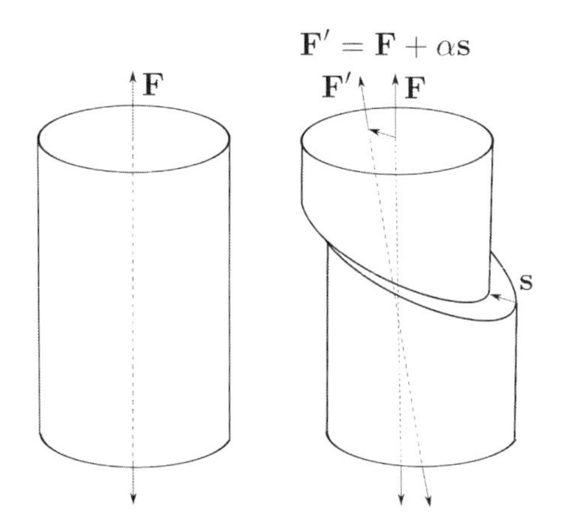

FIGURE 6.2
Slip of an unconstrained single-crystal. Notice how the axis along which the force is applied rotates as a consequence of the slip. The amount of shear implemented is written $\alpha\mathbf{s}$. If the force axis is constrained to be vertical then the sample will rotate relative to the testing machine by the extent illustrated.

the tensile lies in the shaded triangle will therefore be $[011](1\bar{1}1)$. The construction illustrated is known as Diehl's rule.

The unconstrained shear deformation of a single crystal as illustrated in Figure 6.2 leads to the rotation of the axis along which the force is applied. The normal to the slip plane, the tensile axis, and the slip direction are all in the same plane and remain so during rotation. The tensile axis rotates toward the slip direction, on this common plane, during the course of deformation. If the axis along which the force is applied is not allowed to change orientation, then it is the slip system which rotates during deformation, Figure 6.2.

Example 6.1: Elongation during single-crystal deformation
Referring to Figure 6.3, derive an equation for the elongation of a single crystal due to shear on a plane with unit normal **n**.

The definition of a shear strain γ is

$$\gamma = \frac{\text{displacement parallel to shear plane}}{\text{height over which displacement acts}} = \frac{|\mathbf{d}|}{\mathbf{x}.\mathbf{n}} \quad \text{so the magnitude of } \mathbf{d} = \gamma \mathbf{x}.\mathbf{n}.$$

It follows therefore that

$$\mathbf{y} = \mathbf{x} + \mathbf{d} = \mathbf{x} + \gamma(\mathbf{x}.\mathbf{n})\mathbf{s} \qquad \text{so the elongation is } = \frac{|\mathbf{y}|}{|\mathbf{x}|}.$$

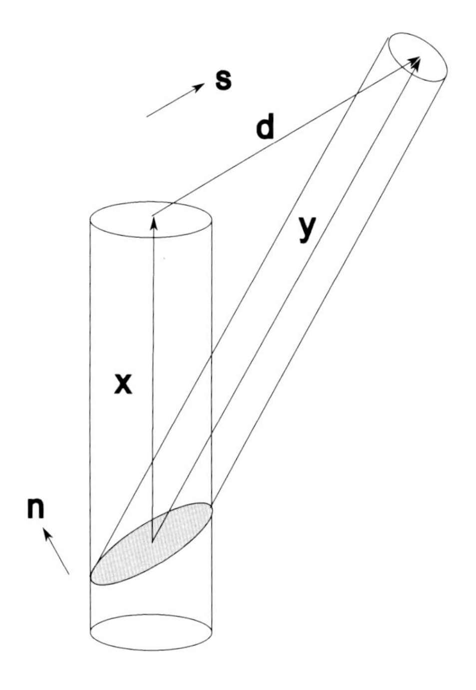

FIGURE 6.3
Schematic diagram showing the shear of a single crystal. **n** is a unit normal vector to the plane on which the shear occurs, **d** is a vector describing the displacement of the far end of the crystal due to the shear, **x** is its initial length above the slip plane, and **y** the corresponding length following the deformation. **s** is a unit vector parallel to the shear direction.

A more general method for calculating the elongation during single-crystal deformation is described on page 157.

Example 6.2: ε martensite

A simple shear can transform the cubic-F structure of austenite into that of the hexagonal close-packed ε martensite. This consists of a displacement of $\frac{a}{6}\langle 112\rangle$ on every second close-packed plane. Calculate the shear strain associated with the transformation. (a is the lattice parameter of austenite.)

How many different crystallographic variants of ε-martensite can in principle form within a single austenite grain?

Figure 6.4 shows a stereographic projection of austenite showing the ε habit planes of the form $\{111\}$ and shear directions of the form $\langle 112\rangle$. Which is the most favored crystallographic variant that is stimulated to grow by applying a tensile stress along the direction [123]?

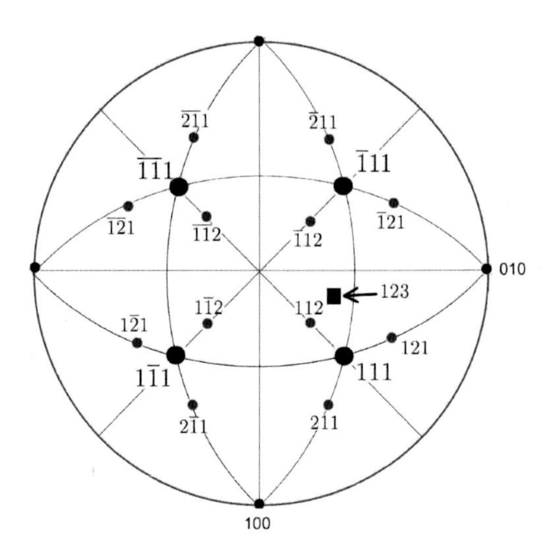

FIGURE 6.4
Stereographic projection of austenite showing the planes and shear directions corresponding to the deformation that leads to the formation of ε martensite.

The shear strain is given by the displacement divided by the height. The height is the d-spacing of two close-packed planes of austenite, i.e., $\frac{2a}{\sqrt{3}}$. The displacement is the magnitude of $\frac{a}{6}\langle 112\rangle$, i.e., $\frac{a}{\sqrt{6}}$. Therefore, the shear strain is $\frac{1}{2\sqrt{2}}$.

The habit plane is $\{111\}$ (four variants) and there are three directions of the form $\langle 11\bar{2}\rangle$ in each of these planes. Therefore, 12 variants in all.

The strain is a shear deformation on a close-packed plane, but with the shear direction along $\langle 11\bar{2}\rangle$. Therefore, either using Diehl's rule, or by calculation, the plane stimulated would be $(\bar{1}11)$. In that plane, there are three shear directions,

$[211]$, $[1\bar{1}2]$, and $[\bar{2}\bar{1}1]$. Of these, the [211] direction gives the largest Schmid factor.

6.2 Texture

Imagine that there is a polycrystalline substance in which the individual crystals are randomly oriented relative to an *external frame of reference*.[1] Suppose that this polycrystalline substance is subjected to rolling deformation. The individual crystals will then tend to rotate such that the slip planes comply with the external forces in a manner akin to that illustrated in Figure 6.2b. This means that the original random distribution of crystals in the undeformed sample becomes non-random. The deformed material is said to become *crystallographically textured* [4].

Most polycrystalline materials show texture due to processing. Texture can arise during solidification when those crystals which have a fast growth direction parallel to that of the heat flow will dominate the final structure. It can arise during recrystallization and phase transformation when selective nucleation leads to the formation of a biased distribution of crystals.

A convenient method for communicating texture caused by deformation, for example, in rolled sheet, is by stating the set $\{h\ k\ l\}\langle u\ v\ w\rangle$, the planes which lie roughly parallel to the rolling plane, and direction in the rolling plane that tends to be parallel to the rolling direction. The overall texture can be represented as the sum of components:

$$\text{texture} = \sum_i \lambda_i \{h\ k\ l\}_i \langle u\ v\ w\rangle_i \tag{6.2}$$

where λ represents the weighting given to a particular type of texture. The major components of the deformation texture of austenite are $\{1\ 1\ 0\}\langle 1\ \bar{1}\ 2\rangle$ and $\{1\ 1\ 2\}\langle 1\ 1\ \bar{1}\rangle$, the so-called *brass* and *copper* textures, respectively. Recrystallization in a ferritic steel often leads to a *cube* texture $\{1\ 0\ 0\}\langle 0\ 0\ 1\rangle$ *cube* component and special thermomechanical processing of the type used to produce magnetically soft metal is associated with the *Goss texture* $\{1\ 1\ 0\}\langle 0\ 0\ 1\rangle$.

Texture can be plotted on a stereogram. A *pole figure* in this context consists of a stereogram with its axes defined relative to the external frame of reference, and with particular hkl poles from each of the crystals in the polycrystalline aggregate plotted on to it. If the distribution of crystals is random, the pole figure would appear as in Figure 6.5a, and if the distribution is non-random then this would be apparent in the pole figure, as shown in Figure 6.5b.

An *inverse pole figure* is one in which the sample frame is plotted relative to the crystal axes of a reference crystal. Figure 6.6 illustrates an example where the rolling, normal, and transverse directions (RD, ND, and TD, respectively) are plotted relative

[1] This frame may, for example, consist of a set of vectors parallel to the rolling direction, transverse direction, and thickness of a rolled plate; alternatively, the principal axes of an applied system of stresses.

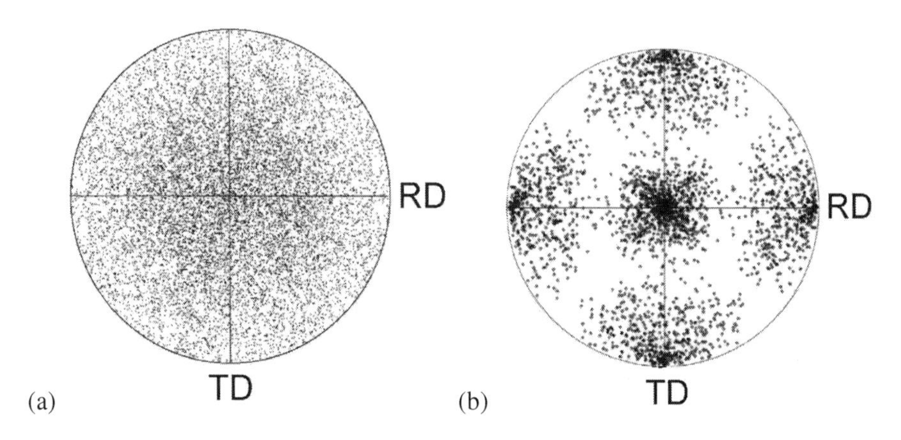

FIGURE 6.5
(a) Figure showing a random distribution of 100 poles from a polycrystalline sample containing 500 crystals, plotted relative to the rolling direction (RD) and transverse direction (TD). Notice that the distribution of poles does not appear random but this is an artifact due to the angular distortion on a stereographic projection. (b) Corresponding pole figure after texture is introduced by phase transformation. Here there is a strong tendency for the 100 poles to cluster around the sample axes. The distribution of poles can be plotted as contours rather than the dots illustrated. Diagrams courtesy of Saurabh Kundu.

to the crystal axes as defined by the stereographic triangle. In a cubic system, the 24 stereographic triangles with $\langle 100 \rangle$, $\langle 101 \rangle$, and $\langle 111 \rangle$ at the corner of each triangle, are crystallographically equivalent. Therefore, the orientation of the rolling direction, for example, from each of the 24 triangles can be plotted on just one of the stereographic triangles in the form of poles, or contour plots representing the design of poles in angular space.

Crystallographic texture can be measured using a variety of techniques such as X-ray diffraction or electron back-scattered diffraction (EBSD) [5]. In the former case, the polycrystalline sample is exposed in a system set to detect X-rays from a particular reflection (Bragg angle) and the sample is systematically tilted and rotated in order to capture intensity at a variety of orientations. In EBSD, an electron beam is rocked about a fixed point on the surface of a grain using a scanning electron microscope. At particular angles, the beam is Bragg diffracted, resulting in a reduction in intensity picked up by the detector, leading to the formation of channelling patterns of the type illustrated in Figure 6.7. These patterns can be interpreted to determine the orientation of the crystal relative to the sample axes.

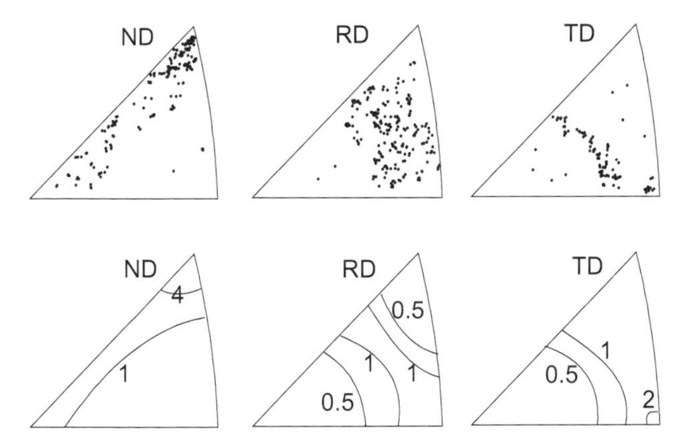

FIGURE 6.6
Inverse pole figures for a cubic crystal. The diagrams at the top have poles corresponding to the rolling, normal, and transverse directions plotted relative to the crystal axes, whereas those at the bottom have the same data represented as contours representing the density of poles.

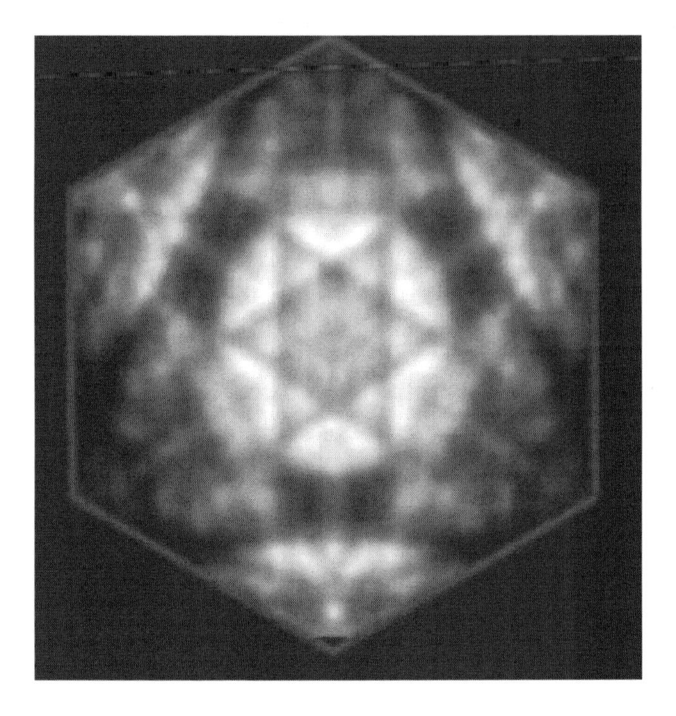

FIGURE 6.7
A simulated electron channelling pattern from silicon carbide. Reproduced from Winkelmann et al. [6], with permission of Elsevier.

6.3 Orientation distribution functions

A pole figure does not give information about the orientation of a particular crystal relative to another, since the poles are not, in the final plot, identified with particular crystals. An orientation distribution function is a more rigorous method of expressing the orientations of crystals relative to the frame of reference. It uses three Euler angles (φ_1, Φ, and φ_2) which define the relative orientation of the crystal and sample frames of reference, as illustrated in Figure 6.8.

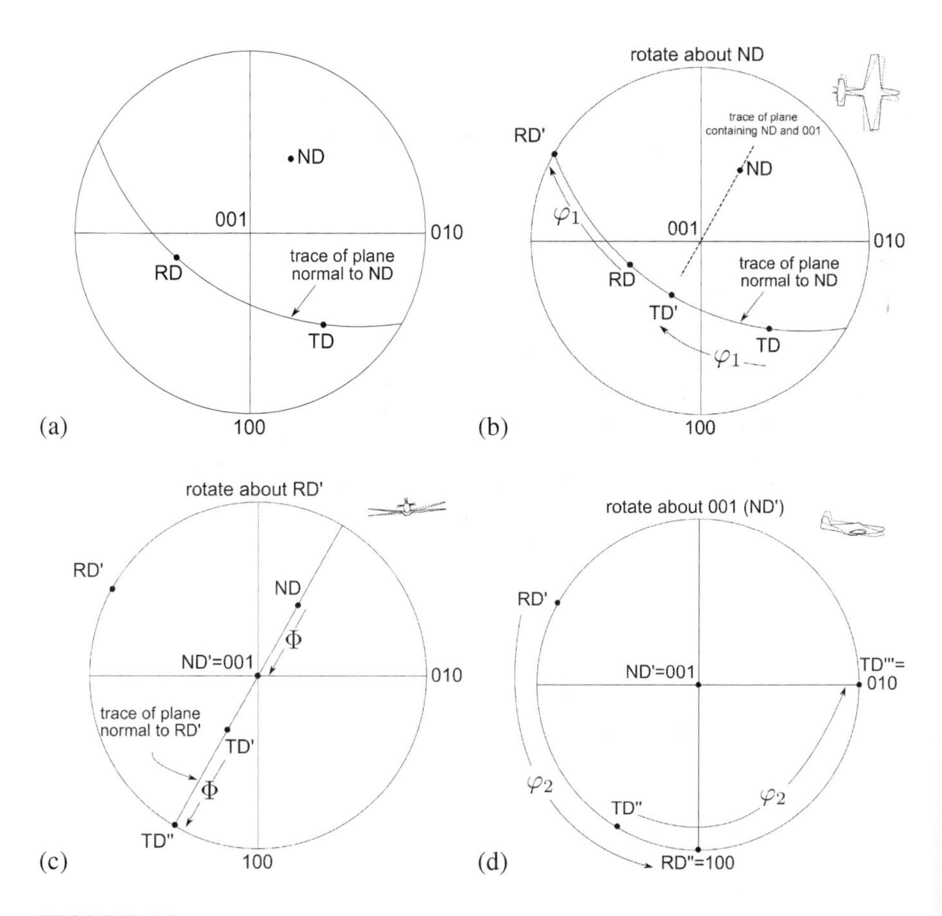

FIGURE 6.8
Operations that define the three (Bunge) Euler angles. (a) Initial relationship between crystal and sample coordinates. (b) Rotation by φ_1 about ND to generate RD' in a position normal to the plane containing ND and 001. (c) Rotate by Φ about RD' to generate ND' parallel to 001. (d) Rotate by φ_2 about 001 to bring crystal and sample axes into coincidence. The insets show the rotations relative to an aircraft.

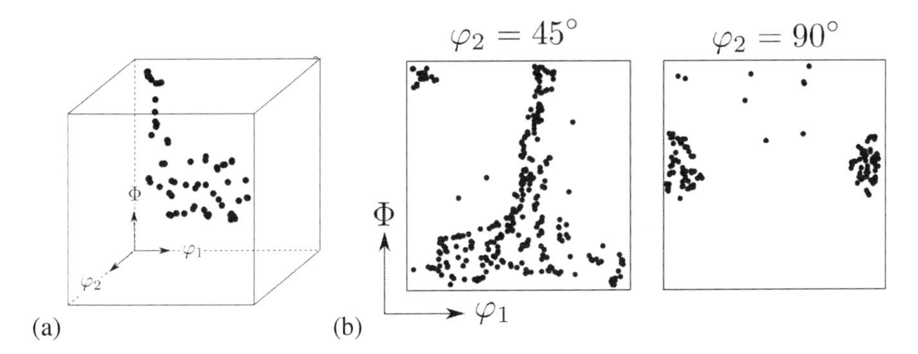

FIGURE 6.9
(a) An orientation distribution plotted on a cube with axes defined by the Euler angles. Each dot represents a crystal orientation relative to sample axes. (b) Sections of the cube to show how the density of orientations varies.

The set of measured Euler angles made by the RD, TD, and ND directions for a particular crystal is plotted as a point in a cube whose axes scale with φ_1, Φ, and φ_2. Each point inside this cube is a single crystal.

There are very many industrial processes which control and exploit texture on a grand scale. Beverage cans are incredibly thin even though their bodies are made without fracture using extreme deformation including deep drawing. The texture of the material prior to drawing is controlled such that plastic instability is avoided in the thickness direction. For similar reasons, the steels used in the manufacture of formed automobile bodies are texture-controlled. Iron alloys used to make electricity transformers have to be magnetically soft. This is achieved by controlling the crystallographic texture to minimize magnetic losses.

Example 6.3: Euler angles relating two frames

By comparison with the operations illustrated in Figure 6.8a, deduce the set φ_1, Φ, and φ_2 relating the bases A and B in Figure 6.10. Represent the relative orientation of the two frames on a section of an orientation distribution function.

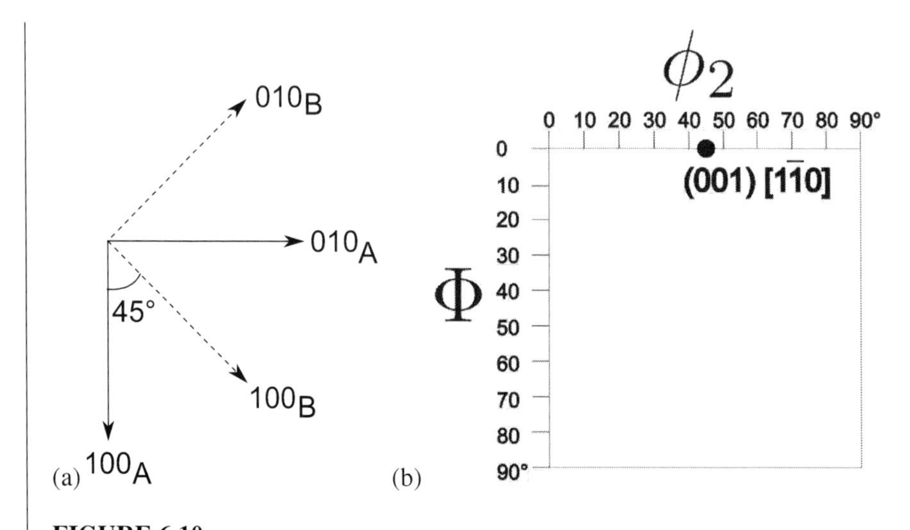

FIGURE 6.10
(a) Note that $[001]_A \parallel [001]_B$. (b) Corresponding ODF with $\phi_1 = 0$.

The Euler angles define the operations needed to bring the two frames of reference into coincidence. Referring to Figure 6.8b, since $[001]_A \parallel [001]_B$, $\varphi_1 = 0°$. For the same reason, $\Phi = 0°$ (Figure 6.8c) and the remaining angle φ_2 is clearly 45°. Figure 6.10b shows the ODF section $\phi_1 = 0$, referred to the basis-B as the reference frame.

6.4 Summary

Crystallographic texture can be understood by a thought experiment in which a large single crystal is broken up into many smaller ones which are then joined together in somewhat different orientations that are not random. The material then behaves as if it has some of the characteristics of a single crystal (such as anisotropy) while other properties are more a feature of random polycrystals (such as the multiple slip systems needed per crystal in order to achieve plasticity without the creation of voids).

We have discussed how texture can arise from plastic deformation, but many kinds of processing can lead to its development. The heat treatment of deformed materials to induce recrystallization can lead to recrystallization textures because not all "nuclei" grow at the same rate [7]. The nuclei have to be sufficiently different in orientation from their surroundings to accomplish rapid growth. Similar selection can occur in phase transformations where the product phase forms at a grain boundary because certain boundaries make it possible for the product to adopt a favorable orientation relationship with the parent phase on either side of the interface [8, 9].

Both magnetic [10] and electrical fields [11] can also lead to orientational order in appropriate materials.

So how does one produce a polycrystalline material with a random arrangement of grains? The short answer is, with enormous difficulty. Powder techniques can partially reach this goal; metallic or ceramic powders that are sintered together tend to undergo very limited deformation or rotation during fabrication, and hence may approach a random orientational distribution.

References

1. E. Schmid, and W. Boas: *Plasticity of crystals* (translated from the 1935 edition of Kristalplastizitaet): London, U.K.: F. A. Hughes and Co., 1950.

2. J. W. Christian: Deformation by moving interfaces, *Metallurgical Transactions A*, 1982, **13**, 509–538.

3. A. Kelly, and K. M. Knowles: *Crystallography and crystal defects*: 2nd ed., New York: John Wiley & Sons, Inc., 2012.

4. D. N. Lee: *Texture and related phenomena*: Seoul, Republic of Korea: Korean Institute of Metals and Materials, 2006.

5. A. F. Gourgues-Lorenzon: Application of electron backscatter diffraction to the study of phase transformations, *International Materials Reviews*, 2007, **52**, 65–128.

6. A. Winkelmann, B. Schröter, and W. Richter: Dynamical simulations of zone axis electron channelling patterns of cubic silicon carbide, *Ultramicroscopy*, 2003, **98**, 1–7.

7. W. B. Hutchinson: Recrystallisation textures in iron resulting from nucleation at grain boundaries, *Acta Metallurgica*, 1989, **37**, 1047–1056.

8. O. Hashimoto, S. Satoh, and T. Tanaka: $\alpha \rightarrow \gamma \rightarrow \alpha$ transformation behavior during heating from the $\alpha + \gamma$ region in 0.18%C steel, *Tetsu-to-Hagané*, 1980, **66**, 102–111.

9. H. K. D. H. Bhadeshia: Diffusional formation of ferrite in iron and its alloys, *Progress in Materials Science*, 1985, **29**, 321–386.

10. A. D. Sheikh-Ali, D. A. Molodov, and H. Garemstani: Magnetically induced texture development in zinc alloy sheet, *Scripta Materialia*, 2002, **46**, 857–862.

11. D. Coates, W. A. Crossland, J. H. Morrissy, and B. Needham: Electrically induced scattering textures in smectic A phases and their electrical reversal, *Journal of Applied Physics D*, 1978, **11**, 2025–2035.

7

Interfaces, Orientation Relationships

Abstract

An interface is a region where the long-range translational order of a crystal is disrupted because there is an abrupt change in orientation. It is in general a very narrow region within which the atoms may not have a strong sense of belonging to the crystalline material on either side. The density of the interfacial region is likely to be smaller than that of the neighboring crystals with consequences on diffusion, impurity concentrations, and cohesion. Nevertheless, we shall see that the structure of an interface can have elements of periodicity, including regularity in the arrangement of the defect structure within.

7.1 Introduction

Much of the science and technology of polycrystalline materials depends on the nature of interfaces between crystals and the relative orientations of adjacent crystals. The corrosion resistance of a boundary depends on how coherent it is [1], the continuity of deformation varies with texture [2] . . . the list is seemingly endless!

Atoms in the boundary between crystals must in general be displaced from positions they would occupy in the undisturbed crystal, but it is now well established that many interfaces have a periodic structure. In such cases, the misfit between the crystals connected by the boundary is not distributed uniformly over every element of the interface; it is localized periodically into discontinuities that separate patches of the boundary where the fit between the two crystals is good or perfect. When these discontinuities are well separated, they may individually be recognized as interface dislocations which separate coherent patches in the boundary, which is macroscopically said to be semi-coherent. The simple example of a symmetrical tilt boundary illustrates this.

7.2 Symmetrical tilt boundary

The structure of an interface can be understood by creating a boundary beginning with a single crystal that is sliced and the two halves then rotated relative to generate two crystals in different orientations. The rotation is illustrated in Figure 7.1 about an axis normal to the diagram through an angle θ. This leaves a gap between the bicrystal which can be filled with edge dislocations that, after all, have extra half-planes. The misorientation θ between the grains can therefore be described in terms of dislocations. Inserting an edge dislocation of Burgers vector **b** is like forcing a wedge into the lattice, so that each dislocation is associated with a small change in the orientation of the lattice on either side of the extra half plane. If the spacing of dislocations is d, then it follows from Figure 7.1b that

$$\tan\frac{\theta}{2} = \frac{b}{2d} \tag{7.1}$$

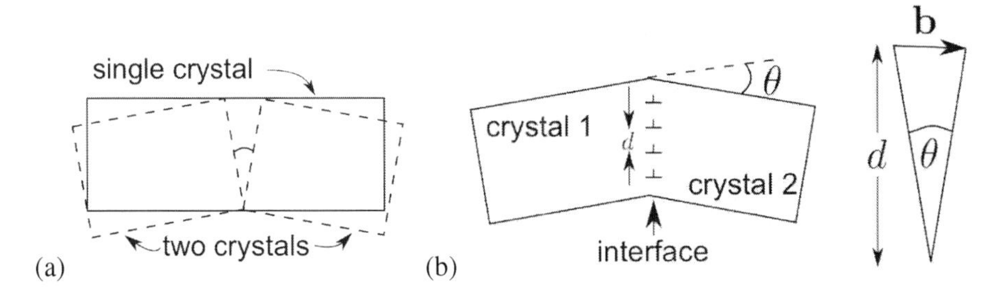

FIGURE 7.1
The creation of an interface. (a) Single crystal cut in two halves which are then tilted through an angle θ about an axis normal to the diagram. (b) The gap within the tilt is filled with edge dislocations, each with a Burgers vector **b** and spacing d to create an interface.

The structure of the interface in terms of dislocations provides a mechanism for calculating the energy per unit area of the interface as a function of the misorientation θ. The energy per unit length W_s of a dislocation is given by [3]

$$\begin{aligned} W_s &= W_c + \frac{1}{2}\int_{r_0}^{r_\infty} \frac{Gb^2}{2\pi(1-\nu)}\frac{\mathrm{d}r}{r} \\ &= W_c + \frac{Gb^2}{4\pi(1-\nu)}\ln\left\{\frac{r_\infty}{r_0}\right\} \end{aligned} \tag{7.2}$$

where W_c is the core energy per unit length, covering the region of the dislocation where elasticity theory fails; G and ν are the shear modulus and Poisson's ratio, respectively; r_0 is the radius of the dislocation core and r_∞ is a cut-off radius beyond which the dislocation strain field can be neglected or is limited by the size of the sample. In arrays of dislocations of the type found in interfaces, the strain fields of individual dislocations are partly compensated by their neighbors so it is a good approximation that $r_\infty \approx d$ in which case $r_\infty \propto 1/\theta$. It follows that Equation 7.2 can be written $W_s = A(B - \ln\theta)$. Therefore, the interfacial energy per unit area σ_i is related to the number of dislocations per unit area $(1/d)$:

$$\begin{aligned}\sigma_i &= \frac{1}{d} \times W_s \\ &= \frac{2}{b}\tan\frac{\theta}{2} \times A(B - \ln\theta).\end{aligned} \tag{7.3}$$

This gives a model for the energy of grain boundaries [4], based on the density of dislocations in the boundary ($\propto d^{-1}$). The variation in energy as a function of misorientation in the symmetrical tilt boundary described above is illustrated in Figure 7.2. The model is not valid for large misorientations where the dislocation spacing becomes comparable with the magnitude of the Burgers vector, since the dislocation cores then begin to overlap.

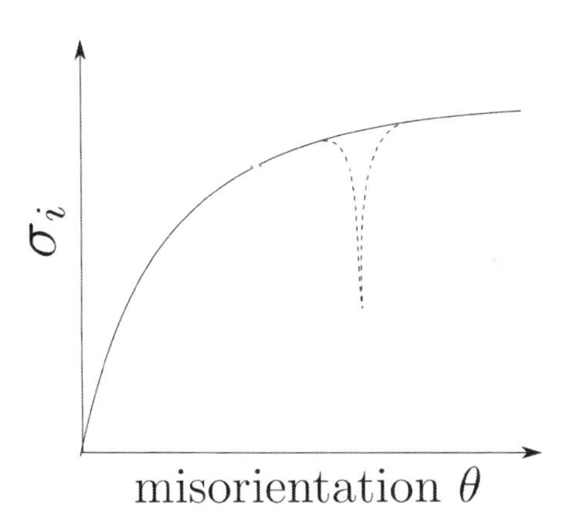

FIGURE 7.2
The interfacial energy as a function of the misorientation for a tilt boundary. The dashed curve represents cusps when the orientation forms a coincidence site lattice.

The symmetrical tilt boundary is an illustration of how interfaces can be described in terms of defect structures with coherent patches in-between the defects. More complex misorientations would require more than one array of dislocations to represent the fit between the crystals, and there will come a point where the defect density becomes so large that it is no longer reasonable to talk of a dislocation structure.

7.3 Coincidence site lattices

For high-misorientation boundaries, the predicted spacings of dislocations may turn out to be so small that the misfit is highly localized with respect to the boundary, and the dislocation model of the interface has only formal significance (it often is said that the dislocations get so close to each other that their cores overlap). The arrangement of atoms in such incoherent boundaries may be haphazard, with little correlation of atomic positions across the boundary.

On the other hand, it is unreasonable to assume that all high-angle boundaries have the disordered structure. There is clear experimental evidence which shows that certain high-angle boundaries exhibit the characteristics of low-energy coherent or semi-coherent interfaces; for example, they exhibit strong faceting, have low mobility in pure materials and the boundary diffusion coefficient may be abnormally small. These observations imply that there are special orientations, which would usually be classified as large mis-orientations, where it is possible to obtain boundaries that have a distinct structure. They contain regions of good fit, which occur at regular intervals in the boundary plane, giving a pattern of good fit points in the interface. It is along these points that the two crystals match exactly at their mutual interface.

The lattice points that are common to the two crystals can be identified by allowing the two lattices, with a common origin, notionally to interpenetrate and fill all space. The set of these coincidence points forms a coincidence site lattice (CSL), and the fraction of lattice points which are also coincidence sites is a rational fraction $1/\Sigma$. Σ is thus the reciprocal density of coincidence sites relative to ordinary lattice sites [5, 6]. The value of Σ is a function of the relative orientation of the two grains and not of the orientation of the boundary plane. The boundary simply intersects the CSL and will contain regions of good fit that have a periodicity of a planar net of the CSL along which the intersection occurs. Boundaries parallel to low-index planes of the CSL are two-dimensionally periodic with relatively small repeat cells. Those boundaries with a high planar coincidence site density may have a relatively low energy per unit area.

Figure 7.3 shows the projection on (100) of two primitive cubic lattices with a common origin, each with a unit lattice parameter, and an orientation relationship described by a rotation of 36.9° about $\langle 100 \rangle$ which results in a $\Sigma 5$ coincidence site lattice. It can be seen that the lattice parameters of the CSL are $\sqrt{5}$, $\sqrt{5}$, and 1 (along the rotation axis) so its volume is exactly five times that of the cubic lattice.

7.4 Representation of orientation relationships

In Figure 7.4a, the choice of basis vectors $\mathbf{a}_i$ is arbitrary though convenient; Figure 7.4b illustrates an alternative basis, a body-centered tetragonal (bct) unit cell describing the same lattice. We label this as basis "B", consisting of the vectors $\mathbf{b}_1$, $\mathbf{b}_2$, and $\mathbf{b}_3$ which define the bct unit cell. It is obvious that $[\mathrm{B}; \mathbf{u}] = [0\ 2\ 1]$, compared

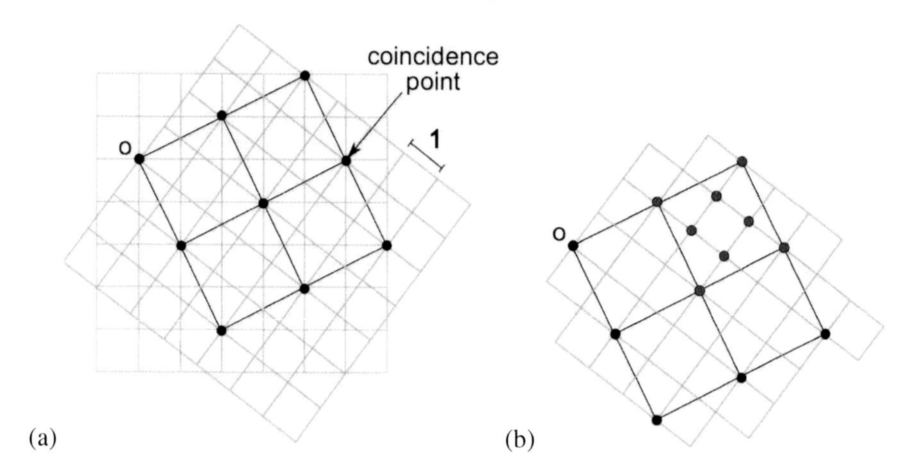

FIGURE 7.3
(a) A rotation of 36.9° about $\langle 100 \rangle$ defines the orientation relationship between two primitive cubic crystals, each of lattice parameter 1, leading to a $\Sigma 5$ coincidence site lattice. The origin is labelled "o". (b) Simplified illustration of the fact that there are only one in five of the lattice points of each of the cubic crystals in coincidence – the four in the middle of the cell are not.

with $[A; \mathbf{u}] = [1\ 1\ 1]$. The following vector equations illustrate the relationships between the basis vectors of A and those of B (Figure 7.4):

$$\begin{aligned} \mathbf{a}_1 &= 1\mathbf{b}_1 + 1\mathbf{b}_2 + 0\mathbf{b}_3 \\ \mathbf{a}_2 &= \bar{1}\mathbf{b}_1 + 1\mathbf{b}_2 + 0\mathbf{b}_3 \\ \mathbf{a}_3 &= 0\mathbf{b}_1 + 0\mathbf{b}_2 + 1\mathbf{b}_3. \end{aligned} \tag{7.4}$$

These equations can also be presented in matrix form as follows:

$$(\mathbf{a}_1\ \mathbf{a}_2\ \mathbf{a}_3) = (\mathbf{b}_1\ \mathbf{b}_2\ \mathbf{b}_3) \times \begin{pmatrix} 1 & \bar{1} & 0 \\ 1 & 1 & 0 \\ 0 & 0 & 1 \end{pmatrix}. \tag{7.5}$$

This 3×3 matrix representing the coordinate transformation is denoted (B J A) and transforms the components of vectors referred to the A basis to those referred to the B basis. The first column of (B J A) represents the components of the basis vector $\mathbf{a}_1$, with respect to the basis B, and so on.

The components of a vector $\mathbf{u}$ can now be transformed between bases using the matrix (B J A) as follows:

$$[B; \mathbf{u}] = (B\ J\ A)[A; \mathbf{u}] \tag{7.6}$$

Notice the juxtapositioning of like basis symbols. Writing (A J B) as the inverse of (B J A):

$$[A; \mathbf{u}] = (A\ J\ B)[B; \mathbf{u}]$$

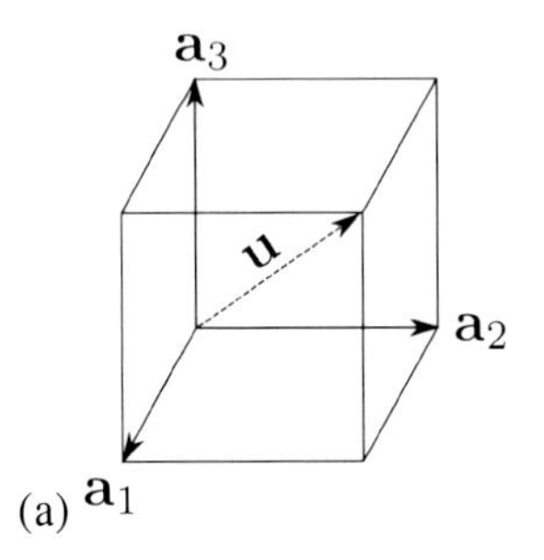

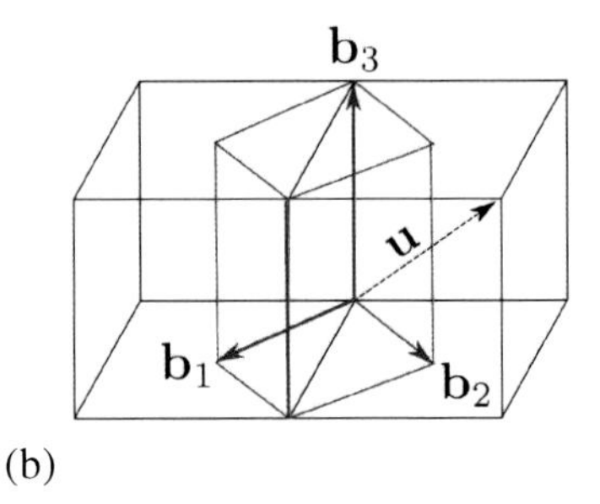

FIGURE 7.4
(a) Cubic coordinate system with orthogonal basis vectors of equal magnitude. (b) Tetragonal coordinate system (red) in which the basis vectors on the basal plane end at the face-centers of the cubic system. The relationship to the cubic cells (black) is also illustrated.

Example 7.1: Coordinate transformation

Two adjacent crystals that have an identical cubic lattice are represented by bases "A" and "B", respectively. The basis vectors $\mathbf{a}_i$ of A and $\mathbf{b}_i$ of B, respectively, define the cubic unit cells of the crystals concerned. The lattice parameter is identical for the two bases, $|\mathbf{a}_i| = |\mathbf{b}_i| = a$. The grains are orientated such that $[0\ 0\ 1]_{\mathrm{A}} \parallel [0\ 0\ 1]_{\mathrm{B}}$, and $[1\ 0\ 0]_{\mathrm{B}}$ makes an angle of 45° with both $[1\ 0\ 0]_{\mathrm{A}}$ and $[0\ 1\ 0]_{\mathrm{A}}$, with $[001]_{\mathrm{A}} \parallel [001]_{\mathrm{B}}$, as illustrated in Figure 7.5. Prove that if $\mathbf{u}$ is a vector such that its components in crystal A are given by $[\mathrm{A}; \mathbf{u}] = [\sqrt{2}\ 2\sqrt{2}\ 0]$, then in the basis B, $[\mathrm{B}; \mathbf{u}] = [3\ 1\ 0]$. Show that the magnitude of $\mathbf{u}$ (i.e., $|\mathbf{u}|$) does not depend on the choice of the basis.

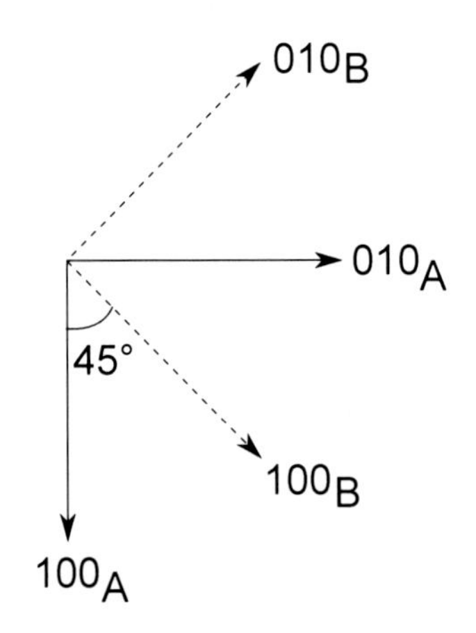

FIGURE 7.5
Diagram illustrating the relation between the bases A and B. Note that $[001]_{\mathrm{A}} \parallel [001]_{\mathrm{B}}$.

Referring to Figure 7.5, and recalling that the matrix (B J A) consists of three columns, each column being the components of one of the basis vectors of A, with respect to B, we have

$$\begin{aligned}[\mathrm{B};\mathbf{a}_1] &= [\cos 45 \quad -\sin 45 \quad 0]\\ [\mathrm{B};\mathbf{a}_2] &= [\sin 45 \quad \cos 45 \quad 0]\\ [\mathrm{B};\mathbf{a}_3] &= [0 \quad 0 \quad 1]\end{aligned}$$

so that

$$(\mathrm{B\ J\ A}) = \begin{pmatrix} \cos 45 & \sin 45 & 0\\ -\sin 45 & \cos 45 & 0\\ 0 & 0 & 1\end{pmatrix}$$

Now, $[\mathrm{B};\mathbf{u}] = (\mathrm{B\ J\ A})[\mathrm{A};\mathbf{u}]$. On substituting for $[\mathrm{A};\mathbf{u}] = [\sqrt{2}\ 2\sqrt{2}\ 0]$, we get $[\mathrm{B};\mathbf{u}] = [3\ 1\ 0]$. Both the bases A and B are orthogonal so that the magnitude of **u** can be obtained using the Pythagoras theorem. Hence, choosing components referred to the basis B, we get:

$$|\mathbf{u}|^2 = (3|\mathbf{b}_1|)^2 + (|\mathbf{b}_2|)^2 = 10a^2 \tag{7.7}$$

With respect to basis A,

$$|\mathbf{u}|^2 = (\sqrt{2}|\mathbf{a}_1|)^2 + (2\sqrt{2}|\mathbf{a}_2|)^2 = 10a^2 \tag{7.8}$$

Hence, $|\mathbf{u}|$ is invariant to the coordinate transformation. This is a general result, since a vector is a physical entity, whose magnitude and direction cannot depend on the choice of a reference frame.

7.4.1 General rotation matrix

If the axis of rotation is represented as a unit vector, then only three independent quantities are needed to define a misorientation between grains: two components of the axis of rotation, and an angle of rotation. It follows that a rotation matrix must also have only three independent terms. The components of any rotation matrix can be written in terms of a vector $\mathbf{u} = [u_1\ u_2\ u_3]$ which lies along the axis of rotation (such that $u_1u_1 + u_2u_2 + u_3u_3 = 1$), and in terms of the right-handed angle of rotation θ as follows (see Appendix B for derivation):

$$(\mathrm{Y\ J\ X}) = \begin{pmatrix} u_1u_1(1-m)+m & u_1u_2(1-m)+u_3n & u_1u_3(1-m)-u_2n\\ u_1u_2(1-m)-u_3n & u_2u_2(1-m)+m & u_2u_3(1-m)+u_1n\\ u_1u_3(1-m)+u_2n & u_2u_3(1-m)-u_1n & u_3u_3(1-m)+m\end{pmatrix} \tag{7.9}$$

where $m = \cos\theta$ and $n = \sin\theta$. The right-handed angle of rotation can be obtained from the fact that

$$J_{11} + J_{22} + J_{33} = 1 + 2\cos\theta \tag{7.10}$$

and the components of the vector **u** along the axis of rotation are given by

$$\begin{aligned} u_1 &= (J_{23} - J_{32})/2\sin\theta \\ u_2 &= (J_{31} - J_{13})/2\sin\theta \\ u_3 &= (J_{12} - J_{21})/2\sin\theta. \end{aligned} \tag{7.11}$$

If the rotation axis in Equation 7.9 is set to be $[001]_{\mathrm{X}}$ then

$$(\mathrm{Y\ J\ X}) = \begin{pmatrix} \cos\theta & \sin\theta & 0 \\ -\sin\theta & \cos\theta & 0 \\ 0 & 0 & 1 \end{pmatrix} \tag{7.12}$$

Referring again to Equation 7.9, if the rotation is a symmetry operation then it should map lattice points onto lattice points, in which case all components of the rotation matrix should be integers. It follows that the trace of the matrix should also be an integer, and since $-1 \leq \cos\theta \leq 1$, the set of possible integers is limited:

$$J_{11} + J_{22} + J_{33} = 1 + 2\cos\theta = -1,\ 0,\ 1,\ 2,\ 3 \tag{7.13}$$

or

$$\theta = 2\pi,\ \frac{2\pi}{2},\ \frac{2\pi}{3},\ \frac{2\pi}{4},\ \frac{2\pi}{6}, \tag{7.14}$$

another way of showing that a five-fold rotation axis is not permitted when translational symmetry is to be maintained.

7.5 Mathematical method for determining Σ

We now consider a mathematical method of determining the CSL formed by allowing the lattices of crystals A and B notionally to interpenetrate; A and B are assumed to be related by a transformation (A S A) which deforms the A lattice into that of B; A and B need not have the same crystal structure or lattice parameters, so that (A S A) need not be a rigid body rotation. Consider a vector **u** which is a lattice vector whose integral components do not have a common factor. As a result of the transformation (A S A), **u** becomes a new vector **x** such that

$$[\mathrm{A}; \mathbf{x}] = (\mathrm{A\ S\ A})[\mathrm{A}; \mathbf{u}] \tag{7.15}$$

x does not necessarily have integral components in the A basis (i.e., it need not be a lattice vector of A). CSL vectors, on the other hand, identify lattice points which are common to both A and B, and therefore are lattice vectors of both crystals. It follows that CSL vectors have integral indices when referred to either crystal. Hence, **x** is only a CSL vector if it has integral components in the basis A. We note that **x** always has integral components in B, because a lattice vector of A (such as **u**) always deforms into a lattice vector of B.

The meaning of Σ is that $1/\Sigma$ of the lattice sites of A or B are common to both A and B. It follows that any primitive lattice vector of A or B, when multiplied by Σ, must give a CSL vector. $\Sigma\mathbf{x}$ must therefore always be a CSL vector and if Equation 7.15 is multiplied by Σ, then we obtain an equation in which the vector **u** always transforms into a CSL vector:

$$\Sigma[\mathrm{A};\mathbf{x}] = \Sigma(\mathrm{A\ S\ A})[\mathrm{A};\mathbf{u}] \tag{7.16}$$

i.e., given that **u** is a lattice vector of A, whose components have no common factor, $\Sigma\mathbf{x}$ is a CSL vector with integral components in either basis. This can only be true if the matrix $\Sigma(\mathrm{A\ S\ A})$ has elements which are all integral since it is only then that $\Sigma[\mathrm{A};\mathbf{x}]$ has elements which are all integral.

It follows that if an integer H can be found such that all the elements of the matrix $H(\mathrm{A\ S\ A})$ are integers (without a common factor), then H is the Σ value relating A and B.

Applying this to the problem considered in Section 7.3, the rotation matrix corresponding to the rotation $180°$ about $[1\ 1\ 2]_\mathrm{A}$ is given by (Equation 7.9)

$$(\mathrm{A\ J\ A}) = \frac{1}{3}\begin{pmatrix} \overline{2} & 1 & \overline{2} \\ 1 & \overline{2} & 2 \\ 2 & 2 & 1 \end{pmatrix} \tag{7.17}$$

and since 3 is the integer which when multiplied with (A J A) gives a matrix of integral elements (without a common factor), the Σ value for this orientation is given by $\Sigma = 3$.

Coincidence site lattices have in recent years become popular because of the advent of orientation imaging in the scanning electron microscope. The microscope is usually associated with software that enables the estimation of the Σ value at every junction between two grains. The accuracy of the technique is limited so such Σ maps should be taken with a pinch of salt, and large Σ values are not very meaningful as representations of low-energy boundaries.

7.6 Summary

Interfaces clearly have a structure that can be periodic. Simple boundaries between misoriented crystals that have identical structures can be described in terms of arrays of dislocations which become visible during transmission electron microscopy. The spacing of these boundary dislocations can be related to the degree of misorientation whereas their line vectors, in the case of a tilt boundary, lie along the tilt axis. The dislocation structure can be used also to estimate the interfacial energy per unit area as long as the dislocation description remains valid.

It is possible that at certain values of the misorientations, a significant fraction of the lattice points between the two crystals that form a boundary, are coincident. Furthermore, the coincident sites form a lattice that is periodic, known as a coincidence site lattice. It is important to realize that this is a reflection of the orientation relationships between the crystals rather than the boundary plane. The properties of a boundary change significantly when the fraction of coincident points, $1/\Sigma$, becomes large. For example, the interfacial energy per unit area declines dramatically as might the boundary diffusion coefficient.

References

1. T. Watanabe: Grain boundary engineering: historical perspective and future prospects, *Journal of Materials Science*, 2011, **46**, 4095–4115.

2. A. F. Gourgues, H. M. Flower, and T. C. Lindley: Electron backscattering diffraction study of acicular ferrite, bainite, and martensite steel microstructures, *Materials Science and Technology*, 2000, **16**, 26–40.

3. J. S. Koehler: On the dislocation theory of plastic deformation, *Physical Review*, 1941, **60**, 397–410.

4. W. T. Read, and W. Schockley: Dislocation models of crystal grain boundaries, *Physical Review*, 1950, **78**, 275–289.

5. M. L. Kronberg, and F. H. Wilson: Secondary recrystallization in copper, *Trans. A.I.M.E.*, 1949, **185**, 501–514.

6. S. Ranganathan: On the geometry of coincidence-site lattices, *Acta Crystallographica*, 1966, **21**, 197–199.

8

Crystallography of Martensitic Transformations

Abstract

Martensitic transformations are exceptionally simple and precise. This is because they do not involve diffusion, which has a tendency to create confusion in the arrangement of atoms. With martensite, the atoms move in a disciplined manner that leads to strong crystallographic predictability, including that of irrational quantities and exact changes in shape. This chapter contains a basic explanation of the theory needed to explain the fine details of the crystallography of martensite and associated consequences.

8.1 Introduction

Martensitic transformations, and displacive transformations in general, remain the only mechanism for the manufacture of bulk nanostructured materials for engineering applications [1, 2]. The transformation is diffusionless and hence can occur at incredibly low temperatures [3], or can propagate at the speed of sound in the material (a few thousand $\mathrm{m\,s^{-1}}$) [4]. There is no composition change when it grows to consume the parent phase.

There are, however, a number of peculiarities associated with martensitic transformations, which are important in the design of alloys. The interface plane between austenite and martensite is known as the *habit plane* and measurements show that the crystallographic indices of that plane are irrational and strange (Table 8.1). Another difficulty is that the martensite-austenite (α/γ) interface has to be glissile, that is, it must be able to move without diffusion. This requires that the interface must contain no more than one set of dislocations. Any more than one array of line vectors can lead to interference between the dislocations which renders the interface sessile. In crystallographic terms, this means that there must be at least one line fully coherent in the α/γ interface. The transformation strain relating the two lattices must therefore, as a minimum, be an *invariant-line strain*.

TABLE 8.1
Habit plane indices for martensite.

Composition /wt.%	Approximate habit plane indices
Low-alloy steels, Fe-28Ni	$\{1\,1\,1\}_\gamma$
Plate martensite in Fe-1.8C	$\{2\,9\,5\}_\gamma$
Fe-30Ni-0.3C	$\{3\,15\,10\}_\gamma$
Fe-8Cr-1C	$\{2\,5\,2\}_\gamma$

Note: The quoted indices are approximate because the habit planes are in general irrational. There is an interesting study on the regularity of the habit plane indices as a function of, for example, strain in the austenite [5].

8.2 Shape deformation

During martensitic transformation, the pattern in which the atoms in the parent crystal are arranged is *deformed* into that appropriate for martensite, there must be a corresponding change in the macroscopic shape of the crystal undergoing transformation. The dislocations responsible for the deformation are in the α/γ interface, with Burgers vectors such that in addition to deformation they cause the change in crystal structure. The deformation is such that an initially flat surface becomes uniformly tilted about the line formed by the intersection of the interface plane with the free surface (Figure 8.1). Any scratch traversing the transformed region is similarly deflected though the scratch remains connected at the α/γ interface.

8.3 Bain strain

We now consider the nature of the strain necessary to transform the fcc lattice of γ into the bcc lattice of α. Such a strain was proposed by Bain in 1924 [6] and hence is known as the "Bain Strain" (Figure 8.2). There is a compression along the z axis and a uniform expansion along the x and y axes.

The Bain strain implies the following *rational* orientation relationship between the parent and product lattices:

$$[0\,0\,1]_{\rm fcc}|\,|[0\,0\,1]_{\rm bcc} \qquad [1\,\bar{1}\,0]_{\rm fcc}|\,|[1\,0\,0]_{\rm bcc} \qquad [1\,1\,0]_{\rm fcc}|\,|[0\,1\,0]_{\rm bcc} \tag{8.1}$$

and although the orientations actually observed are consistent with good matching

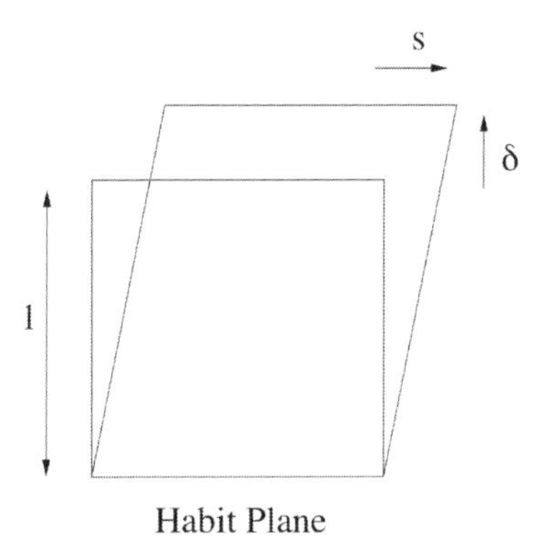

FIGURE 8.1
The measured shape deformation due to martensitic transformation is an invariant-plane strain with a large shear component ($\simeq 0.22$) and a small dilatational strain ($\simeq 0.03$) directed normal to the habit plane in steels.

between the two lattices (Figure 8.3) they are inconsistent with the Bain relationship. Typical observed orientation relationships are irrational but are represented approximately by:

Kurdjumov–Sachs orientation relationship:

$$\{1\,1\,1\}_\gamma \| \;\{0\,1\,1\}_\alpha,$$

$$\langle 1\,0\,\bar{1}\rangle_\gamma \| \;\langle 1\,1\,\bar{1}\rangle_\alpha,$$

Nishiyama–Wasserman orientation relationship:

$$\{1\,1\,1\}_\gamma \;\|\; \{0\,1\,1\}_\alpha,$$

$$\langle 1\,0\,\bar{1}\rangle_\gamma \quad \text{about } 5.3^\circ \text{ from} \langle 1\,1\,\bar{1}\rangle_\alpha \text{ towards } \langle \bar{1}\,1\,\bar{1}\rangle_\alpha,$$

Greninger–Troiano orientation relationship:

$$\{1\,1\,1\}_\gamma \quad \text{about } 0.2^\circ \text{ from } \{0\,1\,1\}_\alpha,$$

$$\langle 1\,0\,\bar{1}\rangle_\gamma \quad \text{about } 2.7^\circ \text{ from } \langle 1\,1\,\bar{1}\rangle_\alpha \text{ towards } \langle \bar{1}\,1\,\bar{1}\rangle_\alpha.$$

A conceptual difficulty therefore is to explain why the observed orientation relations differ from that implied by the Bain strain. The second inconsistency with the Bain strain does not leave any line invariant, thus violating the essential condition

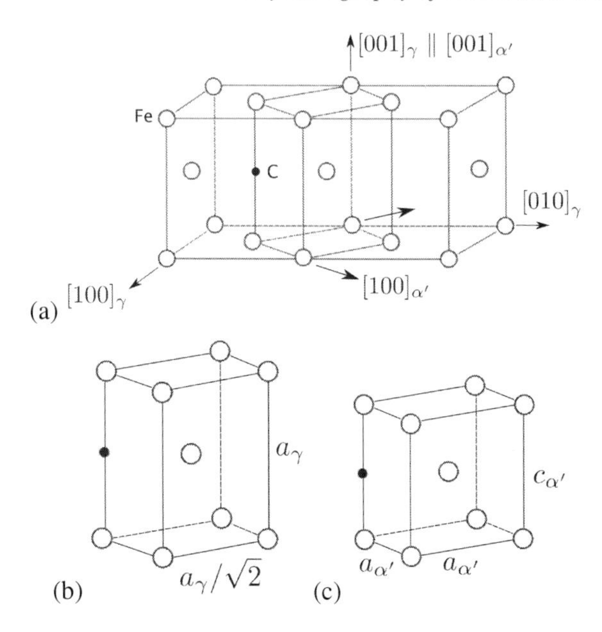

FIGURE 8.2
The Bain strain. The iron atoms at the centers of the faces of the austenite cell parallel to the plane of the diagram are omitted for clarity. (a) The lattice correspondence for formation of martensite from austenite, showing a single carbon atom in an octahedral interstice on the $[001]_\gamma$ axis. (b) Tetragonal unit cell outlined in austenite. (c) Lattice deformation (compression along c-axis) to form martensite with an appropriate c/a ratio. Note that the martensite in iron alloys can be body-centered tetragonal (bct) or body-centered cubic (bcc); none of the descriptions in this chapter are sensitive to this.

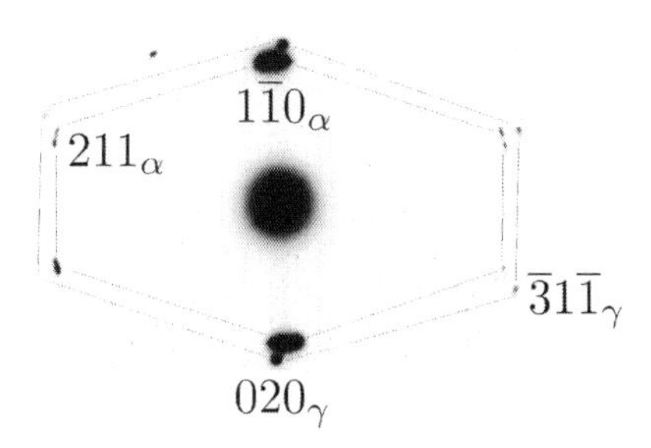

FIGURE 8.3
Electron diffraction pattern from bcc martensite and fcc austenite lattices in steels.

for the existence of a glissile interface that facilitated martensitic transformation. In Figure 8.4a,b, the austenite is represented as a sphere which, as a result of the Bain strain **B**, is deformed into an ellipsoid of revolution which represents the martensite. There are no lines which are left undistorted or unrotated by **B**. There are no lines in the $(0\ 0\ 1)_{\mathrm{fcc}}$ plane which are undistorted. The lines wx and yz are undistorted but are rotated to the new positions $w'x'$ and $y'z'$. Such rotated lines are not invariant. However, the combined effect of the Bain strain **B** and the rigid body rotation **R** is indeed an invariant-line strain (ILS) because it brings yz and $y'z'$ into coincidence (Figure 8.4c). This is the reason why the observed irrational orientation relationship differs from that implied by the Bain strain. The rotation required to convert **B** into an ILS precisely corrects the Bain orientation into that which is observed experimentally.

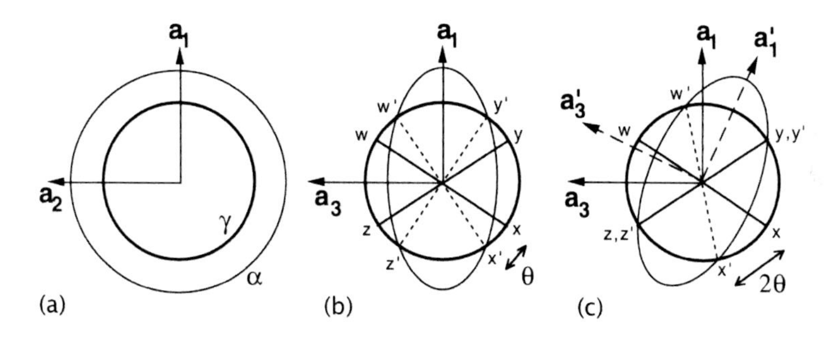

FIGURE 8.4
(a) and (b) show the effect of the Bain strain on austenite, which when undeformed is represented as a sphere of diameter $wx = yz$ in three dimensions. The strain transforms it into an ellipsoid of revolution. (c) Shows the ILS obtained by combining the Bain strain with a rigid body rotation through an angle θ. $\mathbf{a}_1$, $\mathbf{a}_2$, and $\mathbf{a}_3$ refer to $[100]_\gamma$, $[010]_\gamma$, and $[001]_\gamma$ axes, respectively.

As can be seen from Figure 8.4c, there is no rotation which can make **B** into an invariant-plane strain since this would require two non-parallel invariant-lines. Thus, for the fcc→bcc transformation, austenite cannot be transformed into martensite by a homogeneous strain which is an invariant-plane strain. And yet, the observed shape deformation leaves the habit plane undistorted and unrotated, i.e., it is an invariant-plane strain.

The phenomenological theory of martensite crystallography solves this remaining problem (Figure 8.5) [7–10]. The Bain strain converts the structure of the parent phase into that of the product phase. When combined with an appropriate rigid body rotation, the net homogeneous lattice deformation **RB** is an invariant-line strain (step a to c in Figure 8.5). However, the observed shape deformation is an invariant-plane strain $\mathbf{P}_1$ (step a to b in Figure 8.5), but this gives the wrong crystal structure. If a second homogeneous shear $\mathbf{P}_2$ is combined with $\mathbf{P}_1$ (step b to c), then the correct structure is obtained but the wrong shape since

$$\mathbf{P}_1\mathbf{P}_2 = \mathbf{RB}.$$

These discrepancies are all resolved if the shape changing effect of $\mathbf{P_2}$ is cancelled macroscopically by an inhomogeneous lattice-invariant deformation, which may be slip or twinning as illustrated in Figure 8.5.

The theory explains all the observed features of the martensite crystallography. The orientation relationship is predicted by deducing the rotation needed to change the Bain strain into an invariant-line strain. The habit plane does not have rational indices because the amount of lattice-invariant deformation needed to recover the correct macroscopic shape is not usually rational. The theory predicts a substructure in plates of martensite (either twins or slip steps) as is observed experimentally. The transformation goes to all the trouble of ensuring that the shape deformation is macroscopically an invariant-plane strain because this reduces the strain energy when compared with the case where the shape deformation might be an invariant-line strain.

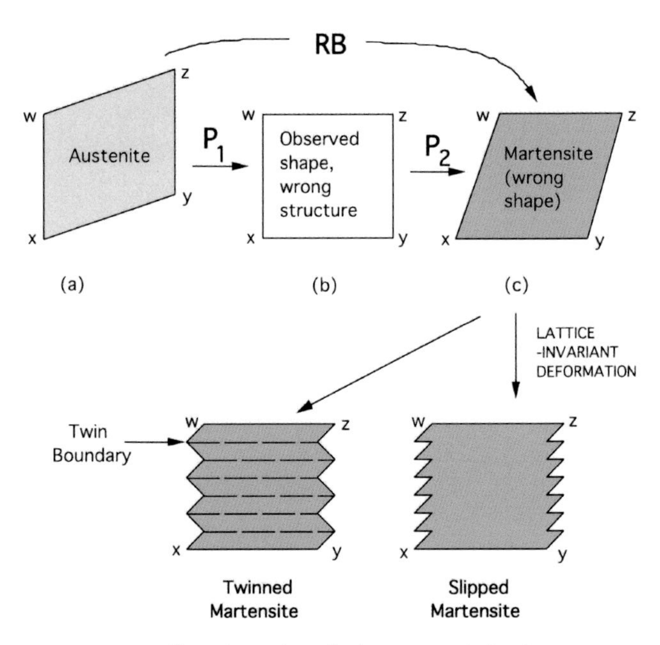

FIGURE 8.5
The phenomenological theory of martensite crystallography.

When austenite transforms martensitically into ε (hcp), the entire change is achieved by a deformation that is an invariant-plane strain as illustrated in Figure 8.6. As a consequence, the habit plane of the martensite is rational, i.e., precisely the close-packed plane, the orientation relationship is rational:

$$(111)_\gamma \parallel (0001)_\mathrm{H} \qquad [10\bar{1}]_\gamma \parallel [10\bar{1}0]_\mathrm{H}.$$

The strain not only transforms the lattice but the resulting shape of the final transformation product is as observed, so there is no need for any lattice-invariant deformation (slip or twinning).

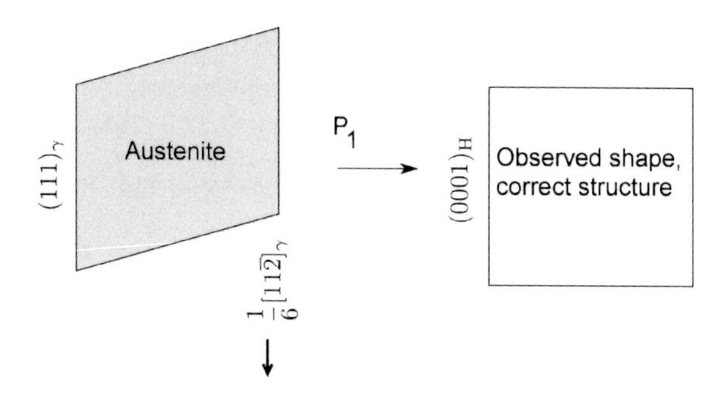

FIGURE 8.6
The fcc→hcp transformation involving a shear of $\frac{a_\gamma}{6}[11\overline{2}]$ on alternate $(111)_\gamma$ planes to generate the hcp structure.

8.4 Summary

It is hard to exaggerate the success of the crystallographic theory of martensitic transformations, both in terms of the conceptual leap in explaining away the apparent difficulties accumulated via careful experiments, and in the quantitative closure between experiment and theory. This has not, in many respects, been achieved for any other solid-state transformations, particularly those that involve diffusion and composition change. The theory has generic value because it is not restricted to changes between particular crystal systems.

References

1. R. Ueji, N. Tsuji, Y. Minamino, and Y. Koizumi: Ultragrain refinement of plain low carbon steel by cold-rolling and annealing of martensite, *Acta Materialia*, 2002, **50**, 4177–4139.

2. H. K. D. H. Bhadeshia: The first bulk nanostructured metal, *Science and Technology of Advanced Materials*, 2013, **14**, 014202.

3. L. Kaufman, and M. Cohen: The martensitic transformation in iron-nickel system, *Trans. AIME Journal of Metals*, 1956, **206**, 1393–1401.

4. R. F. Bunshah, and R. F. Mehl: Rate of propagation of martensite, *Trans. A.I.M.E.*, 1953, **193**, 1251–1258.

5. D. P. Dunne, and N. F. Kennon: The martensite interface — how regular is the habit?, *Materials Science Forum*, 1994, **189-190**, 273–278.

6. E. C. Bain: The nature of martensite, *Trans. AIME*, 1924, **70**, 25–46.

7. J. S. Bowles, and J. K. Mackenzie: The crystallography of martensite transformations, part I, *Acta Metallurgica*, 1954, **2**, 129–137.

8. J. K. Mackenzie, and J. S. Bowles: The crystallography of martensite transformations II, *Acta Metallurgica*, 1954, **2**, 138–147.

9. J. K. Mackenzie, and J. S. Bowles: The crystallography of martensite transformations III FCC to BCT transformations, *Acta Metallurgica*, 1954, **2**, 224–234.

10. M. S. Wechsler, D. S. Lieberman, and T. A. Read: On the theory of the formation of martensite, *Trans. AIME Journal of Metals*, 1953, **197**, 1503–1515.

Part II

A Few Advanced Methods

9

Orientation Relations

Abstract

Macroscopic materials consist of aggregates of crystals. The order that typifies single crystals is disrupted at boundaries so the properties of polycrystalline samples can be sensitive to the relative orientations of neighboring crystals. Or even to the averaged disposition of the myriads of crystals relative to some macroscopic frame of reference. Orientation relationships between crystals are useful in understanding the structure of the interfaces between crystals and the evolution of the shapes of precipitates as they grow in a solid matrix. An entire subject now exists on grain boundary engineering, where, for example, the corrosion resistance of a stainless steel can be improved dramatically by ensuring a preponderance of good-fit boundaries. And there are remarkable experimental techniques that facilitate the crystallographic characterization of millions of crystals in a long blink of an eye.

9.1 Introduction

A substantial part of research on polycrystalline materials is concerned with the accurate determination, assessment, and theoretical prediction of orientation relationships between adjacent crystals. There are obvious practical applications, such as in the study of anisotropy and texture, or in defining the role of interfaces in controlling the mechanical properties. The existence of a reproducible orientation relation between the parent and product phases might determine the ultimate morphology of any precipitates, by allowing the development of low interfacial-energy facets. It is possible that vital clues about nucleation in the solid state might emerge from a detailed examination of orientation relationships, even though these can usually only be measured when the crystals concerned are well into the growth stage. Needless to say, the properties of interfaces depend critically on the relative dispositions of the crystals that they connect.

Perhaps the most interesting experimental observation is that the orientation relationships that are found to develop during phase transformations (or during deformation or recrystallisation experiments) usually are not random [1–3]. The frequency of occurrence of any particular orientation relation usually far exceeds the probability of obtaining it simply by taking the two separate crystals and joining them up in an arbitrary way.

This indicates that there are favored orientation relations, perhaps because it is these which allow the best fit at the interface between the two crystals. This would in turn reduce the interfacial energy, and hence the activation energy for nucleation. Nuclei which, by chance, happen to be orientated in this manner would find it relatively easy to grow, giving rise to the non-random distribution mentioned earlier.

On the other hand, it could be the case that nuclei actually form by the homogeneous deformation of a small region of the parent lattice. The deformation, which transforms the parent structure to that of the product (e.g., Bain strain), would have to be of the kind which minimizes strain energy. Of all the possible ways of accomplishing the lattice change by homogeneous deformation, only a few might satisfy the minimum strain energy criterion – this again would lead to the observed non-random distribution of orientation relationships. It is a problem to understand which of these factors really determine the existence of rational orientation relations. In this chapter we deal with the methods for adequately describing the relationships between crystals.

9.2 Cementite in steels

Cementite, Fe_3C, referred to as θ, is the most common precipitate to be found in steels. It has an orthorhombic crystal structure and can precipitate from supersaturated ferrite or austenite.[1] It sometimes can precipitate at temperatures below 400 K in times too short to allow any substantial diffusion of iron atoms [4–6] although the long range diffusion of carbon atoms is clearly necessary and indeed possible. It has therefore been suggested that the cementite lattice may sometimes be generated by the deformation of the ferrite lattice, combined with the necessary diffusion of carbon into the appropriate sites [4, 7].

Shackleton and Kelly [8] showed that the plane of precipitation of cementite from ferrite is $\{1\ 0\ 1\}_\theta \parallel \{1\ 1\ 2\}_\alpha$. This is consistent with the habit plane containing the direction of reasonable fit between the θ and α lattices [4], i.e., $\langle 0\ 1\ 0\rangle_\theta \parallel \langle 1\ 1\ \bar{1}\rangle_\alpha$. Cementite formed during the tempering of martensite adopts many crystallographic variants of this habit plane in any given plate of martensite; in lower bainite it is usual to find just one such variant, with the habit plane inclined at some 60° to the plate axis. The problem is discussed in detail in [9]. Cementite which forms from austenite usually exhibits the Pitsch [10] orientation relation with $[0\ 0\ 1]_\theta \parallel [\bar{2}\ 2\ 5]_\gamma$ and $[1\ 0\ 0]_\theta \parallel [5\ \bar{5}\ 4]_\gamma$ and a mechanism which involves the intermediate formation of ferrite has been postulated [4] to explain this orientation relationship.

[1] The structure of cementite is described on page 61 but we will assume in what follows that the lattice parameters increase in the order $a > b > c$.

Example 9.1: Bagaryatski orientation relationship

Cementite (θ) has an orthorhombic crystal structure, with lattice parameters $a = 4.5241$, $b = 5.0883$, and $c = 6.7416$ Å along the [1 0 0], [0 1 0] and [0 0 1] directions, respectively. This differs from the convention used on page 61, but is equally valid although we note that the space group symbol for the structure now becomes $Pbnm$ instead of $Pnma$ (the original solution of the crystal structure of cementite by Lipson and Petch used the convention described here [11]). The reason some prefer the convention that leads to the space group $Pnma$ may have something to do with the fact that it is the most abundant space group of known inorganic crystals and minerals [12].

When cementite precipitates from ferrite (α, bcc structure, lattice parameter $a_\alpha = 2.8662$ Å), the lattices are related by the Bagaryatski orientation relationship, given by:

$$[1\ 0\ 0]_\theta \parallel [0\ \bar{1}\ 1]_\alpha,\ [0\ 1\ 0]_\theta \parallel [1\ \bar{1}\ \bar{1}]_\alpha,\ [0\ 0\ 1]_\theta \parallel [2\ 1\ 1]_\alpha \tag{9.1}$$

(i) Derive the coordinate transformation matrix $(\alpha\ J\ \theta)$ representing this orientation relationship, given that the basis vectors of θ and α define the orthorhombic and bcc unit cells of the cementite and ferrite, respectively.

(ii) Published stereograms of this orientation relation have the $(0\ \bar{2}\ 3)_\theta$ plane exactly parallel to the $(1\ 3\ 3)_\alpha$ plane. Show that this can only be correct if the ratio $b/c = 8\sqrt{2}/15$.

The information given indicates parallelisms between vectors in the two lattices. In order to find $(\alpha\ J\ \theta)$, it is necessary to ensure that the magnitudes of the parallel vectors are also equal, since the magnitude of a vector must remain invariant to a coordinate transformation. If the constants k, g, and m are defined as

$$\begin{aligned} k &= \frac{|[1\ 0\ 0]_\theta|}{|[0\ \bar{1}\ 1]_\alpha|} = \frac{a}{a_\alpha\sqrt{2}} = 1.116120 \\ g &= \frac{|[0\ 1\ 0]_\theta|}{|[1\ \bar{1}\ \bar{1}]_\alpha|} = \frac{b}{a_\alpha\sqrt{3}} = 1.024957 \\ m &= \frac{|[0\ 0\ 1]_\theta|}{|[2\ 1\ 1]_\alpha|} = \frac{c}{a_\alpha\sqrt{6}} = 0.960242 \end{aligned} \tag{9.2}$$

then multiplying $[0\ \bar{1}\ 1]_\alpha$ by k makes its magnitude equal to that of $[1\ 0\ 0]_\theta$; the constants g and m can similarly be used for the other two α vectors.

Recalling now the definition of a coordinate transformation matrix, each column of $(\alpha\ J\ \theta)$ represents the components of a basis vector of θ in the α basis. For example, the first column of $(\alpha\ J\ \theta)$ consists of the components of $[1\ 0\ 0]_\theta$ in the α basis, $[0\ \bar{k}\ k]_\alpha$. It follows that $(\alpha\ J\ \theta)$ can be derived simply by inspection of the relations 9.1, 9.2, so that

$$(\alpha\ J\ \theta) = \begin{pmatrix} 0.000000 & 1.024957 & 1.920485 \\ -1.116120 & -1.024957 & 0.960242 \\ 1.116120 & -1.024957 & 0.960242 \end{pmatrix} \tag{9.3}$$

The transformation matrix can therefore be deduced by inspection when the orientation relationship 9.1 is stated in terms of the *basis vectors* of one of the crystals concerned (in this case, the basis vectors of θ are specified in relationship 9.1). On the other hand, orientation relationships can, and often are, specified in terms of vectors other than the basis vectors. Furthermore, electron diffraction patterns may not include direct information about basis vectors. A more general method of solving for $(\alpha\ \mathrm{J}\ \theta)$ is now presented, a method independent of the vectors used in specifying the orientation relationship:

From the relations 9.1 and 9.2 we see that

$$\begin{aligned}
\begin{bmatrix}0 & \overline{k} & k\end{bmatrix}_{\alpha} &= (\alpha\ \mathrm{J}\ \theta)[1\ 0\ 0]_{\theta} \\
\begin{bmatrix}g & \overline{g} & \overline{g}\end{bmatrix}_{\alpha} &= (\alpha\ \mathrm{J}\ \theta)[0\ 1\ 0]_{\theta} \\
\begin{bmatrix}2m & m & m\end{bmatrix}_{\alpha} &= (\alpha\ \mathrm{J}\ \theta)[0\ 0\ 1]_{\theta}.
\end{aligned} \tag{9.4}$$

These equations can be written as:

$$\begin{pmatrix} 0 & g & 2m \\ \overline{k} & \overline{g} & m \\ k & \overline{g} & m \end{pmatrix} = \begin{pmatrix} J_{11} & J_{12} & J_{13} \\ J_{21} & J_{22} & J_{23} \\ J_{31} & J_{32} & J_{33} \end{pmatrix} \begin{pmatrix} 1 & 0 & 0 \\ 0 & 1 & 0 \\ 0 & 0 & 1 \end{pmatrix} \tag{9.5}$$

where the J_{ij} ($i = 1, 2, 3$ and $j = 1, 2, 3$) are the elements of the matrix $(\alpha\ \mathrm{J}\ \theta)$. From Equation 9.5, it follows that

$$(\alpha\ \mathrm{J}\ \theta) = \begin{pmatrix} 0 & g & 2m \\ \overline{k} & \overline{g} & m \\ k & \overline{g} & m \end{pmatrix} = \begin{pmatrix} 0 & 1.024957 & 1.920485 \\ -1.116120 & -1.024957 & 0.960242 \\ 1.116120 & -1.024957 & 0.960242 \end{pmatrix}. \tag{9.6}$$

It is possible to accumulate rounding off errors in such calculations so the use of at least six figures after the decimal point is advisable.

To consider the relationships between *plane normals* (rather than directions) in the two lattices, we have to discover how vectors representing plane normals (always referred to a reciprocal basis) transform. From Equation 5.8, if **u** is any vector,

$$\begin{aligned}
& |\mathbf{u}|^2 = (\mathbf{u}; \alpha^*)[\alpha; \mathbf{u}] = (\mathbf{u}; \theta^*)[\theta; \mathbf{u}] \\
\text{or}\quad & (\mathbf{u}; \alpha^*)(\alpha\ \mathrm{J}\ \theta)[\theta; \mathbf{u}] = (\mathbf{u}; \theta^*)[\theta; \mathbf{u}] \\
\text{giving}\quad & (\mathbf{u}; \alpha^*)(\alpha\ \mathrm{J}\ \theta) = (\mathbf{u}; \theta^*) \\
& (\mathbf{u}; \alpha^*) = (\mathbf{u}; \theta^*)(\theta\ \mathrm{J}\ \alpha) \\
\text{where}\quad & (\theta\ \mathrm{J}\ \alpha) = (\alpha\ \mathrm{J}\ \theta)^{-1}.
\end{aligned} \tag{9.7}$$

$$(\theta \text{ J } \alpha) = \frac{1}{6gmk}\begin{pmatrix} 0 & -3gm & 3gm \\ 2mk & -2mk & -2mk \\ 2gk & gk & gk \end{pmatrix}$$

$$= \begin{pmatrix} 0 & -0.447981 & 0.447981 \\ 0.325217 & -0.325217 & -0.325217 \\ 0.347135 & 0.173567 & 0.173567 \end{pmatrix}$$

If $(\mathbf{u}; \theta^*) = (0\ \overline{2}\ 3)$ is now substituted into Equation 9.8, the corresponding vector

$$(\mathbf{u}; \alpha^*) = \frac{1}{6gmk}(6gk - 4mk\ \ 3gk + 4mk\ \ 3gk + 4mk).$$

For this to be parallel to a (1 3 3) plane normal in the ferrite, the second and third components must equal three times the first; i.e., $3(6gk - 4mk) = (3gk + 4mk)$, which is equivalent to $b/c = 8\sqrt{2}/15$, as required.

It should be noted that the determinant of $(\alpha \text{ J } \theta)$ gives the ratio of the volume of the θ unit cell to that of the ferrite α unit cell. If the coordinate transformation simply involves a rotation of bases, e.g., when it describes the relation between two grains of identical structure, then the matrix is orthogonal and its determinant has a value of unity for all proper rotations (i.e., not involving inversion operations). Such coordinate transformation matrices are called rotation matrices.

A stereographic representation of the Bagaryatski orientation is presented in Figure 9.1, to visualize the angular relationships between poles (plane normals) of crystal planes and give some indication of symmetry; the picture is of course distorted because distance on the stereogram does not scale directly with angle. Angular measurements on stereograms are in general made using Wulff nets and may typically be in error by a few degrees, depending on the size of the stereogram. Space and aesthetic considerations usually limit the number of poles plotted on stereograms, and those plotted usually have rational indices. Separate plots are needed for cases where directions and plane normals of the same indices have a different orientation in space. A coordinate transformation matrix is, by comparison, a precise way of representing orientation relationships; angles between any plane normals or directions can be calculated to any desired degree of accuracy and information on both plane normals and directions can be derived from just one matrix. With a little nurturing, it is possible to picture the meaning of the elements of a coordinate transformation matrix: each column of $(\alpha \text{ J } \theta)$ represents the components of a basis vector of θ in the basis α, and the determinant of this matrix gives the ratio of volumes of the two unit cells.

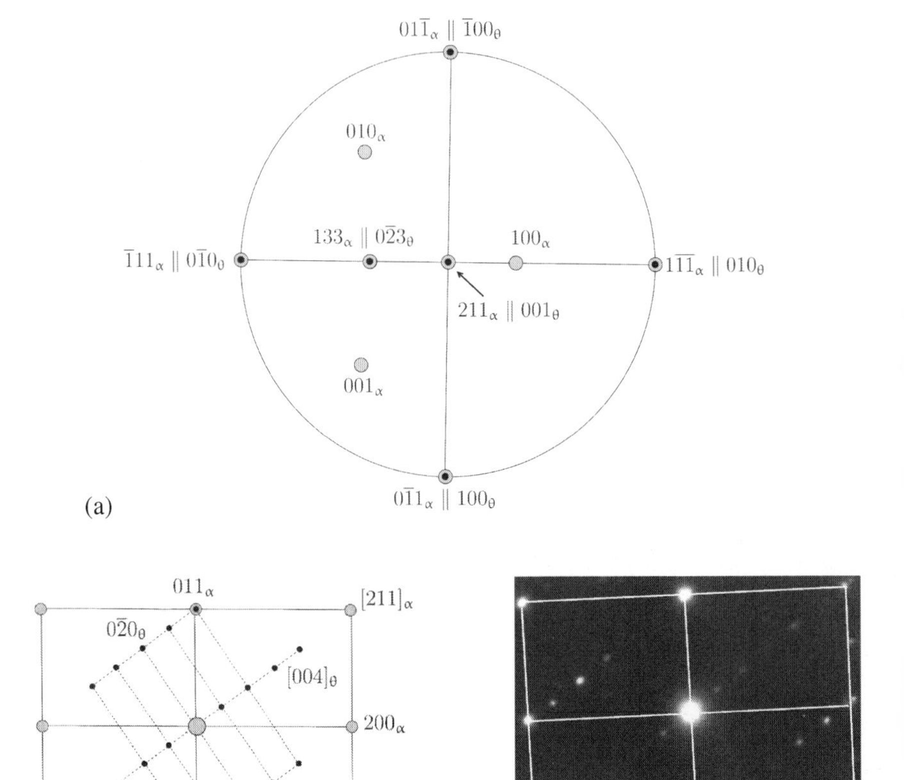

FIGURE 9.1
(a) Stereographic representation of the Bagaryatski orientation relationship between ferrite and cementite in steels, where $[1\ 0\ 0]_\theta \parallel [0\ \bar{1}\ 1]_\alpha$, $[0\ 1\ 0]_\theta \parallel [1\ \bar{1}\ \bar{1}]_\alpha$, and $[0\ 0\ 1]_\theta \parallel [2\ 1\ 1]_\alpha$. All the poles plotted are in the upper hemisphere. (b) Typical electron diffraction pattern consistent with the Bagaryatski orientation relation with $[1\,0\,0]_\theta \parallel [0\,\bar{1}\,1]_\alpha$ normal to the plane of the pattern. (c) An actual electron diffraction pattern corresponding to (b).

9.3 Relations between fcc and bcc crystals

The ratio of the lattice parameters of austenite and ferrite in steels is about 1.24, and there are several other alloys (e.g., Cu-Zn brasses, Cu-Cr alloys) where fcc and bcc precipitates of similar lattice parameter ratios coexist. The orientation relations between these phases vary within a well-defined range, but it is usually the case that a close-packed $\{1\ 1\ 1\}_{\text{fcc}}$ plane is approximately parallel to a $\{0\ 1\ 1\}_{\text{bcc}}$ plane

(Figure 9.2). Within these planes, there can be a significant variation in orientation, with $\langle 1\ 0\ 1\rangle_{\text{fcc}} \parallel \langle 1\ 1\ 1\rangle_{\text{bcc}}$ for the Kurdjumov–Sachs orientation relation [13], and $\langle 1\ 0\ 1\rangle_{\text{fcc}}$ about 5.3° towards $\langle 1\ 1\ 1\rangle_{\text{bcc}}$ for the Nishiyama-Wasserman relation [14, 15]. It is experimentally difficult to distinguish between these relations using ordinary electron diffraction or X-ray techniques, but accurate work, such as that of Crosky et al. [16], shows a spread of fcc-bcc orientation relationships (roughly near the Kurdjumov–Sachs and Nishiyama–Wasserman orientations) within the same alloy. When discussing these small variations in the orientation relationship, it is important to realize that the samples studied are at an advanced stage of transformation. The state of the austenite actually changes as transformation evolves, particularly with respect to local plastic strains. As a result, the reference frame from which the martensite forms is no longer identical to that of the undeformed austenite, which may be the real cause of the spread in orientations. Ideally, a comparison needs to be made of the observed orientation relationship as a function of the number of martensite plates per unit volume. Example 9.3 deals with the exact Kurdjumov–Sachs orientation relationship.

Example 9.2: Kurdjumov–Sachs orientation relationship

Two plates of Widmanstätten ferrite (α ferrite, basis symbols X and Y, respectively) growing in the same grain of austenite (basis symbol γ) are found to exhibit two different variants of the Kurdjumov–Sachs orientation relationship with the austenite; the data below show the sets of parallel vectors of the three lattices:

$$\begin{array}{llll}
[1\,1\,1]_\gamma \parallel [0\,1\,1]_{\text{X}} & & [1\,1\,\bar{1}]_\gamma \parallel [0\,\bar{1}\,1]_{\text{Y}} \\
[\bar{1}\,0\,1]_\gamma \parallel [\bar{1}\,\bar{1}\,1]_{\text{X}} & & [1\,0\,1]_\gamma \parallel [1\,\bar{1}\,1]_{\text{Y}}. \\
[1\,\bar{2}\,1]_\gamma \parallel [2\,\bar{1}\,1]_{\text{X}} & & [1\,\bar{2}\,\bar{1}]_\gamma \parallel [2\,1\,\bar{1}]_{\text{Y}}
\end{array}$$

Derive the matrices (X J γ) and (Y J γ). Hence obtain the rotation matrix (X J Y) describing the orientation relationship between the two Widmanstätten ferrite plates.

The information given relates to parallelisms between vectors in different lattices. It is necessary to equalize the magnitudes of parallel vectors in order to solve for the various coordinate transformation matrices. Defining the constants k, g, and m as

$$k = \frac{a_\gamma\sqrt{3}}{a_\alpha\sqrt{2}} \qquad g = \frac{a_\gamma\sqrt{2}}{a_\alpha\sqrt{3}} \qquad m = \frac{\sqrt{6}a_\gamma}{\sqrt{6}a_\alpha} = \frac{a_\gamma}{a_\alpha},$$

we obtain:

$$\begin{aligned}
[0\ k\ k]_{\text{X}} &= (\text{X J }\gamma)[1\,1\,1]_\gamma \\
[\bar{g}\ \bar{g}\ g]_{\text{X}} &= (\text{X J }\gamma)[\bar{1}\,0\,1]_\gamma \\
[2m\ \bar{m}\ m]_{\text{X}} &= (\text{X J }\gamma)[1\,\bar{2}\,1]_\gamma
\end{aligned}$$

or in other words,

$$\begin{pmatrix} 0 & \overline{g} & 2m \\ k & \overline{g} & \overline{m} \\ k & g & m \end{pmatrix} = \begin{pmatrix} J_{11} & J_{12} & J_{13} \\ J_{21} & J_{22} & J_{23} \\ J_{31} & J_{32} & J_{33} \end{pmatrix} \begin{pmatrix} 1 & \overline{1} & 1 \\ 1 & 0 & \overline{2} \\ 1 & 1 & 1 \end{pmatrix}$$

$$\begin{aligned} (\mathrm{X\ J\ }\gamma) &= \begin{pmatrix} 0 & \overline{g} & 2m \\ k & \overline{g} & \overline{m} \\ k & g & m \end{pmatrix} \begin{pmatrix} 2/6 & 2/6 & 2/6 \\ \overline{3}/6 & 0 & 3/6 \\ 1/6 & \overline{2}/6 & 1/6 \end{pmatrix} \\ &= \frac{1}{6} \begin{pmatrix} 3g+2m & \overline{4}m & \overline{3}g+2m \\ 2k+3g-m & 2k+2m & 2k-3g-m \\ 2k-3g+m & 2k-2m & 2k+3g+m \end{pmatrix} \\ &= \frac{a_\gamma}{a_\alpha} \begin{pmatrix} 0.741582 & -0.666667 & -0.074915 \\ 0.649830 & 0.741582 & -0.166667 \\ 0.166667 & 0.074915 & 0.983163 \end{pmatrix}. \end{aligned}$$

In an similar manner,

$$(\mathrm{Y\ J\ }\gamma) = \frac{a_\gamma}{a_\alpha} \begin{pmatrix} 0.741582 & -0.666667 & 0.074915 \\ 0.166667 & 0.074915 & -0.983163 \\ 0.649830 & 0.741582 & 0.166667 \end{pmatrix}. \tag{9.8}$$

To find the rotation matrix relating X and Y, we proceed as follows:

$[\mathrm{X};\mathbf{u}] = (\mathrm{X\ J\ }\gamma)[\gamma;\mathbf{u}]$ and $[Y;\mathbf{u}] = (\mathrm{Y\ J\ }\gamma)[\gamma;\mathbf{u}]$ and $[\mathrm{X};\mathbf{u}] = (\mathrm{X\ J\ Y})[\mathrm{Y};\mathbf{u}]$

it follows that

$$(\mathrm{X\ J\ }\gamma)[\gamma;\mathbf{u}] = (\mathrm{X\ J\ Y})[\mathrm{Y};\mathbf{u}]$$

substituting for [Y;**u**], we get

$$(\mathrm{X\ J\ }\gamma)[\gamma;\mathbf{u}] = (\mathrm{X\ J\ Y})(\mathrm{Y\ J\ }\gamma)[\gamma;\mathbf{u}]$$

so that

$$(\mathrm{X\ J\ Y}) = (\mathrm{X\ J\ }\gamma)(\gamma\ \mathrm{J\ Y})$$

carrying out this operation, the required X-Y orientation relation is obtained as follows:

$$(\mathrm{X\ J\ Y}) = \begin{pmatrix} 0.988776 & 0.147308 & -0.024972 \\ -0.024972 & 0.327722 & 0.944445 \\ 0.147308 & -0.933219 & 0.327722 \end{pmatrix}.$$

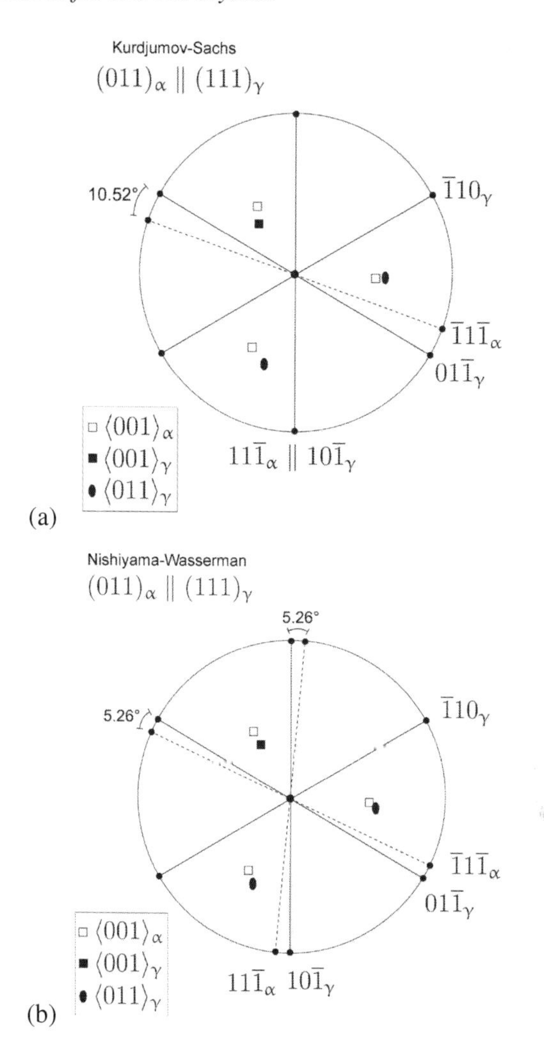

FIGURE 9.2
(a) Stereographic representation of the Kurdjumov–Sachs orientation relationship. Note that the positions of the base vectors of the γ lattice are consistent with the matrix (X J γ) derived in Example 9.3, Equation 9.8, where each column of (X J γ) represents the components of a basis vector of γ in the basis X, so that $[1\ 0\ 0]_\gamma$, $[0\ 1\ 0]_\gamma$, and $[0\ 0\ 1]_\gamma$ are approximately parallel to $[1\ 1\ 0]_\alpha$, $[\bar{1}\ 1\ 0]_\alpha$, and $[0\ 0\ 1]_\alpha$ respectively, as seen in the stereographic representation. (b) Stereographic representation of the Nishiyama–Wasserman orientation relationship which can be generated from the Kurdjumov–Sachs orientation by a rotation of 5.26° about $[0\ 1\ 1]_\alpha$. The necessary rotation makes $[\bar{1}\ \bar{1}\ 2]_\gamma$ exactly parallel to $[0\ \bar{1}\ 1]_\alpha$. The Nishiyama–Wasserman relation is midway between the two variants of the Kurdjumov–Sachs relation which share the same parallel close-packed plane. The stereograms also show that the Kurdjumov–Sachs and Nishiyama–Wasserman orientation relationships do not differ much from the γ/α orientation relationship implied by the Bain strain illustrated in Figure 8.2.

We see that the matrix (X J Y) is orthogonal because it represents an axis transformation between like orthogonal bases. In fact, (X J γ) and (Y J γ) each equal an orthogonal matrix times a scalar factor a_γ/a_α; this is because the bases X, Y, and γ are themselves orthogonal.

In this example, we chose to represent the Kurdjumov–Sachs orientation relationship by a coordinate transformation matrix (X J γ). *Named* orientation relationships like this usually assume the parallelism of particular low index planes and directions. In the example, these parallelisms are independent of the lattice parameters of the fcc and bcc structures concerned. In such cases, the orientation relationship may be represented by a pure rotation matrix, relating the orthogonal, but not necessarily orthonormal, bases of the two structures. Orientation relationships are indeed often specified in this way, or in some equivalent manner such as an axis-angle pair or a set of three Euler angles. This provides an economic way of representing orientation relations, but it should be emphasized that there is a loss of information in doing this. For example, a coordinate transformation matrix like (X J γ) not only gives information about vectors which are parallel, but also gives a ratio of the volumes of the two unit cells. Axis-angle pairs and Euler angles assume orthonormal axes.

9.4 Relationships between grains of identical structure

The relationship between two crystals which are of identical structure but which are misoriented with respect to each other is described in terms of a rotation matrix representing the rigid body rotation which can be imagined to generate one crystal from the other. As discussed below, any rotation of this kind, which leaves the common origin of the two crystals fixed, can also be described in terms of a rotation of 180° or less about an axis passing through that origin.

Example 9.3: Axis-angle pairs and rotation matrices

Two ferrite grains X and Y can be related by a rotation matrix

$$(\mathrm{Y\ J\ X}) = \frac{1}{3}\begin{pmatrix} 2 & 2 & \bar{1} \\ \bar{1} & 2 & 2 \\ 2 & \bar{1} & 2 \end{pmatrix}$$

where the basis vectors of X and Y define the respective bcc unit cells. Show that the crystal Y can be generated from X by a right-handed rotation of 60° about an axis parallel to the $[1\ 1\ 1]_\mathrm{X}$ direction.

A rigid body rotation leaves the magnitudes and relative positions of all vectors in that body unchanged. For example, an initial vector **u** with components $[u_1\ u_2\ u_3]_\mathrm{X}$ relative to the X basis, due to rotation becomes a new vector **x**, with the *same* components $[u_1\ u_2\ u_3]_\mathrm{Y}$, but with respect to the rotated basis Y. The fact that **x** represents a different direction than **u** (due to the rotation operation) means that its components in the X basis, $[w_1\ w_2\ w_3]_X$ must differ from $[u_1\ u_2\ u_3]_X$. The components of **x** in either basis are obviously related by

$$[\mathrm{Y}; \mathbf{x}] = (\mathrm{Y\ J\ X})[\mathrm{X}; \mathbf{x}]$$

in other words,

$$[u_1\ u_2\ u_3] = (\mathrm{Y\ J\ X})[w_1\ w_2\ w_3]. \tag{9.9}$$

However, if **u** happens to lie along the axis of rotation relating X and Y, then not only will [X;**u**] = [Y;**x**] as before, but its *direction* also remains invariant to the rotation operation, so that [X;**x**] = [Y;**x**] and

$$(\mathrm{Y\ J\ X})[\mathrm{X}; \mathbf{x}] = [\mathrm{Y}; \mathbf{x}]$$

so that

$$(\mathrm{Y\ J\ X})[\mathrm{X}; \mathbf{u}] = [\mathrm{X}; \mathbf{u}]$$

and hence

$$\{(\mathrm{Y\ J\ X}) - \mathbf{I}\}[\mathrm{X}; \mathbf{u}] = 0 \tag{9.10}$$

where **I** is a 3×3 identity matrix. Any rotation axis must satisfy an equation of this form; expanding Equation 9.10, we get

$$\begin{aligned} -\frac{1}{3}u_1 + \frac{2}{3}u_2 - \frac{1}{3}u_3 &= 0 \\ -\frac{1}{3}u_1 - \frac{1}{3}u_2 + \frac{2}{3}u_3 &= 0 \\ \frac{2}{3}u_1 - \frac{1}{3}u_2 - \frac{1}{3}u_3 &= 0. \end{aligned}$$

Solving these simultaneously gives $u_1 = u_2 = u_3$, proving that the required rotation axis lies along the $[1\ 1\ 1]_X$ direction, which also is the $[1\ 1\ 1]_Y$ direction.

The angle and sense of rotation can be determined by examining a vector **v** which lies at 90° to **u**. If, say, $\mathbf{v} = [\bar{1}\ 0\ 1]_X$, then as a result of the rotation operation it becomes $\mathbf{z} = [\bar{1}\ 0\ 1]_Y = [0\ \bar{1}\ 1]_X$, making an angle of 60° with **v**, giving the required angle of rotation. Since $\mathbf{v} \wedge \mathbf{z}$ gives $[1\ 1\ 1]_X$, it is also a rotation in the right-handed sense.

9.4.1 Comments

(i) The problem illustrates the fact that the orientation relation between two grains can be represented by a matrix such as (Y J X), or by an axis-angle pair such as $[1\ 1\ 1]_X$ and 60°. Obviously, the often used practice of stating a misorientation between grains in terms of just an angle of rotation is inadequate and incorrect.

(ii) The Euler theorem for rigid rotations [17] states that "regardless of the way a coordinate system is rotated from its original orientation, it always is possible to find a fixed axis in space about which a single rotation of the initial coordinates ends at the final orientation" (the original theorem is in Latin – the English quote is from [18]). If the axis of rotation is constrained to be a unit vector, then only three independent quantities are needed to define a misorientation between grains: two components of the axis of rotation, and an angle of rotation. It follows that a rotation matrix must also have only three independent terms. The relationship between the axis-angle pair and rotation matrix is given in Equation 7.9.

From the definition of a coordinate transformation matrix, each column of (Y J X) gives the components of a basis vector of X in the basis Y. It follows that

$$[1\,0\,0]_{\mathrm{X}} \parallel [2\,\bar{1}\,2]_{\mathrm{Y}} \qquad [0\,1\,0]_{\mathrm{X}} \parallel [2\,2\,\bar{1}]_{\mathrm{Y}} \qquad [0\,0\,1]_{\mathrm{X}} \parallel [\bar{1}\,2\,2\,]_{\mathrm{Y}}.$$

Suppose now that there exists another ferrite crystal (basis symbol Z), such that

$$[0\,\bar{1}\,0]_{\mathrm{Z}} \parallel [2\,\bar{1}\,2]_{\mathrm{Y}} \qquad [1\,0\,0]_{\mathrm{Z}} \parallel [2\,2\,\bar{1}]_{\mathrm{Y}} \qquad [0\,0\,1]_{\mathrm{Z}} \parallel [\bar{1}\,2\,2]_{\mathrm{Y}}$$

$$(\mathrm{Y\;J\;Z}) = \frac{1}{3}\begin{pmatrix} 2 & \bar{2} & \bar{1} \\ 2 & 1 & 2 \\ \bar{1} & \bar{2} & 2 \end{pmatrix}$$

with the crystal Y being generated from Z by a right-handed rotation of 70.52° about $[1\,0\,\bar{1}]_{\mathrm{Z}}$ direction. It can easily be demonstrated that

$$(\mathrm{Z\;J\;X}) = \begin{pmatrix} 0 & 1 & 0 \\ \bar{1} & 0 & 0 \\ 0 & 0 & 1 \end{pmatrix} \qquad \text{from} \qquad (\mathrm{Z\;J\;X}) = (\mathrm{Z\;J\;Y})(\mathrm{Y\;J\;X})$$

so that crystal X can be generated from Z by a rotation of 90° about $[0\;0\;1]_{\mathrm{X}}$ axis. However, this clearly is a symmetry operation of a cubic crystal, and it follows that crystal X can never be distinguished experimentally from crystal Z, so that the matrices (Y J X) and (Y J Z) are crystallographically equivalent, as are the corresponding axis-angle pair descriptions. In other words, Y can be generated from X either by a rotation of 60° about $[1\;1\;1]_{\mathrm{X}}$, or by a rotation of 70.52° about $[\bar{1}\;0\;1]_{\mathrm{X}}$. The two axis-angle pair representations are equivalent. There are actually 24 matrices like (Z J X) which represent symmetry rotations in cubic systems. It follows that a cubic bicrystal can be represented in 24 equivalent ways, with 24 axis-angle pairs. Any rotation matrix like (Y J X) when multiplied by rotation matrices representing symmetry operations (e.g., (Z J X)) will lead to the 24 axis-angle pair representations. The degeneracy of other structures is as follows [19, 20]: cubic (24), hexagonal (12), hexagonal close-packed (6), tetragonal (8), trigonal (6), orthorhombic (4), monoclinic (2), and triclinic (1). In general, the number N of axis-angle pairs is given by

$$N = 1 + N_2 + 2N_3 + 3N_4 + 5N_6$$

where N_2, N_3, N_4, and N_6 refer to the number of diads, triads, tetrads, and hexads in the symmetry elements of the structure concerned.

Figure 9.3 is an electron diffraction pattern taken from an internally twinned martensite plate in a Fe-4Ni-0.4C wt% steel. It contains two $\langle 0\ 1\ 1\rangle$ bcc zones, one from the parent plate (m) and the other from the twin (t). The pattern clearly illustrates how symmetry makes it possible to represent the same bi-crystal in terms of more than one axis-angle pair. This particular pattern shows that the twin crystal can be generated from the parent in at least three different ways: (i) rotation of 70.52° about the $\langle 0\ 1\ 1\rangle$ zone axes of the patterns, (ii) rotation of 180° about the $\{1\ 1\ \overline{1}\}$ plane normal, and (iii) rotation of 180° about the $\{2\ \overline{1}\ 1\}$ plane normal. It is apparent that these three operations would lead to the same orientation relation between the twin and the parent lattices.

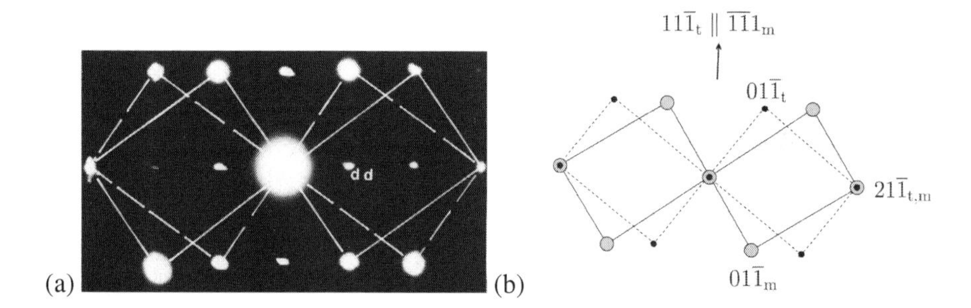

FIGURE 9.3
(a) Electron diffraction pattern from a martensite plate (m) and its twin (t). Spots not connected by lines (e.g., "dd") arise from double diffraction. (b) Interpretation of the diffraction pattern.

Example 9.4: "Double" twinning

Plates of bcc martensite in Fe-30.4Ni wt% contain $\{1\ 1\ \overline{2}\}$ transformation twins, the two twin orientations X and Y being related by a rotation of 60° about a $\langle 1\ 1\ 1\rangle$ axis. Deformation of this martensite at low temperatures leads to the formation of twins on $\{5\ 8\ 11\}$ planes, the direction of the twinning shear being $\langle \overline{5}\ \overline{1}\ 3\rangle$. This is a very rare mode of twinning deformation in bcc systems; show that its occurrence may be related to the fact that such twins can propagate without any deviation, across the already existing transformation twins in the martensite [21].

The orientation relationship between the transformation twins is clearly the same as the matrix (Y J X) of Example 9.4. It is easy to show therefore, that:

$$[\overline{5}\,\overline{1}\,3]_X \parallel \langle\overline{5}\,3\,\overline{1}\rangle_Y$$

$$(5\,8\,11)_X \parallel \{5\,11\,8\}_Y.$$

It follows that $\{5\ 8\ 11\}$ deformation twins can propagate without deviation across the transformation twins, since the above planes and directions, respectively, are crystallographically equivalent and indeed parallel. This may explain the occurrence of such an unusual deformation twinning mode.

9.5 The metric

For cubic crystals, it is a familiar result that if the indices $[u_1\ u_2\ u_3]$ of a direction **u** in the lattice are numerically identical to the Miller indices $(h_1\ h_2\ h_3)$ of a plane in the lattice, then the normal to this plane (**h**) is parallel to the direction mentioned. In other words, $\mathbf{u} \parallel \mathbf{h}$, and since $[\mathrm{A}; \mathbf{u}] = [u_1\ u_2\ u_3]$ and $(\mathbf{h}; \mathrm{A}^*) = (h_1\ h_2\ h_3)$, we have $[u_1\ u_2\ u_3] = [h_1\ h_2\ h_3]$. ("A" represents the basis of the cubic system concerned, and "A*" the corresponding reciprocal basis, in the usual way).

However, this is a special case reflecting the high symmetry of cubic systems, and it is generally not true that if $u_i = h_i$, then $\mathbf{u} \parallel \mathbf{h}$. For example, the $[1\ 2\ 3]$ direction in cementite is not parallel to the $(1\ 2\ 3)$ plane normal.

Consider an arbitrary crystal system, defined by a basis A (basis vectors $\mathbf{a}_i$), and by the corresponding reciprocal basis A* consisting of the basis vectors $\mathbf{a}_i^*$. To find the angle between the direction **u** and the plane normal **h**, it would be useful to have a matrix (A* G A), which allows the transformation of the components of a vector referred to the basis A, to those referred to its reciprocal basis A*. (The symbol G is used, rather than J, to follow convention). This matrix, called the metric, with components G_{ij} can be determined in exactly the same manner as any other coordinate transformation matrix. Each column of (A* G A) thus consists of the components of one of the basis vectors of A, when referred to the basis A*. For example,

$$\mathbf{a}_1 = G_{11}\mathbf{a}_1^* + G_{21}\mathbf{a}_2^* + G_{31}\mathbf{a}_3^*$$

Taking successive scalar dot products with $\mathbf{a}_1$, $\mathbf{a}_2$, and $\mathbf{a}_3$, respectively, on both sides yields:

$$G_{11} = \mathbf{a}_1.\mathbf{a}_1, \qquad G_{21} = \mathbf{a}_1.\mathbf{a}_2 \qquad G_{31} = \mathbf{a}_1.\mathbf{a}_3$$

since $\mathbf{a}_i.\mathbf{a}_j^* = 0$ when $i \neq j$. The rest of the elements of (A* G A) can be determined in a similar way, so that

$$(\mathrm{A}^*\ \mathrm{G}\ \mathrm{A}) = \begin{pmatrix} \mathbf{a}_1.\mathbf{a}_1 & \mathbf{a}_2.\mathbf{a}_1 & \mathbf{a}_3.\mathbf{a}_1 \\ \mathbf{a}_1.\mathbf{a}_2 & \mathbf{a}_2.\mathbf{a}_2 & \mathbf{a}_3.\mathbf{a}_2 \\ \mathbf{a}_1.\mathbf{a}_3 & \mathbf{a}_2.\mathbf{a}_3 & \mathbf{a}_3.\mathbf{a}_3 \end{pmatrix}. \qquad (9.11)$$

It is easily demonstrated that the determinant of (A* G A) equals the square of the volume of the cell formed by the basis vectors of A. We note in passing that for orthonormal coordinates, (Z* G Z) =**I**. A list of metric tensors is given in Table 9.1.

Example 9.5: Plane normals and directions in an orthorhombic structure

A crystal with an orthorhombic structure has lattice parameters a, b, and c. If the edges of the orthorhombic unit cell are taken to define the basis θ, determine the metric (θ* G θ). Hence derive the equation giving the angle ϕ between a plane normal $(\mathbf{h}; \theta^*) = (h_1\ h_2\ h_3)$ and any direction $[\theta; \mathbf{u}] = [u_1\ u_2\ u_3]$.

From the definition of a scalar dot product, $\mathbf{h}.\mathbf{u}/|\mathbf{h}||\mathbf{u}| = \cos\phi$. Now,

$$(\theta^*\ \mathrm{G}\ \theta) = \begin{pmatrix} a^2 & 0 & 0 \\ 0 & b^2 & 0 \\ 0 & 0 & c^2 \end{pmatrix} \qquad (\theta\ \mathrm{G}\ \theta^*) = \begin{pmatrix} a^{-2} & 0 & 0 \\ 0 & b^{-2} & 0 \\ 0 & 0 & c^{-2} \end{pmatrix}.$$

From Equation 5.8,

$$\begin{aligned} |\mathbf{h}|^2 &= \mathbf{h}.\mathbf{h} = (\mathbf{h};\theta^*)[\theta;\mathbf{h}] \\ &= (\mathbf{h};\theta^*)(\theta\ \mathrm{G}\ \theta^*)[\theta^*;\mathbf{h}] \\ &= h_1^2/a^2 + h_2^2/b^2 + h_3^2/c^2. \end{aligned}$$

Similarly,

$$\begin{aligned} |\mathbf{u}|^2 &= \mathbf{u}.\mathbf{u} = (\mathbf{u};\theta)[\theta^*;\mathbf{u}] \\ &= (\mathbf{u};\theta)(\theta^*\ \mathrm{G}\ \theta)[\theta;\mathbf{u}] \\ &= u_1 a^2 + u_2 b^2 + u_3^2 c^2 \end{aligned}$$

Since

$$\mathbf{h}.\mathbf{u} = (\mathbf{h};\theta^*)[\theta;\mathbf{u}] = h_1u_1 + h_2u_2 + h_3u_3$$

it follows that

$$\cos\phi = \frac{(h_1u_1 + h_2u_2 + h_3u_3)}{\sqrt{(h_1^2/\ a^2 + h_2^2/b^2 + h_3^2/c^2)(u_1a^2 + u_2b^2 + u_3^2c^2)}}.$$

9.6 More about the vector cross product

Suppose that the basis vectors **a**, **b**, and **c** of the basis θ define an orthorhombic unit cell, then the cross product between two arbitrary vectors **u** and **v** referred to this basis may be written:

$$\mathbf{u} \wedge \mathbf{v} = (u_1\mathbf{a} + u_2\mathbf{b} + u_3\mathbf{c}) \wedge (v_1\mathbf{a} + v_2\mathbf{b} + v_3\mathbf{c})$$

where $[\theta;\mathbf{u}] = [u_1\ u_2\ u_3]$ and $[\theta;\mathbf{v}] = [v_1\ v_2\ v_3]$. This equation can be expanded to give:

$$\mathbf{u} \wedge \mathbf{v} = (u_2v_3 - u_3v_2)(\mathbf{b} \wedge \mathbf{c}) + (u_3v_1 - u_1v_3)(\mathbf{c} \wedge \mathbf{a}) + (u_1v_2 - u_2v_1)(\mathbf{a} \wedge \mathbf{b})$$

Since $\mathbf{a}.\mathbf{b}\wedge\mathbf{c} = V$, the volume of the orthorhombic unit cell, and since $\mathbf{b}\wedge\mathbf{c} = V\mathbf{a}^*$, it follows that

$$\mathbf{u} \wedge \mathbf{v} = V\Big[(u_2v_3 - u_3v_2)\mathbf{a}^* + (u_3v_1 - u_1v_3)\mathbf{b}^* + (u_1v_2 - u_2v_1)\mathbf{c}^*\Big].$$

TABLE 9.1
Elements of metric tensor (A* G A) and its inverse. The term $\phi = (-2\cos\{\beta\}\cos\{\alpha\}\cos\{\gamma\} + \cos^2\{\beta\} + \cos^2\{\alpha\} + \cos^2\{\gamma\} - 1)^{-1}$

					(A* G A)				
Class	G_{11}	G_{12}	G_{13}	G_{21}	G_{22}	G_{23}	G_{31}	G_{32}	G_{33}
Cubic	a^2	0	0	0	a^2	0	0	0	a^2
Tetragonal	a^2	0	0	0	a^2	0	0	0	c^2
Orthorhombic	a^2	0	0	0	b^2	0	0	0	c^2
Hexagonal	a^2	$-0.5a^2$	0	$-0.5a^2$	a^2	0	0	0	c^2
Monoclinic	a^2	0	$ac\cos\{\beta\}$	0	b^2	0	$ac\cos\{\beta\}$	0	c^2
Triclinic	a^2	$ab\cos\{\gamma\}$	$ac\cos\{\beta\}$	$ab\cos\{\gamma\}$	b^2	$bc\cos\{\alpha\}$	$ac\cos\{\beta\}$	$bc\cos\{\alpha\}$	c^2

					(A G A*)				
Class	G_{11}	G_{12}	G_{13}	G_{21}	G_{22}	G_{23}	G_{31}	G_{32}	G_{33}
Cubic	a^{-2}	0	0	0	a^{-2}	0	0	0	a^{-2}
Tetragonal	a^{-2}	0	0	0	a^{-2}	0	0	0	c^{-2}
Orthorhombic	a^{-2}	0	0	0	b^{-2}	0	0	0	c^{-2}
Hexagonal	$\frac{b^2}{a^2b^2-0.25a^4}$	$-\frac{2}{a^2-4b^2}$	0	$-\frac{2}{a^2-4b^2}$	$-\frac{4}{a^2-4b^2}$	0	0	0	c^{-2}
Monoclinic	$\frac{\csc^2\{\beta\}}{a^2}$	0	$-\frac{\cot\{\beta\}\csc\{\beta\}}{ac}$	0	b^{-2}	0	$-\frac{\cot\{\beta\}\csc\{\beta\}}{ac}$	0	$\frac{\csc^2\{\beta\}}{c^2}$
Triclinic	$-\phi\frac{\sin^2\{\alpha\}}{a^2}$	$-\phi\frac{\cos\{\beta\}\cos\{\alpha\}-\cos\{\gamma\}}{ab}$	$\phi\frac{\cos\{\beta\}-\cos\{\alpha\}\cos\{\gamma\}}{ac}$	$-\phi\frac{\cos\{\beta\}\cos\{\alpha\}-\cos\{\gamma\}}{ab}$	$-\phi\frac{\sin^2\{\beta\}}{b^2}$	$\phi\frac{\cos\{\alpha\}-\cos\{\beta\}\cos\{\gamma\}}{bc}$	$\phi\frac{\cos\{\beta\}-\cos\{\alpha\}\cos\{\gamma\}}{ac}$	$\phi\frac{\cos\{\alpha\}-\cos\{\beta\}\cos\{\gamma\}}{bc}$	$-\phi\frac{\sin^2\{\gamma\}}{c^2}$

Hence, $\mathbf{u} \wedge \mathbf{v}$ gives a vector whose components are expressed in the reciprocal basis. Writing $\mathbf{x} = \mathbf{u} \wedge \mathbf{v}$, with $(\mathbf{x}; \theta^*) = (w_1\ w_2\ w_3)$, it follows that $w_1 = V(u_2v_3 - u_3v_2)$, $w_2 = V(u_3v_1 - u_1v_3)$ and $w_3 = V(u_1v_2 - u_2v_1)$. The cross product of two directions thus gives a normal to the plane containing the two directions. If necessary, the components of $\mathbf{x}$ in the basis θ can easily be obtained using the metric, since $[\theta; \mathbf{x}] = (\theta\ G\ \theta^*)[\theta^*; \mathbf{x}]$. Similarly, the cross product of two vectors $\mathbf{h}$ and $\mathbf{k}$ which are referred to the reciprocal basis θ^*, such that $(\mathbf{h}; \theta^*) = (h_1\ h_2\ h_3)$ and $(\mathbf{k}; \theta^*) = (k_1\ k_2\ k_3)$, can be shown to be:

$$\mathbf{h} \wedge \mathbf{k} = \frac{1}{V}\left[(h_2k_3 - h_3k_2)\mathbf{a} + (h_3k_1 - h_1k_3)\mathbf{b} - (h_2k_1 - h_1k_2)\mathbf{c}\right].$$

Hence, $\mathbf{h} \wedge \mathbf{k}$ gives a vector whose components are expressed in the real basis. The vector cross product of two plane normals gives a direction (zone axis) which is common to the two planes represented by the plane normals.

9.7 Summary

The term "orientation relationship" has more than one meaning. When the relationship is reproducible, for example when precipitates form in a solid matrix away from heterogeneous nucleation sites, the strong implication is that there is something that favors that behavior. The reasons could include:

1. The reduction in interfacial energy between the precipitate and matrix when a specific relationship is adopted. This facilitates nucleation because the activation energy for nucleation scales with the cube of the interfacial energy per unit area.
2. That the nucleus is generated by a homogeneous deformation of the matrix, in which case the deformation that minimizes the strain energy will determine the observed orientation relationship.

The second kind of relationship is not reproducible but arises because deformation or other external influences induce the alignment of crystals in a polycrystalline sample. In this case, the average orientation and the (different) relative orientations of adjacent crystals matter in determining the anisotropy and macroscopic properties of the material as a whole. This subject falls in the domain of crystallographic texture as introduced in Chapter 6, but in geology a different descriptor is used: *fabric*. Ice at ambient pressure has a hexagonal structure with the resistance to slip being some 60 times larger on non-basal planes [22]. Initially isotropic polycrystalline ice then becomes anisotropic when fabric develops because the basal slip-planes of different crystals tend to align during flow [23]. A similar effect happens when hexagonal titanium alloys are friction stir welded [24].

References

1. P. L. Ryder, and W. Pitsch: The crystallographic analysis of grain boundary precipitation, *Acta Metallurgica*, 1966, **14**, 1437–1448.

2. P. L. Ryder, W. Pitsch, and R. F. Mehl: Crystallography of the precipitation of ferrite on austenite grain boundaries in a Co-20%Fe alloy, *Acta Metallurgica*, 1967, **15**, 1431–1440.

3. J. W. Christian: Martensitic transformations: a current assessment, In: *The mechanism of phase transformations in crystalline solids, Monography 33*. London, U.K.: Institute of Metals, 1969:129–142.

4. K. W. Andrews: The structure of cementite and its relation to ferrite, *Acta Metallurgica*, 1963, **11**, 939–946.

5. H. K. D. H. Bhadeshia: *Bainite in steels: theory and practice*: 3rd ed., Leeds, U.K.: Maney Publishing, 2015.

6. A. I. Tyshchenko, W. Theisen, A. Oppenkowski, S. Siebert, O. N. Razumov, A. P. Skoblik, V. A. Sirosh, Y. N. Petrov, and V. G. Gavriljuk: Low-temperature martensitic transformation and deep cryogenic treatment of a tool steel, *Materials Science & Engineering A*, 2010, **527**, 7027–7039.

7. W. Hume-Rothery, G. V. Raynor, and A. T. Little: On the carbide and nitride particles in titanium steels, *Journal of the Iron and Steel Institute*, 1942, **145**, 129–141.

8. D. N. Shackleton, and P. M. Kelly: The crystallography of cementite precipitation in the bainite tranformation, *Acta Metallurgica*, 1967, **15**, 979–992.

9. H. K. D. H. Bhadeshia: The lower bainite transformation and the significance of carbide precipitation, *Acta Metallurgica*, 1980, **28**, 1103–1114.

10. W. Pitsch: Der orientierungszusammenhang zwischen zementit und austenit, *Acta Metallurgica*, 1962, **10**, 897–900.

11. H. Lipson, and N. J. Petch: The crystal structure of cementite Fe_3C, *Journal of the Iron and Steel Institute*, 1940, **142**, 95P–103P.

12. V. S. Urusov, and T. N. Nadezhina: Frequency distribution and selection of space groups in inorganic crystal chemistry, *Journal of Structural Chemistry*, 2009, **50**, S22–S37.

13. G. V. Kurdjumov, and G. Sachs: über den mechanismus der stahlhärtung, *Zietschrift für Physik A Hadrons and Nuclei*, 1930, **64**, 325–343.

14. Z. Nishiyama: X-ray investigation of the mechanism of the transformation from face centered cubic lattice to body centered cubic, *Science Reports of Tohoku Imperial University*, 1934, **23**, 637–634.

15. G. Wassermann: Einflusse der α-γ-umwandlung eines irreversiblen nickelstahls auf kristallorientierung und zugfestigkeit, *Arch. Eisenhüttenwes.*, 1933, **6**, 347–351.

16. A. Crosky, P. G. McDougall, and J. S. Bowles: The crystallography of the precipitation of alpha rods from beta Cu Zn alloys, *Acta Metllurgica*, 1980, **28**, 1495–1504.

17. L. Euler: Formulae generales pro trandlatione quacunque corporum rigidorum, *Novi commentarii Academiae Scientiarum Imperialis Petropolitanae*, 1975, **20**, 189–207.

18. B. Palais, and R. Palais: Euler's fixed point theorem: The axis of a rotation, *Journal of Fixed Point Theory and Applications*, 2007, **2**, 215–220.

19. F. F. Lange: Mathematical characterization of a general bicrystal, *Acta Metallurgica*, 1967, **15**, 311–318.

20. C. Goux: Structure of joined grains-crystallographic consideration and methods of structure estimation, *Canadian Metallurgical Quarterly*, 1974, **13**, 9–31.

21. P. C. Rowlands, E. O. Fearon, and M. Bevis: Deformation twinning in Fe-Ni and Fe-Ni-C martensites, *Trans. A.I.M.E.*, 1968, **242**, 1559–1562.

22. P. Duval, M. F. Ashby, and I. Andereman: Rate-controlling processes in the creep of polycrystalline ice, *The Journal of Physical Chemistry*, 1983, **87**, 4066–4074.

23. O. Castelnau, and P. Duval: Viscoplastic modelling of texture development in polycrystalline ice with a self-consistent approach: comparison with bound estimates, *Journal of Geophysical Research*, 1996, **101**, 13851–13868.

24. R. Fonda: Texture development in friction stir welds, *Science and Technology of Welding and Joining*, 2011, **16**, 288–294.

10

Homogeneous deformations

Abstract

Many operations in crystallography involve physical deformations that lead to an observable change in shape of the affected region of the crystal. Such deformations can in general be factorized mathematically into components that involve only a rigid body rotation, and another that is a pure strain which involves the stretching of the body along three orthogonal directions. The sign and magnitude of the stretch can vary according to the nature of the deformation. Homogeneous deformations help represent changes in crystal structure or the reorientation of the crystal. They can also be used to define the nature of interfaces between phases.

10.1 Introduction

A homogeneous deformation is one in which points that initially are collinear remain so after the deformation. It follows that a plane will similarly remain a plane in such circumstances. Each volume element in the solid therefore experiences the same change in shape. Figure 10.1 illustrates the difference between a homogeneous and heterogeneous deformation. In this chapter, we describe deformations that alter the length and/or direction of vectors, i.e., physical deformations, with the initial and resultant vectors referred to a fixed reference frame.

Elastic strain energy calculations often are based on the assumption that the deformations are homogeneous [1] and this usually is justified given that the elastic strain fields may extend over large distances whereas atomic perturbations are just that. Similarly, the stability of crystal lattices has been studied by subjecting them to homogeneous deformation along the direction of one of its axes, and to another small deformation along an arbitrary direction [2].

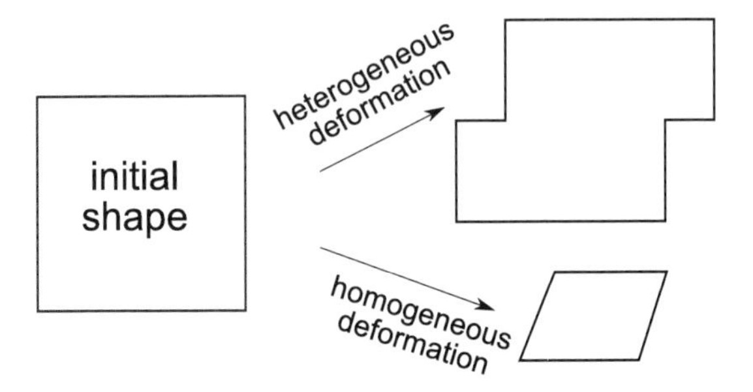

FIGURE 10.1
An illustration of the difference between homogeneous deformations and those that are heterogeneous. Examples of the former include shear, hydrostatic expansion and uniaxial expansion, whereas slip caused by the motion of a dislocation, or the formation of a crack, or the motion of rigid continental plates, involves heterogeneous deformation.

10.2 Homogeneous deformations

We can now return to the question of martensite, and how the homogeneous deformation known as the Bain strain [3] might transform the austenite lattice (parameter a_γ) to a bcc or bct martensite (parameters $a_{\alpha'}$, $c_{\alpha'}$ where $a_{\alpha'} \leq c_{\alpha'}$). Referring to Figure 10.2, the basis "A" is defined by the basis vectors $\mathbf{a}_i$, each of magnitude a_γ, and basis "B" is defined by basis vectors $\mathbf{b}_i$ as illustrated in Figure 10.2b. Focusing attention on the bct representation of the austenite unit cell, it is evident that a compression along the $[0\ 0\ 1]_\mathrm{B}$ axis, coupled with expansions along $[1\ 0\ 0]_\mathrm{B}$ and $[0\ 1\ 0]_\mathrm{B}$ would accomplish the transformation of the bct austenite unit cell into a bcc or bct α' cell. This deformation, referred to the basis B, can be written as:

$$\eta_1 = \eta_2 = \sqrt{2}(a_{\alpha'}/a_\gamma)$$

along $[1\ 0\ 0]_\mathrm{B}$ and $[0\ 1\ 0]_\mathrm{B}$, respectively, and

$$\eta_3 = c_{\alpha'}/a_\gamma$$

along the $[0\ 0\ 1]_\mathrm{B}$ axis.

The deformation just described can be written as a 3×3 matrix, referred to the austenite lattice. In other words, we imagine that a part of a single crystal of austenite undergoes the prescribed deformation, allowing us to describe the strain in terms of the remaining (and undeformed) region, which forms a fixed reference basis. Hence, the deformation matrix does not involve any change of basis, and this point is emphasised by writing it as (A S A), with the same basis symbol on both sides of S:

$$[\mathrm{A}; \mathbf{v}] = (\mathrm{A\ S\ A})[\mathrm{A}; \mathbf{u}].$$

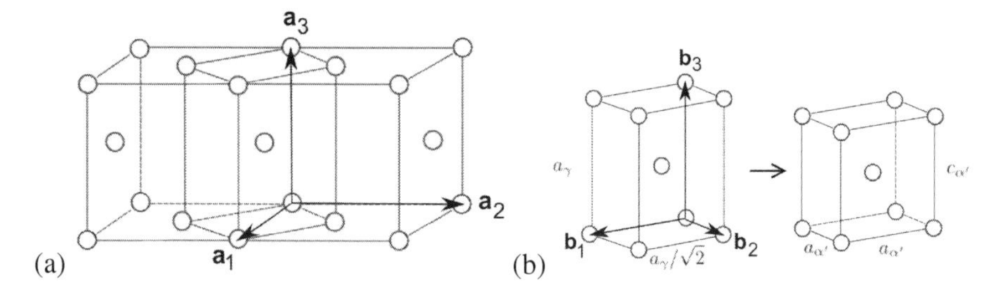

FIGURE 10.2
(a) Two austenite unit cells, one the conventional cubic-F (basis A) and the other a body-centered tetragonal cell (basis B). Some of the atoms in face-centering positions on the vertical faces of the the cubic cell of austenite are not illustrated for the sake of clarity. (b) How the body centered tetragonal cell can be deformed by the Bain strain into that of body-centered cubic or body-centered tetragonal cell of martensite.

(A S A) converts the vector [A;**u**] into a new vector [A;**v**], with **v** still referred to the basis A.

The difference between a coordinate transformation (B J A) and a deformation matrix (A S A) is illustrated in Figure 10.3, where $\mathbf{a}_i$ and $\mathbf{b}_i$ are the basis vectors of the bases A and B, respectively.

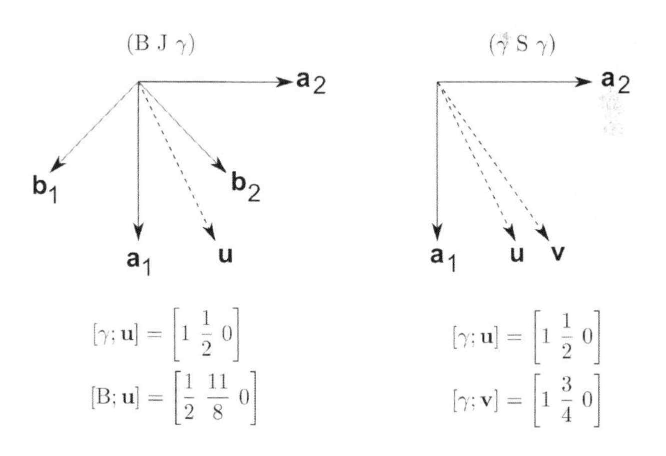

FIGURE 10.3
Difference between coordinate transformation $(\mathrm{B\ J}\ \gamma)$ and deformation matrix $(\gamma\ \mathrm{S}\ \gamma)$. The latter is defined with respect to a single reference frame whereas the former deals with a change in the reference frame.

We see that an advantage of the Mackenzie–Bowles notation is that it enables a clear distinction to be made between 3×3 matrices which represent changes of axes and those which represent physical deformations referred to a single axis system. The following additional information can be deduced from Figure 10.2:

Vector components before Bain strain	Vector components after Bain strain
$[1\ 0\ 0]_{\mathrm{A}}$	$[\eta_1\ 0\ \ 0]_{\mathrm{A}}$
$[0\ 1\ 0]_{\mathrm{A}}$	$[0\ \ \eta_2\ 0]_{\mathrm{A}}$
$[0\ 0\ 1]_{\mathrm{A}}$	$[0\ \ 0\ \ \eta_3]_{\mathrm{A}}$

so the matrix (A S A) can be written as follows, with each column consisting of the components of a vector equal to the basis vector of A, following deformation:

$$(\mathrm{A\ S\ A}) = \begin{pmatrix} \eta_1 & 0 & 0 \\ 0 & \eta_2 & 0 \\ 0 & 0 & \eta_3 \end{pmatrix}. \tag{10.1}$$

Each column of the deformation matrix represents the components of the new vector formed as a result of the deformation of a vector equal to one of the basis vectors of A.

The strain represented by (A S A) is called a pure strain because it involves stretching along three orthogonal axes; for this reason, the matrix representation (A S A) is symmetrical. Stretching means extensions or contractions but no rotations, and the three initially orthogonal directions are referred to as the principal axes which obviously remain orthogonal and unrotated during the deformation.

The ratio of its final length to initial length is the *principal deformation* associated with that principal axis. The directions $[1\ 0\ 0]_{\mathrm{B}}$, $[0\ 1\ 0]_{\mathrm{B}}$, and $[0\ 0\ 1]_{\mathrm{B}}$ are therefore the principal axes of the Bain strain, and η_i the respective principal deformations. In the particular example under consideration, $\eta_1 = \eta_2$ so that any two perpendicular lines in the $(0\ 0\ 1)_{\mathrm{B}}$ plane could form two of the principal axes. Since $\mathbf{a}_3 \parallel \mathbf{b}_3$ and since $\mathbf{a}_1$ and $\mathbf{a}_2$ lie in $(0\ 0\ 1)_{\mathrm{B}}$, it is clear that the vectors $\mathbf{a}_i$ must also be the principal axes of the Bain deformation.

Since the deformation matrix (A S A) is referred to a basis system which coincides with the principal axes, the off-diagonal components of this matrix must be zero. Column 1 of the matrix (A S A) represents the new coordinates of the vector [1 0 0], after the latter has undergone the deformation described by (A S A), and a similar rationale applies to the other columns. (A S A) deforms the fcc γ lattice into a bcc or bct α' lattice, and the ratio of the final to initial volume of the material is simply $\eta_1\eta_2\eta_3$; the ratio is more generally given by the determinant of the deformation matrix. Finally, it should be noted that any tetragonality in the martensite can readily be taken into account by writing $\eta_3 = c_{\alpha'}/a_\gamma$, where $c_{\alpha'}/a_\alpha$ is the aspect ratio of the bct martensite unit cell.

Example 10.1: Bain strain: undistorted vectors

Given that the principal distortions of the Bain strain (A S A), referred to the crystallographic axes of the fcc γ lattice (lattice parameter a_γ), are $\eta_1 = \eta_2 =$

1.123883, and $\eta_3 = 0.794705$, show that the vector

$$[\mathrm{A};\mathbf{x}] = [-0.645452\ 0.408391\ 0.645452]$$

remains undistorted, though not unrotated as a result of the operation of the Bain strain. Furthermore, show that for **x** to remain unextended as a result of the Bain strain, its components x_i must satisfy the equation

$$(\eta_1^2 - 1)x_1^2 + (\eta_2^2 - 1)x_2^2 + (\eta_3^2 - 1)x_3^2 = 0. \tag{10.2}$$

As a result of the deformation (A S A), the vector **x** becomes a new vector **y**, according to the equation

$$(\mathrm{A\ S\ A})[\mathrm{A};\mathbf{x}] = [\mathrm{A};\mathbf{y}] = [\eta_1 x_1\ \eta_2 x_2\ \eta_3 x_3] = [-0.723412\ 0.458983\ 0.512944]$$

Now,

$$|\mathbf{x}|^2 = (\mathbf{x};\mathrm{A}^*)[\mathrm{A};\mathbf{x}] = a_\gamma^2(x_1^2 + x_2^2 + x_3^2) \tag{10.3}$$

and,

$$|\mathbf{y}|^2 = (\mathbf{y};\mathrm{A}^*)[\mathrm{A};\mathbf{y}] = a_\gamma^2(y_1^2 + y_2^2 + y_3^2). \tag{10.4}$$

Using these equations, and the numerical values of x_i and y_i obtained above, it is easy to show that $|\mathbf{x}| = |\mathbf{y}|$. It should be noted that although **x** remains unextended, it is rotated by the strain (A S A), since $x_i \neq y_i$. Using Equations 10.3 and 10.4 with $y_i = \eta_i x_i$, yields the required Equation 10.2. Since η_1 and η_2 are equal and greater than 1, and since η_3 is less than unity, Equation 10.2 amounts to the equation of a right-circular cone, the axis of which coincides with $[0\ 0\ 1]_\mathrm{A}$ Any vector initially lying on this cone will remain unextended as a result of the Bain strain.

This process was illustrated in Figure 8.4 where a spherical volume of austenite was deformed into an ellipsoid of revolution. Notice that the principal axes ($\mathbf{a}_i$) remain unrotated by the deformation, and that lines such as xw and yz which become $x'w'$ and $y'z'$, respectively, remain unextended by the deformation (since they are all diameters of the original sphere), although rotated through the angle θ. The lines xw and yz of course lie on the initial cone described by Equation 10.2. Suppose now that the ellipsoid resulting from the Bain strain is rotated through a right-handed angle of θ, about the axis $\mathbf{a}_2$, then Figure 8.4c illustrates that this rotation will cause the initial and final cones of unextended lines to touch along yz, bringing yz and $y'z'$ into coincidence. If the total deformation can therefore be described as (A S A) combined with the above rigid body rotation, then such a deformation would leave the line cd both unrotated and unextended; such a deformation is called an *invariant-line strain*. Notice that the total deformation, consisting of (A S A) and a rigid body rotation, is no longer a pure strain since the vectors parallel to the principal axes of (A S A) are rotated into the new positions $\mathbf{a}'_i$ Figure 8.4c.

It will later be shown that the lattice deformation in a martensitic transformation must contain an invariant line. Therefore, the Bain strain must be combined with a suitable rigid body rotation in order to define the total lattice deformation. This explains why the experimentally observed orientation relationship between martensite and austenite does not correspond to that implied by Figure 10.2. The need to have an invariant line in the martensite-austenite interface means that the Bain strain does not in itself constitute the total transformation strain, which can be factorized into the Bain strain and a rigid body rotation. It is this total strain which determines the final orientation relationship although the Bain strain accomplishes the total fcc to bcc lattice change.

10.3 Eigenvectors and eigenvalues

It is useful to establish a method of determining the directions which remain *unrotated*, though not necessarily undistorted, as a consequence of the deformation concerned. Vectors lying along such unrotated directions are called *eigenvectors* of the deformation matrix, and the ratios of their final to initial lengths are the corresponding *eigenvalues* of the matrix. Considering the deformation matrix (A S A), the unrotated directions may be determined by solving the equations

$$(\mathrm{A\ S\ A})[\mathrm{A};\mathbf{u}] = \lambda[\mathrm{A};\mathbf{u}] \tag{10.5}$$

where **u** is a unit vector lying along an eigenvector, A is a convenient orthonormal basis, and λ is a positive scalar quantity. This equation shows that the vector **u** does not change in direction as a result of (A S A), although its length changes by the ratio λ. If **I** is a 3×3 identity matrix, then on rearranging Equation 10.5, we obtain

$$\{(\mathrm{A\ S\ A}) - \lambda\mathbf{I}\}[\mathrm{A};\mathbf{u}] = 0 \tag{10.6}$$

which can be written more fully as:

$$\begin{pmatrix} S_{11}-\lambda & S_{12} & S_{13} \\ S_{21} & S_{22}-\lambda & S_{23} \\ S_{31} & S_{32} & S_{33}-\lambda \end{pmatrix} \begin{pmatrix} u_1 \\ u_2 \\ u_3 \end{pmatrix} = 0 \tag{10.7}$$

where $[\mathrm{A};\mathbf{u}] = [u_1\ u_2\ u_3]$. This system of homogeneous equations has non-trivial solutions if the determinant

$$\begin{vmatrix} S_{11}-\lambda & S_{12} & S_{13} \\ S_{21} & S_{22}-\lambda & S_{23} \\ S_{31} & S_{32} & S_{33}-\lambda \end{vmatrix} = 0. \tag{10.8}$$

The expansion of this determinant yields an equation which is in general cubic in λ; the roots of this equation are the three eigenvalues λ_i. Associated with each of the eigenvalues is a corresponding eigenvector whose components may be obtained by substituting each eigenvalue, in turn, into Equation 10.7. Of course, since every vector which lies along the unrotated direction is an eigenvector, if **u** is a solution of Equation 10.7 then $k\mathbf{u}$ must also satisfy Equation 10.7, k being a scalar constant. If the matrix (A S A) is real, then there must exist three eigenvalues, at least one of which is necessarily real. If (A S A) is symmetrical, then all three of its eigenvalues are real; the existence of three real eigenvalues does not however imply that the deformation matrix is symmetrical. Every real eigenvalue implies the existence of a corresponding vector which remains unchanged in direction as a result of the operation of (A S A).

Example 10.2: Eigenvectors and eigenvalues

Find the eigenvalues and eigenvectors of

$$(\mathrm{A\ S\ A}) = \begin{pmatrix} 18 & -6 & -6 \\ -6 & 21 & 3 \\ -6 & 3 & 21 \end{pmatrix}$$

To solve for the eigenvalues, we use Equation 10.8 to form the determinant

$$\begin{vmatrix} 18-\lambda & -6 & -6 \\ 6 & 21-\lambda & 3 \\ -6 & 3 & 21-\lambda \end{vmatrix} = 0$$

which on expansion gives the cubic equation

$$(12-\lambda)(\lambda-30)(\lambda-18) = 0$$

with the roots

$$\lambda_1 = 12, \qquad \lambda_2 = 30 \qquad \text{and} \qquad \lambda_3 = 18.$$

To find the eigenvector $\mathbf{u} = [\mathrm{A}; \mathbf{u}] = [u_1\ u_2\ u_3]$ corresponding to λ_1, we substitute λ_1 into Equation 10.7 to obtain

$$\begin{aligned} 6u_1 - 6u_2 - 6u_3 &= 0 \\ -6u_1 + 9u_2 + 3u_3 &= 0 \\ -6u_1 + 3u_2 + 9u_3 &= 0. \end{aligned}$$

These equations can be solved simultaneously to show that $u_1 = 2u_2 = 2u_3$. The other two eigenvectors, **v** and **x**, corresponding to λ_2 and λ_3, respectively, can be determined in a similar way. Hence, it is found that:

$$\begin{aligned} [\mathrm{A}; \mathbf{u}] &= (6^{-\frac{1}{2}})[2\ 1\ 1] \\ [\mathrm{A}; \mathbf{v}] &= (3^{-\frac{1}{2}})[\bar{1}\ 1\ 1] \\ [\mathrm{A}; \mathbf{x}] &= (2^{-\frac{1}{2}})[0\ 1\ \bar{1}]. \end{aligned}$$

All vectors parallel to **u**, **v**, or **x** remain unchanged in direction, though not in magnitude, due to the deformation (A S A).

10.3.1 Comments

(i) Since the matrix (A S A) is symmetrical, we find three real eigenvectors, which form an orthogonal set.

(ii) A negative eigenvalue implies that a vector initially parallel to the corresponding eigenvector becomes antiparallel (changes sign) on deformation. A deformation like this is physically impossible.

(iii) If a new orthonormal basis B is defined, consisting of unit basis vectors parallel to **u**, **v**, and **x**, respectively, then the deformation (A S A) can be expressed in the new basis with the help of a similarity transformation. Keeping like basis symbols adjacent, it follows that

$$(\mathrm{B\ S\ B}) = (\mathrm{B\ J\ A})(\mathrm{A\ S\ A})(\mathrm{A\ J\ B}) \tag{10.9}$$

where the columns of (A J B) consist of the components (referred to as the basis A) of the eigenvectors **u**, **v**, and **x**, respectively, so that

$$\begin{aligned}(\mathrm{B\ S\ B}) &= \begin{pmatrix} u_1 & u_2 & u_3 \\ v_1 & v_2 & v_3 \\ w_1 & w_2 & w_3 \end{pmatrix} \begin{pmatrix} 18 & -6 & -6 \\ 6 & 21 & 3 \\ -6 & 3 & 21 \end{pmatrix} \begin{pmatrix} u_1 & v_1 & w_1 \\ u_2 & v_2 & w_2 \\ u_3 & v_3 & w_3 \end{pmatrix} \\ &= \begin{pmatrix} 18 & 0 & 0 \\ 0 & 30 & 0 \\ 0 & 0 & 12 \end{pmatrix}.\end{aligned}$$

Notice that the off diagonal terms of (B S B) are zero because it is referred to a basis formed by the principal axes of the deformation – i.e., the three orthogonal eigenvectors. The matrix representing the Bain strain (page 136) is also diagonal because it too is referred to the principal axes of the strain. Any real symmetrical matrix can be diagonalized using the procedure illustrated above. (B S B) is called the *diagonal* representation of the deformation and this special representation will henceforth be identified by placing a bar over the matrix symbol: $(\mathrm{B}\ \overline{\mathrm{S}}\ \mathrm{B})$.

10.4 Stretch and rotation

A *pure strain* consists of simple extensions or contractions along the principal axes. The ratios of the final to initial lengths of vectors parallel to the principal axes are called the principal deformations, and the change in length per unit length the principal strains. It has a symmetrical matrix representation irrespective of the choice of basis. It is possible to find three real and orthogonal eigenvectors.

On the other hand, the shear (Z P_2 Z) illustrated in Figure 10.4a is not a pure deformation because it is possible to identify just two mutually perpendicular eigenvectors, both of which must lie in the invariant-plane. All other vectors are rotated by the shearing action. The original object, represented as a sphere, is sheared into an ellipsoid. The invariant-plane of the deformation contains the $\mathbf{z}_1$ and $\mathbf{z}_2$ axes. The deformation can be *imagined* to occur in two stages, the first one involving simple extensions and contractions along the $\mathbf{y}_1$ and $\mathbf{y}_3$ directions, respectively (Figure 10.4b) and the second involving a rigid body rotation of the ellipsoid, about the axis $\mathbf{z}_2 \parallel \mathbf{y}_2$, through a right-handed angle ϕ.

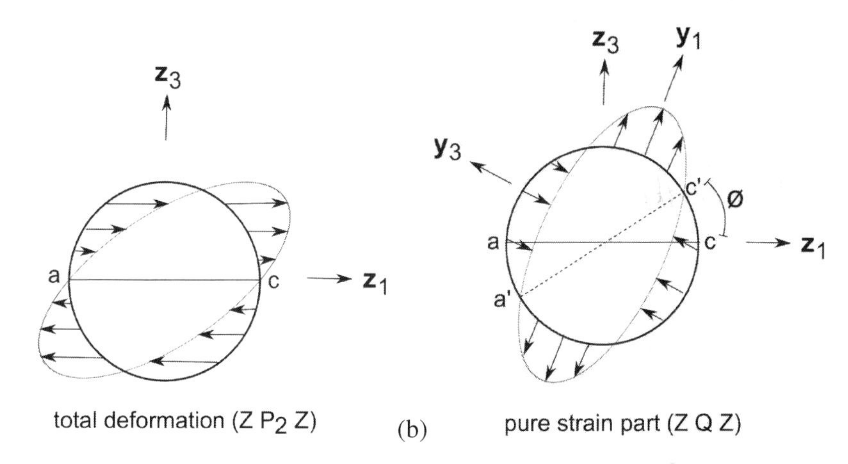

FIGURE 10.4
Factorization of a simple shear (Z P_2 Z) into a pure deformation (Z Q Z) and a right-handed rigid body rotation of ϕ about $\mathbf{z}_2$. In (a), ac is the trace of the invariant plane. (Z Q Z) leaves ac undistorted but rotated to $a'c'$ and rigid body rotation brings $a'c'$ into coincidence with ac. The axes $\mathbf{y}_1$, $\mathbf{y}_2$, and $\mathbf{y}_3$ are the principal axes of the pure deformation. The undeformed shape is represented as a sphere in three dimensions.

In essence, we have just carried out an imaginary factorization of the impure strain (Z P_2 Z) into a pure strain (Figure 10.4b) and a rigid body rotation. If the pure strain part is referred to as (Z Q Z) and the rotation part as (Z J Z), then

$$(\mathrm{Z\ P_2\ Z}) = (\mathrm{Z\ J\ Z})(\mathrm{Z\ Q\ Z}). \tag{10.10}$$

It was arbitrarily chosen that the pure strain would occur first and be followed by the rigid body rotation, but the reverse order is equally acceptable,

$$(\mathrm{Z\ P_2\ Z}) = (\mathrm{Z\ Q_2\ Z})(\mathrm{Z\ J_2\ Z})$$

where in general,

$$(\mathrm{Z\ Q_2\ Z}) \neq (\mathrm{Z\ Q\ Z}) \quad \text{and} \quad (\mathrm{Z\ J_2\ Z}) \neq (\mathrm{Z\ J\ Z}).$$

Any real deformation can in general be factorized into a pure strain and a rigid body rotation, but it is important to realize that the factorization is simply a mathematical convenience and the deformation does not actually occur in the two stages. The factorization in no way indicates the path by which the initial state reaches the final state and is merely phenomenological.

The actual factorization can be considered in terms of the arbitrary deformation (Z S Z), referred to as an orthonormal basis Z. Bearing in mind that (Z S′ Z) is the transpose of (Z S Z),

$$\begin{aligned} (\mathrm{Z\ S'\ Z})(\mathrm{Z\ S\ Z}) &= (\mathrm{Z\ Q'\ Z})(\mathrm{Z\ J'Z})(\mathrm{Z\ J\ Z})(\mathrm{Z\ Q\ Z}) \\ \text{or} \quad (\mathrm{Z\ S'\ Z})(\mathrm{Z\ S\ Z}) &= (\mathrm{Z\ Q\ Z})^2 \end{aligned} \tag{10.11}$$

since $(\mathrm{Z\ J'Z})(\mathrm{Z\ J\ Z}) = \mathbf{I}$ and $(\mathrm{Z\ Q'\ Z}) = (\mathrm{Z\ Q\ Z})$, since (Z Q Z) is a pure deformation with a symmetrical matrix representation. If the product (Z S′ Z)(Z S Z) is written as the symmetrical matrix (Z T Z), then the eigenvalues λ_i of (Z T Z) are also the eigenvalues of $(\mathrm{Z\ Q\ Z})^2$, so that the eigenvalues of (Z Q Z) are $\sqrt{\lambda_i}$. If the eigenvectors of (Z T Z) are **u**, **v**, and **w** (with corresponding eigenvalues λ_1, λ_2, and λ_3, respectively), then (Z T Z) can be diagonalized by similarity transforming it to another orthonormal basis Y formed by the vectors **u**,**v**, and **w**. From Equation 10.9,

$$(\mathrm{Y\ \overline{T}\ Y}) = \begin{pmatrix} \lambda_1 & 0 & 0 \\ 0 & \lambda_2 & 0 \\ 0 & 0 & \lambda_3 \end{pmatrix} = (\mathrm{Y\ J\ Z})(\mathrm{Z\ T\ Z})(\mathrm{Z\ J\ Y})$$

where the columns of (Z J Y) consist of the components of **u**,**v**, and **w**, respectively, when the latter are referred to as the Z basis. It follows that since $(\mathrm{Y\ \overline{Q}\ Y})^2 = (\mathrm{Y\ \overline{T}\ Y})$,

$$(\mathrm{Z\ Q\ Z}) = (\mathrm{Z\ J\ Y})(\mathrm{Y\ \overline{T}\ Y})^{\frac{1}{2}}(\mathrm{Y\ J\ Z})$$

where the square root of a diagonal matrix $(\mathrm{Y\ \overline{T}\ Y})$ is such that $(\mathrm{Y\ \overline{T}\ Y})^{\frac{1}{2}}(\mathrm{Y\ \overline{T}\ Y})^{\frac{1}{2}} = (\mathrm{Y\ \overline{T}\ Y})$. It follows that:

$$(\mathrm{Z\ Q\ Z}) = \begin{pmatrix} u_1 & v_1 & w_1 \\ u_2 & v_2 & w_2 \\ u_3 & v_3 & w_3 \end{pmatrix} \begin{pmatrix} \sqrt{\lambda_1} & 0 & 0 \\ 0 & \sqrt{\lambda_2} & 0 \\ 0 & 0 & \sqrt{\lambda_3} \end{pmatrix} \begin{pmatrix} u_1 & u_2 & u_3 \\ v_1 & v_2 & v_3 \\ w_1 & w_2 & w_3 \end{pmatrix}. \tag{10.12}$$

It is worth repeating that in Equation 10.12, λ_i are the eigenvalues of the matrix (Z S′ Z)(Z S Z) and u_i, v_i, and w_i are the components, in the basis Z of the eigenvectors of (Z S′ Z)(Z S Z). The rotation part of the strain (Z S Z) is simply

$$(\mathrm{Z\ J\ Z}) = (\mathrm{Z\ S\ Z})(\mathrm{Z\ Q\ Z})^{-1}. \tag{10.13}$$

10.5 Interfaces

A vector parallel to a principal axis of a pure deformation may become extended but is not changed in direction by the deformation. The ratio η of its final to initial length is called a principal deformation associated with that principal axis and the corresponding quantity $(\eta - 1)$ is called a principal strain. We have seen that when two of the principal strains of the pure deformation differ in sign from the third, all three being non-zero, it is possible to obtain a total strain which leaves one line invariant (page 107). It intuitively seems advantageous to have the invariant-line in the interface connecting the two crystals, since their lattices would then match exactly along that line.

A completely undistorted interface would have to contain two non-parallel directions which are invariant to the total transformation strain. The following example illustrates the characteristics of such a transformation strain, called an invariant-plane strain, which allows the existence of a plane which remains unrotated and undistorted during the deformation.

Example 10.3: Deformations and interfaces

A pure strain (A Q A), referred to an orthonormal basis "A" whose basis vectors are parallel to the principal axes of the deformation, has the principal deformations $\eta_1 = 1.192281$, $\eta_2 = 1$, and $\eta_3 = 0.838728$. Show that (A Q A) combined with a rigid body rotation gives a total strain which leaves a plane unrotated and undistorted.

Because (A Q A) is a pure strain referred to a basis composed of unit vectors parallel to its principal axes, it consists of simple extensions or contractions along the basis vectors $\mathbf{a}_1$, $\mathbf{a}_2$, and $\mathbf{a}_3$. Hence, Figure 10.5 can be constructed by analogy for the Bain strain illustrated on page 107. Since $\eta_2 = 1$, $ef \parallel \mathbf{a}_2$ remains unextended and unrotated by (A Q A), and if a rigid body rotation (about fe as the axis of rotation) is added to bring yz into coincidence with $y'z'$, then the two vectors ef and ab remain invariant to the total deformation. Any combination of ef and ab will also remain invariant, and hence all lines in the plane containing ef and $y'z'$ are invariant, giving an invariant plane. Thus, a pure strain when combined with a rigid body rotation can only generate an invariant-plane strain if two of its principal strains have opposite signs, the third being zero. Since it is the pure strain which actually accomplishes the lattice change (the rigid body rotation causes no further lattice change), any two lattices related by a pure strain with these characteristics may be joined by a fully coherent interface.

(A Q A) actually represents the pure strain part of the total transformation strain required to change an fcc lattice to an hcp (hexagonal close-packed) lattice, assuming that there is no volume change, by shearing on the $\{1\ 1\ 1\}_\gamma$ plane, in the $\langle 1\ 1\ \overline{2}\rangle_\gamma$ direction, the magnitude of the shear being equal to half the twinning shear (see Chapter 11). Consistent with the proof given above, a fully coherent interface is observed experimentally when hcp martensite is formed in this manner.

It is worth noting that the total deformation, i.e., the combination of the pure and rotational strains, completely defines the set of crystallographic features of the hcp martensite [4–6]. The set consists of the habit plane, orientation relationship, and shape deformation associated with the transformation.

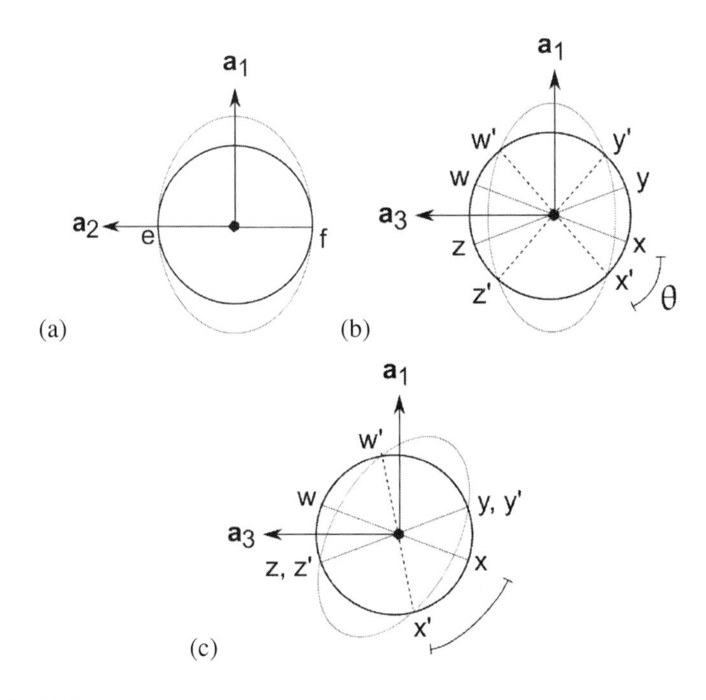

FIGURE 10.5
(a,b) Illustration of the strain (A Q A), the undeformed crystal represented initially as a sphere of diameter ef which also is an invariant line since $\eta_2 = 1$. The strain (c) illustrates that a combination of (A Q A) with a rigid body rotation about the axis ef generates another invariant line $z'y'$, so that the net deformation becomes an invariant-plane strain.

10.6 Topology of grain deformation

Metallic alloys are produced in very large quantities using plastic-deformation in order to achieve particular shapes and properties. The microstructure changes during deformation, with an increase in the defect density and in the amount of grain boundary area per unit volume (S_V) and grain edge length per unit volume (L_V). All of these changes are important in determining the course of phase transformations and recrystallization processes in general.

The equiaxed grain structure typical of most undeformed materials can be represented using Kelvin tetrakaidecahedra which accomplish the "division of space with a minimum of partitional area" [7]. A tetrakaidecahedron has 8 hexagonal and 6 square faces, Figure 10.6, with 36 edges, each of length a. All of the edges can be described in terms of just six vectors, as listed in Table 10.1. It then becomes possible to operate on these vectors by a deformation matrix in order to calculate consequential changes in grain parameters.

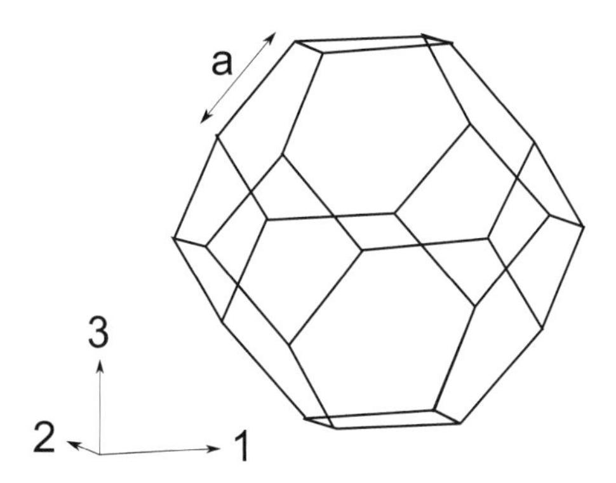

FIGURE 10.6
The Kelvin tetrakaidecahedron.

TABLE 10.1
Vectors defining the edges of a tetrakaidecahedron.

Vector	Components
1	$[a \quad 0 \quad 0]$
2	$[0 \quad a \quad 0]$
3	$[-\frac{a}{2} \quad -\frac{a}{2} \quad \frac{a}{\sqrt{2}}]$
4	$[\frac{a}{2} \quad -\frac{a}{2} \quad \frac{a}{\sqrt{2}}]$
5	$[\frac{a}{2} \quad \frac{a}{2} \quad \frac{a}{\sqrt{2}}]$
6	$[-\frac{a}{2} \quad \frac{a}{2} \quad \frac{a}{\sqrt{2}}]$

10.6.1 Plane strain deformation

A general deformation matrix **S** acts on a vector **u** to give a new vector **v** as follows [8–10]:

$$\begin{pmatrix} S_{11} & S_{12} & S_{13} \\ S_{21} & S_{22} & S_{23} \\ S_{31} & S_{32} & S_{33} \end{pmatrix} \begin{pmatrix} u_1 \\ u_2 \\ u_3 \end{pmatrix} = \begin{pmatrix} v_1 \\ v_2 \\ v_3 \end{pmatrix} \tag{10.14}$$

Consider first the orientation of the tetrakaidecahedron as illustrated in Figure 10.6. The tetrakaidecahedron is completely specified by the six initial vectors listed in Table 10.1. For plane strain deformation, all S_{ij} in Equation 10.14 are zero except that $S_{11} \times S_{22} \times S_{33} = 1$ to conserve volume, and $S_{11} \times S_{33} = 1$ since $S_{22} = 1$. For a diagonal matrix, the terms S_{11}, S_{22}, and S_{33} represent the principal distortions, i.e., the ratios of the final to initial lengths of unit vectors along the principal axes. It follows that for a diagonal **S**, the true strains are given by $\epsilon_{11} = \ln\{S_{11}\}$, $\epsilon_{22} = \ln\{S_{22}\}$, and $\epsilon_{33} = \ln\{S_{33}\}$.

The application of the deformation to the initial set of vectors results in the new set of vectors listed in Table 10.2. The latter are used to calculate the area and edge-lengths of the deformed object. Using Equation 10.14 and the conditions for plane strain deformation, it can be shown that the final to initial area (A/A_0) and edge-length (L/L_0) ratios for the deformed tetrakaidecahedron are given by:

$$\frac{A}{A_0} \equiv \frac{S_V}{S_{V_0}} = \frac{S_{11} + 3(S_{11}\sqrt{1+2S_{33}^2} + \sqrt{S_{11}^2 + 2S_{33}^2}) + S_{33}\sqrt{2(1+S_{11}^2)}}{3(2\sqrt{3}+1)} \tag{10.15}$$

$$\frac{L}{L_0} \equiv \frac{L_V}{L_{V_0}} = \frac{1 + S_{11} + 2\sqrt{1 + S_{11}^2 + 2S_{33}^2}}{6} \tag{10.16}$$

Here S_{V_0} and L_{V_0} are the values at zero strain, of grain surface area and edge-length per unit volume. These equations apply strictly to the grain orientation illustrated in Figure 10.6, relative to **S**. From stereology [11], $S_{V_0} = 2/\overline{L}$ so that $\frac{S_V}{S_{V_0}} \equiv 2S_V\overline{L}$, and $L_{V_0} = 9.088/\overline{L}^2$, where $\overline{L}$ is the mean linear intercept commonly used to define the grain size [10, 12]. It follows that Equations 10.15 and 10.16 implicitly contain the grain size as an input variable.

The grain-orientation illustrated in Figure 10.6 may not be representative. Suppose that we wish to orient the tetrakaidecahedron randomly with respect to the deformation. A rotation matrix **R** can be generated using random numbers to rotate the object relative to the axes defining **S**. Equation 10.14 then becomes

$$\begin{pmatrix} S_{11} & S_{12} & S_{13} \\ S_{21} & S_{22} & S_{23} \\ S_{31} & S_{32} & S_{33} \end{pmatrix} \begin{pmatrix} R_{11} & R_{12} & R_{13} \\ R_{21} & R_{22} & R_{23} \\ R_{31} & R_{32} & R_{33} \end{pmatrix} \begin{pmatrix} u_1 \\ u_2 \\ u_3 \end{pmatrix} = \begin{pmatrix} v_1 \\ v_2 \\ v_3 \end{pmatrix}. \tag{10.17}$$

The results are illustrated in Figure 10.7. For comparison purposes, the results

TABLE 10.2
Components of the six vectors listed in Table 10.1, following plane strain or axisymmetric deformation.

Deformed Vector	Components
1	$[aS_{11} \quad 0 \quad 0]$
2	$[0 \quad aS_{22} \quad 0]$
3	$[-\frac{aS_{11}}{2} \quad -\frac{aS_{22}}{2} \quad \frac{aS_{33}}{\sqrt{2}}]$
4	$[\frac{aS_{11}}{2} \quad -\frac{aS_{22}}{2} \quad \frac{aS_{33}}{\sqrt{2}}]$
5	$[\frac{aS_{11}}{2} \quad \frac{aS_{22}}{2} \quad \frac{aS_{33}}{\sqrt{2}}]$
6	$[-\frac{aS_{11}}{2} \quad \frac{aS_{22}}{2} \quad \frac{aS_{33}}{\sqrt{2}}]$

are plotted against the equivalent strain:

$$\epsilon = \left(\frac{2}{3}\right)^{\frac{1}{2}} \left(\epsilon_{11}^2 + \epsilon_{22}^2 + \epsilon_{33}^2 + \frac{1}{2}\gamma_{13}^2 + \frac{1}{2}\gamma_{12}^2 + \frac{1}{2}\gamma_{23}^2\right)^{\frac{1}{2}} \tag{10.18}$$

where ϵ_{11}, ϵ_{22}, and ϵ_{33} are the normal components and γ_{13}, γ_{12}, and γ_{23} are the shear components of strain (the tangents of the shear angles). For homogeneous plane strain compression, $\epsilon = (2/\sqrt{3})\epsilon_{11}$. In Figure 10.7, the dashed line represents the outcome for the orientation illustrated in Figure 10.6 and the points are for the 99 other results of randomly oriented tetrakaidecahedra. It is clear that the orientation of the tetrakaidecahedron does not make much of a difference to the outcome as far as the surface and edge-lengths per unit volume are concerned. This is probably because the tetrakaidecahedron is almost isotropic in shape.

10.6.2 Sequential deformations

The method here is general – all that is needed is to define the matrix **S** for the appropriate circumstances. There are cases where two or more different kinds of deformation are used in sequence, for example, cross-rolling in which the plate is rotated through 90° after a degree of reduction. This is readily tackled by generalizing Equation 10.14. Rotation through 90° about the compression axis [0 0 1] is given by [10]:

$$\mathbf{R} = \begin{pmatrix} 0 & 1 & 0 \\ -1 & 0 & 0 \\ 0 & 0 & 1 \end{pmatrix} \tag{10.19}$$

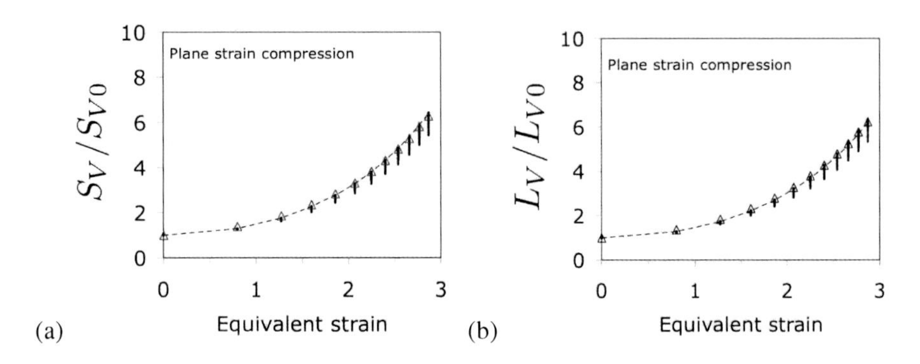

FIGURE 10.7
Calculations for plane strain deformation. The curve represents the data for the tetrakaidecahedron oriented as illustrated in Figure 10.6. The small-points are 99 other cases where the tetrakaidecahedron is randomly oriented relative to **S**.

Writing the first rolling pass as **S** and the cross-rolling pass as **T**, the net deformation **U** is given by $\mathbf{T} \times \mathbf{R} \times \mathbf{S}$:

$$\mathbf{U} = \begin{pmatrix} T_{11} & 0 & 0 \\ 0 & 1 & 0 \\ 0 & 0 & 1/T_{11} \end{pmatrix} \begin{pmatrix} 0 & 1 & 0 \\ -1 & 0 & 0 \\ 0 & 0 & 1 \end{pmatrix} \begin{pmatrix} S_{11} & 0 & 0 \\ 0 & 1 & 0 \\ 0 & 0 & 1/S_{11} \end{pmatrix} = \begin{pmatrix} 0 & T_{11} & 0 \\ -S_{11} & 0 & 0 \\ 0 & 0 & \frac{1}{T_{11}S_{11}} \end{pmatrix}. \tag{10.20}$$

10.6.3 Complex deformations

Flat product rolling is approximated as plane strain compression, but friction with the rolls leads to shears. The plane strain condition is strictly satisfied only at the center of the rolled material. The matrix **S** can be used to deal with the simultaneous actions of plane strain compression and simple shear (on the rolling plane and in the rolling direction) by generalizing the deformation matrix Equation 10.14 as follows:

$$\begin{pmatrix} S_{11} & 0 & S'_{13} \\ 0 & 1 & 0 \\ 0 & 0 & 1/S_{11} \end{pmatrix}. \tag{10.21}$$

The final shear strain S'_{13} arises from the imposed shear strain, S_{13}, modified by the compression S_{11} and is represented by $S_{13} \times S_{11}$.

10.6.4 Multidirectional deformations

The effect of sequential deformations in different directions, as for example in the cross-rolling of plate, is illustrated in Figure 10.8 which represents calculations done

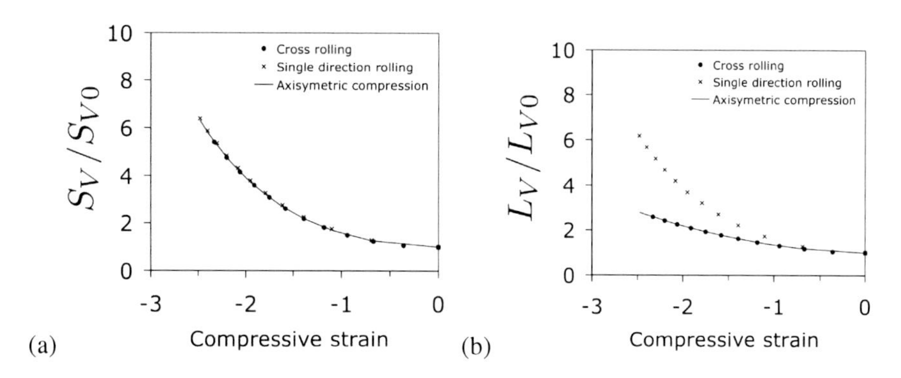

FIGURE 10.8
Comparison of cross-rolling (identical strains in the two rolling directions), axisymmetric compression and single-direction rolling.

using Equation 10.20. In this graph, the rolling strains are the same in both directions ($T_{11} = S_{11}$) and data for axisymmetric compression are also included. To allow a comparison between these deformation modes, the data are plotted as a function of the compressive strain. Notice that the axisymmetric compression and cross-rolling with equal strains in both directions give exactly identical results, illustrating that the results depend on the final strain components and not on the strain path to reach them.

The area ratio for single-direction rolling is only slightly larger than for the cross-rolling when plotted against the compressive strain. This is not surprising given that for the same rolling reduction, the length along the rolling direction for single-direction rolling will be much larger than obtained by cross-rolling. In contrast, the edge ratio becomes much larger for the single-direction scenario, reflecting its greater microstructural anisotropy.

10.7 Summary

Solid state phase transformations often are limited by the fact that the diffusion of atoms becomes difficult at low homologous temperatures. Nevertheless, there may be a tendency for transformation to occur to a lower free energy state, but by a mechanism in which the structure of the parent phase is changed into that of the product by a deformation which is driven not by external forces, but by the favorable change in free energy. The deformation is real in the sense that the shape of the crystal will change in order to reflect the difference between the atomic patterns of the parent and product crystals.

Transformations of this kind are described generically as "displasive" because the structural change is accomplished by a coordinated shift of atoms [13]. The deformations described in this chapter are homogeneous; in the case illustrated in Figure 10.9a, the atoms themselves are placed in the correct final positions by the homogeneous deformation of the lattice. However, the shuffle transformation in Figure 10.9 cannot be described using a homogeneous deformation matrix. The term *shuffle* refers here to small displacements within the unit cell. There are cases, as we shall see in Chapter 11 where the transformation is a mixture of lattice distortion and shuffles, so the homogeneous deformation places some of the atoms in the correct position whereas others have to move small distances so that overall structure of the product phase is correct.

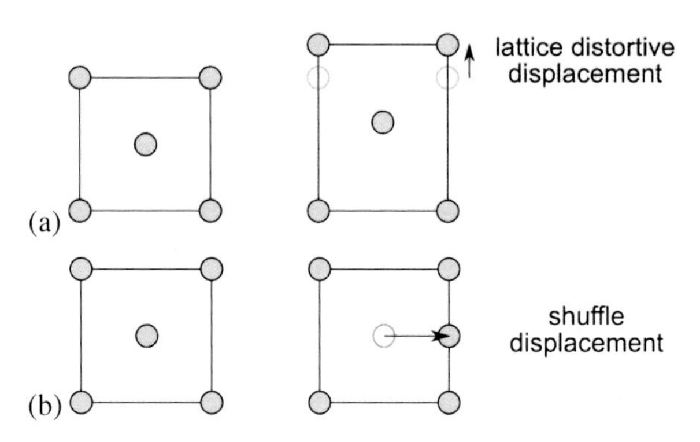

FIGURE 10.9
An illustration of a lattice distortive and shuffle transformation [13].

References

1. M. Mooney: A theory of large elastic deformation, *Journal of Applied Physics*, 1940, **11**, 582–592.

2. M. Born, and R. Fürth: The stability of crystal lattices III, *Mathematical Proceedings of the Cambridge Philosophical Society*, 1940, **36**, 454–465.

3. E. C. Bain: The nature of martensite, *Trans. AIME*, 1924, **70**, 25–46.

4. J. G. Parr: The crystallographic relationship between the phases γ and ε in the system iron-manganese, *Acta Crystallographica*, 1952, **5**, 842–843.

5. T. Anantharaman, and J. Christian: The existence of a macroscopic shear in the transformation in cobalt, *The London, Edinburgh, and Dublin Philosophical Magazine and Journal of Science*, 1952, **43**, 1338–1342.

6. J. S. Bowles, and C. S. Barrett: Crystallography of transformations, *Progress in Metal Physics*, 1952, **3**, 1–41.

7. W. Thomson: On the division of space with minimum partitional area, *Acta mathematica*, 1887, **11**, 121–134.

8. I. Czinege, and T. Reti: Determination of local deformation in cold formed products by a measurement of the geometric characteristics of the crystallites, In: *Eighteenth International Machine Tool Design and Research Conference, Forming*, vol. 1. 1977:159–163.

9. S. B. Singh, and H. K. D. H. Bhadeshia: Topology of grain deformation, *Materials Science and Technology*, 1998, **15**, 832–834.

10. H. K. D. H. Bhadeshia: *Geometry of crystals*: 2nd edition, Institute of Materials, 2001.

11. E. E. Underwood: *Quantitative stereology*: Addison-Wesley Publication Company, 1970.

12. C. Mack: On clumps formed when convex laminae or bodies are placed at random in two or three dimensions, *Proceedings of the Cambridge Philosophical Society*, 1956, **52**, 246–250.

13. M. Cohen, G. B. Olson, and P. C. Clapp: On the classification of displacive transformations (what is martensite?), In: G. B. Olson, and M. Cohen, eds. *International Conference on Martensitic Transformations ICOMAT '79*. Masschusetts: Alpine Press, 1979:1–11.

11

Invariant-plane strains

Abstract

An invariant-plane strain is a deformation that leaves a plane unchanged and in its original orientation. A simple shear deformation applied to a solid is an example, with all the displacements parallel to the shear plane which itself is unaffected. The same deformation can be imagined as a combination of a pure strain that involves just expansions and contractions along orthogonal axes, and a rigid body rotation. There is much to be learned from such factorizations, which although phenomenological, reveal the degeneracies in the combinations of pure strain and rotation that can lead to similar outcomes. If a strain leaves a plane invariant, then two such strains on different planes leave just a line undistorted and unrotated. Both kinds of deformations produced shape changes but are also important in defining the structure of interfaces, as we shall see in the chapters that follow.

11.1 Introduction

The deformation of crystals by the conservative glide of dislocations on a single set of crystallographic planes causes shear in the direction of the resultant Burgers vector of the dislocations concerned, a direction which lies in the slip plane. The slip plane and slip direction constitute a slip system. The material in the slip plane remains crystalline during slip and since there is no reconstruction of this material during slip (e.g., by localized melting followed by resolidification), there can be no change in the relative positions of atoms in the slip plane. The atomic arrangement on the slip plane is thus completely unaffected by the deformation.

Another mode of conservative plastic deformation is mechanical twinning, in which the parent lattice is homogeneously sheared into the twin orientation. The plane on which the twinning shear occurs is again unaffected by the deformation and forms a coherent boundary between the parent and its twin.

If a material which has a Poisson's ratio equal to zero is uniaxially stressed below its elastic limit, then the plane that is normal to the stress axis is unaffected by the deformation since the only non-zero strain is that parallel to the stress axis. Polycrystalline beryllium has a Poisson's ratio close to zero [1] so that stretching would lead to little transverse strain.

All these strains belong to a class of deformations called *invariant-plane strains*. The operation of an invariant-plane strain (IPS) always leaves one plane of the parent crystal completely undistorted and unrotated; this plane is the invariant plane. The condition for a strain to leave a plane undistorted is, as described on page 143, is that the principal deformations of its pure strain component, η_1, η_2, and η_3 are greater than, equal to, and less than unity, respectively. However, as seen in Figures. 10.5, this does not ensure that the undistorted plane is also unrotated; it is necessary to combine the pure strain with an appropriate rotation to generate the invariant plane.

A homogeneous deformation (A S A) strains a vector **u** into another vector **v** which in general may have a different direction and magnitude:

$$[\mathrm{A}; \mathbf{v}] = (\mathrm{A\ S\ A})[\mathrm{A}; \mathbf{u}] \tag{11.1}$$

However, the deformation could also have been defined with respect to another basis, say "B" (basis vectors $\mathbf{b}_i$), to give the deformation matrix (B S B), using a similarity transformation:

$$(\mathrm{B\ S\ B}) = (\mathrm{B\ J\ A})(\mathrm{A\ S\ A})(\mathrm{A\ J\ B}) \tag{11.2}$$

The nature of the physical deformation does not, of course, change; the strain is simply referred to as another coordinate system [2]. We can now proceed to examine the nature of invariant-plane strains. Figure 11.1 illustrates three such strains, defined with respect to a right-handed orthonormal basis Z, such that $\mathbf{z}_3$ is parallel to the unit normal **p** of the invariant plane; $\mathbf{z}_1$ and $\mathbf{z}_2$ lie within the invariant plane, $\mathbf{z}_1$ being parallel to the shear component of the strain concerned. Figure 11.1a illustrates an invariant-plane strain which is purely dilatational, and is of the type to be expected when a plate-shaped precipitate grows diffusionally [3, 4]. The change of shape, as illustrated in Figure 11.1a, due to the growth of this precipitate then reflects the volume change accompanying transformation.

In Figure 11.1b, the invariant-plane strain corresponds to a simple shear, involving no change of volume, as in the macroscopically homogeneous deformation of crystals by slip. The shape of the parent crystal alters in a way which reflects the shear character of the deformation.

The most general invariant-plane strain (Figure 11.1c) involves both a volume change and a shear; if **d** is a unit vector in the direction of the displacements involved, then $m\mathbf{d}$ represents the displacement vector, where m is a scalar giving the magnitude of the displacements. $m\mathbf{d}$ may be factorized as $m\mathbf{d} = s\mathbf{z}_1 + \delta\mathbf{z}_3$, where s and δ are the shear and dilatational components, respectively, of the invariant-plane strain. The strain illustrated in Figure 11.1c is of the type associated with the martensitic transformation of γ iron into hcp iron. This involves a shear on the $\{1\ 1\ 1\}_\gamma$ planes in $\langle 1\ 1\ \overline{2}\rangle_\gamma$ direction, the magnitude of the shear being $1/\sqrt{8}$. However, there also is a dilatational component to the strain, since hcp iron is more dense than fcc iron, consistent with the fact that the former is the stable form at high pressures [5]. There is therefore a volume contraction on martensitic transformation, an additional displacement δ normal to the $\{1\ 1\ 1\}$ austenite planes [6, 26].

It has often been suggested that the passage of a single Shockley partial dislocation on a close-packed plane of austenite leads to the formation of a 3-layer thick region of hcp, since this region contains the correct stacking sequence of close-packed

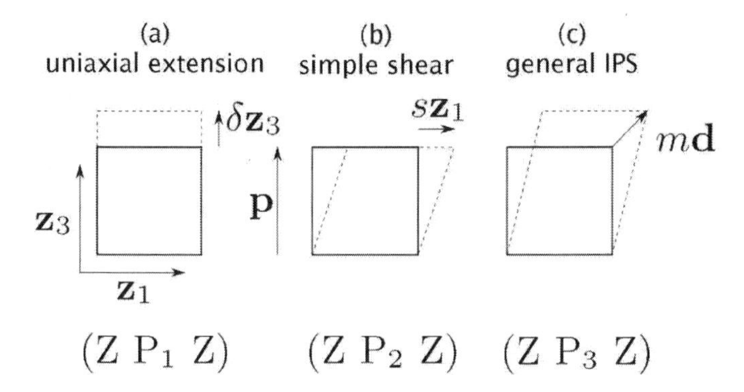

FIGURE 11.1
Three kinds of invariant-plane strains. The cubes indicate the shape before deformation. δ, s and m represent the magnitudes of the dilatational strain, shear strain and general displacement respectively. **p** is a unit vector, the shear strain s is parallel to $\mathbf{z}_1$, whereas δ is parallel to $\mathbf{z}_3$.

planes for the hcp lattice. It has not been possible until recently to prove this because such a small region of hcp material gives very diffuse and unconvincing hcp reflections in electron diffraction experiments. However, the δ component of the fcc-hcp martensite transformation strain has now been detected to be present for single stacking faults, proving the hcp model of such faults [6]. In Figure 11.2, the residual contrast that is visible in part (b) in spite of the fact that the diffraction vector is normal to the displacement in the plane of the fault, is indicative of a change in interplanar spacing, i.e., a volume change normal to the fault plane, consistent with the presence of the hcp phase. The contrast from a fault should otherwise be invisible in the absence of a change in interplanar spacing.

Turning now to the description of the strains illustrated in Figure 11.1, we follow the procedure of Chapter 8, to find the matrices (Z P Z); the symbol "P" in the matrix representation is used to identify specifically an invariant-plane strain, the symbol "S" being the representation of any general deformation. Each column of such a matrix represents the components of a new vector generated by deformation of a vector equal to one of the basis vectors of Z. It follows that the three matrices representing the deformations of Figure 11.1a–c are, respectively,

$$(\mathrm{Z\ P_1\ Z}) = \begin{pmatrix} 1 & 0 & 0 \\ 0 & 1 & 0 \\ 0 & 0 & 1+\delta \end{pmatrix}, \qquad (\mathrm{Z\ P_2\ Z}) = \begin{pmatrix} 1 & 0 & \mathrm{s} \\ 0 & 1 & 0 \\ 0 & 0 & 1 \end{pmatrix},$$

$$(\mathrm{Z\ P_3\ Z}) = \begin{pmatrix} 1 & 0 & \mathrm{s} \\ 0 & 1 & 0 \\ 0 & 0 & 1+\delta \end{pmatrix}.$$

These have a particularly simple form because the basis Z has been chosen carefully, such that $\mathbf{p} \parallel \mathbf{z}_3$ and the direction of the shear is parallel to $\mathbf{z}_1$. However, it is often

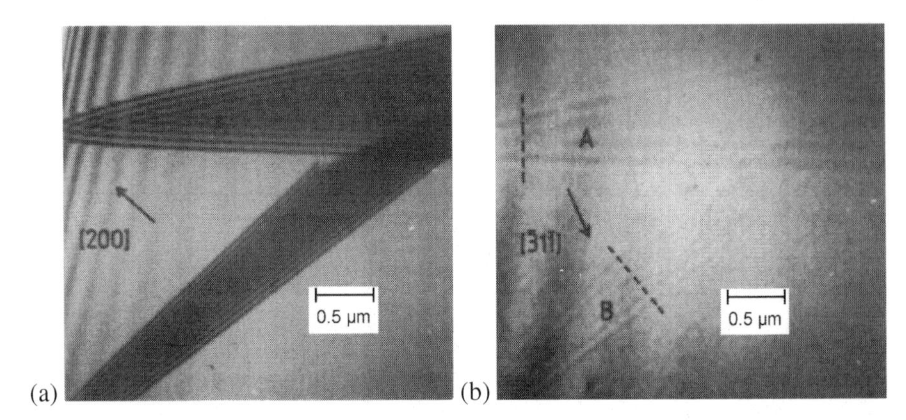

FIGURE 11.2
Stacking faults in austenitic stainless steel. (a) Bright field image of faults. (b) Image with **g**.**R** equal to an integer for both of the faults illustrated, where **g** is the diffraction vector and **R** is the displacement vector. After Brooks et al. [6], reproduced with permission of Elsevier.

necessary to represent invariant-plane strains in a crystallographic basis, or in some other basis X. This can be achieved with the similarity transformation law, Equation 11.2. If (X J Z) represents the coordinate transformation from the basis Z to X, we have

$$(\mathrm{X\ P\ X}) = (\mathrm{X\ J\ Z})(\mathrm{Z\ P\ Z})(\mathrm{Z\ J\ X}).$$

Expansion of this equation gives [7]

$$(\mathrm{X\ P\ X}) = \begin{pmatrix} 1 + md_1p_1 & md_1p_2 & md_1p_3 \\ md_2p_1 & 1 + md_2p_2 & md_2p_3 \\ md_3p_1 & md_3p_2 & 1 + md_3p_3 \end{pmatrix} \tag{11.3}$$

where d_i are the components of **d** in the X basis, such that $(\mathbf{d}; \mathrm{X}^*)[\mathrm{X}; \mathbf{d}] = 1$. The vector **d** points in the direction of the displacements involved; a vector which is parallel to **d** remains parallel following deformation, although the ratio of its final to initial length may be changed. The quantities p_i are the components of the invariant-plane normal **p**, referred to the X* basis, normalized to satisfy $(\mathbf{p}; \mathrm{X}^*)[\mathrm{X}; \mathbf{p}] = 1$. Equation 11.3 may be simplified as follows:

$$(\mathrm{X\ P\ X}) = \mathbf{I} + m[\mathrm{X}; \mathbf{d}](\mathbf{p}; \mathrm{X}^*). \tag{11.4}$$

The multiplication of a single-column matrix with a single-row matrix gives a 3×3 matrix, whereas the reverse order of multiplication gives a scalar quantity. The matrix (X P X) can be used to study the way in which vectors representing directions (referred to the X basis) deform. In order to examine the way in which vectors which are plane normals (i.e., referred to the reciprocal basis X*) deform, we proceed in the following manner.

The property of a homogeneous deformation is that points which originally are collinear remain collinear after the deformation [8]. Lines that initially are coplanar remain so following deformation. It follows that an initial direction **u** which lies in a plane whose normal is initially **h**, becomes a new vector **v** within a plane whose normal is **k**, where **v** and **k** result from the deformation of **u** and **h**, respectively. Now, $\mathbf{h}.\mathbf{u} = \mathbf{k}.\mathbf{v} = 0$, so that:

$$(\mathbf{h}; \mathrm{X}^*)[\mathrm{X}; \mathbf{u}] = (\mathbf{k}; \mathrm{X}^*)[\mathrm{X}; \mathbf{v}] = (\mathbf{k}; \mathrm{X}^*)(\mathrm{X\ P\ X})[\mathrm{X}; \mathbf{u}]$$

i.e.,

$$(\mathbf{k}; \mathrm{X}^*) = (\mathbf{h}; \mathrm{X}^*)(\mathrm{X\ P\ X})^{-1} \tag{11.5}$$

Equation 11.5 describes the way in which plane normals are affected by the deformation (X P X). From Equation 11.4, it can be shown that

$$(\mathrm{X\ P\ X})^{-1} = \mathbf{I} - mq[\mathrm{X}; \mathbf{d}](\mathbf{p}; \mathrm{X}^*) \tag{11.6}$$

where $1/q = \det(\mathrm{X\ P\ X}) = 1 + m(\mathbf{p}; \mathrm{X}^*)[\mathbf{d}; \mathrm{X}]$. The inverse of (X P X) is thus another invariant-plane strain in the opposite direction.

Example 11.1: Tensile tests on single-crystals

A thin cylindrical single-crystal specimen of α iron is tensile tested at $-140°$C, the tensile axis being along the [4 4 1] direction (the cylinder axis). On application of a tensile stress, the entire specimen deforms by twinning on the (1 1 2) plane and in the $[1\ 1\ \bar{1}]$ direction, the magnitude of the twinning shear being $2^{-\frac{1}{2}}$. Calculate the plastic strain recorded along the tensile axis, assuming that the ends of the specimen are always maintained in perfect alignment. (Refs. [9–11] contain details on single crystal deformation).

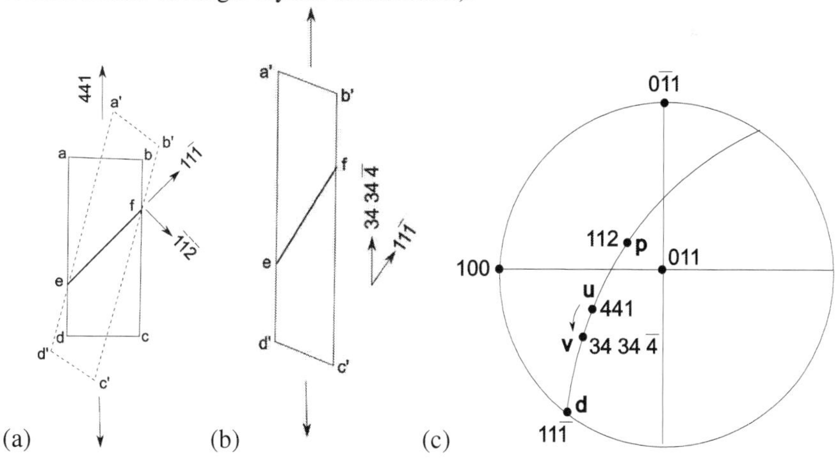

FIGURE 11.3
Longitudinal section of the tensile specimen illustrating the $(1\ \bar{1}\ 0)$ plane. All directions refer to the crystal basis. The tensile axis rotates toward **d**, in the plane containing the original direction of the tensile axis (**u**) and **d**.

Figure 11.3a illustrates the deformation involved; the parent crystal basis α consists of basis vectors which form the conventional bcc unit cell of α-iron. The effect of the mechanical twinning is to alter the original shape $abcd$ to $a'b'c'd'$. ef is a trace of $(1\ 1\ 2)_\alpha$ on which the twinning shear occurs in the $[1\ 1\ \overline{1}]_\alpha$ direction. However, as in most ordinary tensile tests, the ends of the specimen are constrained to be vertically aligned at all times; $a'd'$ must therefore be vertical and the deforming crystal must rotate to comply with this requirement. The configuration illustrated in Figure 11.3c is thus obtained, with ad and $a'd'$ parallel, the tensile strain being $(a'd' - ad)/(ad)$.

As discussed earlier, mechanical twinning is an invariant-plane strain; it involves a homogeneous simple shear on the twinning plane, a plane which is not affected by the deformation and which is common to both the parent and twin regions. Equation 11.3 can be used to find the matrix $(\alpha\ \mathrm{P}\ \alpha)$ describing the mechanical twinning, given that the normal to the invariant-plane is $(\mathbf{p};\alpha^*) = a_\alpha 6^{-\frac{1}{2}}(1\ 1\ 2)$, the displacement direction is $[\alpha;\mathbf{d}] = a_\alpha^{-1}3^{-\frac{1}{2}}[1\ 1\ \overline{1}]$ and $m = 2^{-\frac{1}{2}}$. It should be noted that $\mathbf{p}$ and $\mathbf{d}$ respectively satisfy the conditions $(\mathbf{p};\alpha^*)[\alpha;\mathbf{p}] = 1$ and $(\mathbf{d};\alpha^*)[\alpha;\mathbf{d}] = 1$, as required for Equation 11.3. Hence

$$(\alpha\ \mathrm{P}\ \alpha) = \frac{1}{6}\begin{pmatrix} 7 & 1 & 2 \\ 1 & 7 & 2 \\ \overline{1} & \overline{1} & 4 \end{pmatrix}.$$

Using this, we see that an initial vector $[\alpha;\mathbf{u}] = [4\ 4\ 1]$ becomes a new vector $[\alpha;\mathbf{v}] = (\alpha\ \mathrm{P}\ \alpha)[\alpha;\mathbf{u}] = \frac{1}{6}[34\ 34\ \overline{4}]$ due to the deformation. The need to maintain the specimen ends in alignment means that $\mathbf{v}$ is rotated to be parallel to the tensile axis. Now, $|\mathbf{u}| = 5.745a_\alpha$ where a_α is the lattice parameter of the ferrite, and $|\mathbf{v}| = 8.042a_\alpha$, giving the required tensile strain as $(8.042-5.745)/5.745 = 0.40$.

11.1.1 Comments

(i) From Figure 11.3 it is evident that the end faces of the specimen will also undergo deformation (ab to $a'b'$) and if the specimen gripping mechanism imposes constraints on these ends, then the rod will tend to bend into the form of an "S". For thin specimens this effect may be small.

(ii) The tensile axis at the beginning of the experiment was [4 4 1], but at the end is $\frac{1}{6}[34\ 34\ \overline{4}]$. The tensile direction clearly has rotated during the course of the experiment. The direction in which it has moved is $\frac{1}{6}[34\ 34\ \overline{4}] - [4\ 4\ 1] = \frac{1}{6}[10\ 10\ \overline{10}]$, parallel to $[1\ 1\ \overline{1}]$, the shear direction $\mathbf{d}$. In fact, any initial vector $\mathbf{u}$ will be displaced toward $\mathbf{d}$ to give a new vector $\mathbf{v}$ as a consequence of the IPS. Using Equation 11.4, we see that

$$[\alpha;\mathbf{v}] = (\alpha\ \mathrm{P}\ \alpha)[\alpha;\mathbf{u}] = [\alpha;\mathbf{u}] + m[\alpha;\mathbf{d}](\mathbf{p};\alpha^*)[\alpha;\mathbf{u}] = [\alpha;\mathbf{u}] + \beta[\alpha;\mathbf{d}]$$

where β is a scalar quantity $\beta = m(\mathbf{p};\alpha^*)[\alpha;\mathbf{u}]$.

Evidently, $\mathbf{v} = \mathbf{u}+\beta\mathbf{d}$, with $\beta = 0$ if $\mathbf{u}$ lies in the invariant-plane. All points in the lattice are thus displaced in the direction $\mathbf{d}$, although the extent of displacement depends on β.

(iii) Suppose now that only a volume fraction V_V of the specimen underwent the twinning deformation, the remainder being unaffected by the applied stress. The tensile strain recorded over the whole specimen as the gauge length would simply be $0.40\ V_V$, which is obtained by replacing m in Equation 11.3 by $V_V m$ [10, 12].

(iv) If the shear strain is allowed to vary, as is usually the case during slip deformation, then the position of the tensile axis is still given by $\mathbf{v} = \mathbf{u} + \beta\mathbf{d}$, with β and $\mathbf{v}$ both varying as the test progresses. Since $\mathbf{v}$ is a linear combination of $\mathbf{u}$ and $\beta\mathbf{d}$, it must always lie in the plane containing both $\mathbf{u}$ and $\mathbf{d}$. Hence, the tensile axis rotates in the direction $\mathbf{d}$ within the plane defined by the original tensile axis and the shear direction, as illustrated in Figure 11.3.

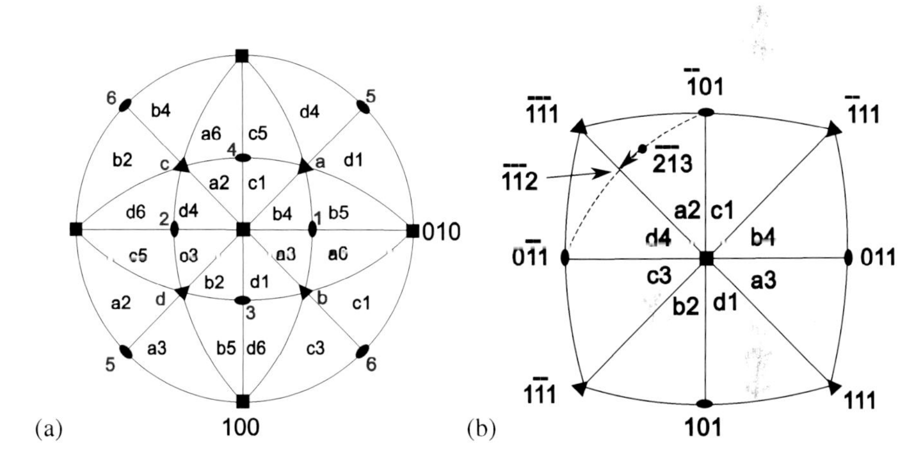

FIGURE 11.4
Stereographic analysis of slip in fcc single-crystals. The dashed curve in (b) is the trace of (111).

Considering further the deformation of single-crystals, an applied stress σ can be resolved into a shear stress τ acting on a slip system. The relationship between σ and τ can be shown (page 81) to be $\tau = \sigma \cos\phi\ cos\lambda$, where ϕ is the angle between the slip plane normal and the tensile axis, and λ is the angle between the slip direction and the tensile axis. Glide will first occur in the particular slip system for which τ exceeds the critical resolved shear stress necessary to initiate dislocation motion on that system. In austenite, glide is easiest on $\{1\ 1\ 1\}\langle 0\ 1\ \bar{1}\rangle$ and the γ standard projection (Figure 11.4a) can be used [9] to determine the particular slip system which has the maximum resolved shear stress due to a tensile stress applied along $\mathbf{u}$. For example, if $\mathbf{u}$ falls within the stereographic triangle labelled a2, then $(\bar{1}\ 1\ 1)[0\ \bar{1}\ 1]$ can be shown to be the most highly stressed system. Hence, when τ reaches a critical value (the critical resolved shear stress), this system alone operates, giving "easy glide" since there is very little work hardening at this stage of deformation; the dislocations which accomplish the shear can simply glide out of the crystal and there is no interference with this glide since none of the other slip systems are active. Of course, the tensile axis is continually rotating toward **d** and may eventually fall on the boundary between two adjacent triangles in Figure 11.4a. If **u** falls on the boundary between triangles a2 and d4, then the slip systems $(\bar{1}\ 1\ 1)[0\ \bar{1}\ 1]$ and $(1\ \bar{1}\ 1)[\bar{1}\ 0\ 1]$ are both equally stressed. This means that both systems can simultaneously operate and *duplex slip* is said to occur; the work hardening rate drastically increases as dislocations moving on different planes interfere with each other in a way which hinders glide and increases the defect density. It follows that a crystal which is initially orientated for single slip eventually deforms by multiple slip.

Example 11.2: Transition from easy glide to duplex slip

A single-crystal of austenite is tensile tested at 25°C, the stress being applied along $[\bar{2}\ \bar{1}\ 3]$ direction; the specimen deforms by easy glide on the $(\bar{1}\ 1\ 1)[0\ \bar{1}\ 1]$ system. If slip can only occur on systems of this form, calculate the tensile strain necessary for the onset of duplex slip. Assume that the ends of the specimen are maintained in alignment throughout the test.

The tensile axis (**v**) is expected to rotate toward the slip direction, its motion being confined to the plane containing the initial tensile axis (**u**) and the slip direction (**d**). In Figure 11.4b, **v** will therefore move on the trace of the (1 1 1) plane. Duplex slip is expected to begin when **v** reaches the great circle which separates the stereographic triangles a2 and d4 of Figure 11.4b, since the $(1\ \bar{1}\ 1)[\bar{1}\ 0\ 1]$ slip system will have a resolved shear stress equal to that on the initial slip system. The tensile axis can be expressed as a function of the shear strain m as in the example on the tensile testing of single crystals (page 157):

$$[\gamma;\mathbf{v}] = [\gamma;\mathbf{u}] + m[\gamma;\mathbf{d}](\mathbf{p};\gamma^*)[\gamma;\mathbf{u}]$$

where $(\mathbf{p};\gamma^*) = \frac{a_\gamma}{\sqrt{3}}(\bar{1}\ 1\ 1)$ and $[\gamma;\mathbf{d}] = \frac{1}{\sqrt{2}a_\gamma}[0\ \bar{1}\ 1]$, so that

$$[\gamma;\mathbf{v}] = [\gamma;\mathbf{u}] + \frac{4m}{\sqrt{6}}[0\ \bar{1}\ 1] = [\bar{2}\ \bar{1}\ 3] + \frac{4m}{\sqrt{6}}[0\ \bar{1}\ 1]$$

When duplex slip occurs, **v** must lie along the intersection of the (1 1 1) and (1 1 0) planes, the former being the plane on which **v** is confined to move and the latter being the boundary between triangles a2 and d4. It follows that $\mathbf{v} \parallel [1\ 1\ \bar{2}]$ and must be of the form $\mathbf{v} = [v\ v\ \overline{2v}]$. Substituting this into the earlier equation gives

$$[v\ v\ 2\overline{v}] = [\bar{2}\ \bar{1}\ 3] + \frac{4m}{\sqrt{6}}[0\ \bar{1}\ 1]$$

and on comparing coefficients from both sides of this equation, we obtain

$$[\gamma;\mathbf{v}] = [\bar{2}\ \bar{2}\ 4]$$

so that the tensile strain required is $(|\mathbf{v}| - |\mathbf{u}|)/|\mathbf{u}| = 0.31$.

11.2 Deformation twins

We can now proceed to study twinning deformations [13, 14], in greater depth, noting that a twin is said to be any region of a parent which has undergone a homogeneous shear to give a reorientated region with the same crystal structure. The example below illustrates some of the important concepts of twinning deformation.

Example 11.3: Twins in fcc crystals

Show that the austenite lattice can be twinned by a shear deformation on the $\{1\,1\,1\}$ plane and in the $\langle 1\,1\,\bar{2}\rangle$ direction. Deduce the magnitude of the twinning shear, and explain why this is the most common twinning mode in fcc crystals. Derive the matrix representing the orientation relationship between the twin and parent lattices.

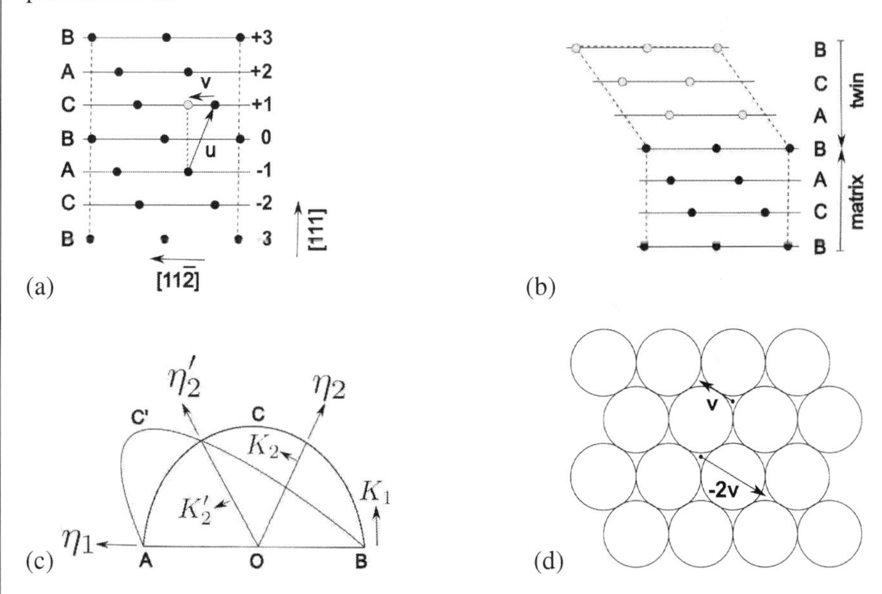

FIGURE 11.5
Twinning in the fcc austenite crystal structure. The diagrams represent sections in the $(1\,\bar{1}\,0)$ plane. The vector **v** is $\frac{a_\gamma}{6}[11\bar{2}]$. In (c), K_2' and η_2' are the final positions of the undistorted plane K_2 and the undistorted direction η_2, respectively. (d) A close-packed $\{111\}$ plane showing that a shear along **v** is not equivalent to one parallel to $-$**v**, since the magnitude of the reverse displacement has to be twice as large.

The $\{1\,1\,1\}$ planes in fcc crystals have a stacking sequence $\ldots ABCABC \ldots$ The region of the parent which becomes reorientated due to the twinning shear can be generated by reflection across the twinning plane; the stacking sequence across the plane **B** which is the coherent twin interface is therefore

$\ldots ABCA\mathbf{B}ACBA \ldots$ Figure 11.5a illustrates how a stack of close-packed planes (stacking sequence $\ldots ABC \ldots$) may be labelled $\ldots - 1, 0, +1 \ldots$ Reflection across 0 can be achieved by shearing atoms in the $+1$ plane into positions which are directly above (i.e., along $\langle 1\ 1\ 1 \rangle$) the atoms in the -1 plane.

Figure 11.5a is a section of the lattice on the $(\bar{1}\ 1\ 0)$ plane; it is evident that a displacement of all the atoms on $+1$ through a distance $v = |\mathbf{v}|$ along $\langle 1\ 1\ \bar{2} \rangle$ gives the required reflection across the twinning plane 0. The twinning shear is given by the equation $s^2 = (v/d)^2$, where d is the spacing of the $(1\ 1\ 1)$ planes. Since $v^2 = u^2 - 4d^2$, we may write

$$s^2 = (u/d)^2 - 4 \tag{11.7}$$

where $u = |\mathbf{u}|$ and $\mathbf{u}$ connects a site on the $+1$ plane to an equivalent site on the -1 plane (Figure 11.5a). Hence, the fcc lattice can be twinned by a shear of magnitude $s = 1/\sqrt{2}$ on $\{1\ 1\ 1\}$.

To answer why a crystal twins in a particular way, it is necessary to make the physically reasonable assumption that the twinning mode should lead to the smallest possible shear. When the twin is forced to form in a constrained environment, such as within a polycrystalline material, the shape change resulting from the shear deformation causes elastic distortions in both the twin and the matrix. The consequent strain energy increase (per unit volume of material) is given approximately by $E = (c/r)\mu s^2$ [15–17], where c and r represent the thickness and length of the twin respectively, and μ is the shear modulus. This also is the reason why mechanical twins have a lenticular morphology, since the small thickness to length ratio of thin-plates minimises E. Annealing twins grow diffusionally so there is no physical deformation involved in their growth. Hence, their shape is not restricted to that of a thin plate, the morphology being governed by the need to minimise interface energy. It is interesting that annealing and mechanical twins are crystallographically equivalent (if we ignore the absence of a shape change in the former) but their mechanisms of growth are very different.

Equation 11.7 indicates that s can be minimised by choosing twinning planes with large d-spacings and by choosing the smallest vector $\mathbf{u}$ connecting a site on the $+1$ plane to an equivalent site on the -1 plane; for the $(1\ 1\ 1)$ plane the smallest $\mathbf{u}$ is $\frac{1}{2}[1\ 1\ \bar{2}]$, as illustrated in Figure 11.5a. Equation 11.7 can also be used to show that none of the planes of slightly smaller spacing than $\{1\ 1\ 1\}$ can lead to twins with $s < 2^{-\frac{1}{2}}$; two of these planes are also mirror planes and thus cannot serve as the invariant-plane (K_1, Figure 11.5b) of the reflection twin.

From Figure 11.5a,d we see that the twin lattice could also have been obtained by displacing the atoms in the $+1$ plane through a distance $2v$ along $[\bar{1}\ \bar{1}\ 2]$ had $\mathbf{u}$ been chosen to equal $[\sqrt{2}\ \sqrt{2}\ \ 0]$, giving $s = \sqrt{2}$. This larger shear is of course inconsistent with the hypothesis that the favored twinning mode involves the smallest shear, and indeed, this mode of twinning is not observed. To obtain the smallest shear, the magnitude of the vector $\mathbf{v}$ must also be minimised; in the example under consideration, the correct choice of $\mathbf{v}$ will connect a lattice site of plane +1 with the projection of its nearest neighbor lattice site on plane -1. The twinning direction is therefore expected to be along $[1\,1\,\overline{2}]$. It follows that the operative twin mode for the fcc lattice should involve a shear of magnitude $s = 2^{-\frac{1}{2}}$ on $\{1\,1\,1\}\langle 1\,1\,\overline{2}\rangle$.

The matrix-twin orientation relationship (M J T) can be deduced from the fact that the twin was generated by a shear which brought atoms in the twin into positions which are related to the parent lattice points by reflection across the twinning plane (the basis vectors of M and T define the fcc unit cells of the matrix and twin crystals respectively). From Figure 11.5 we note that:

$$[1\,1\,\overline{2}]_{\mathrm{M}} \parallel [1\,1\,\overline{2}]_{\mathrm{T}} \qquad [\overline{1}\,1\,0]_{\mathrm{M}} \parallel [\overline{1}\,1\,0]_{\mathrm{T}} \qquad [1\,1\,1]_{\mathrm{M}} \parallel [\overline{1}\,\overline{1}\,\overline{1}]_{\mathrm{T}}$$

It follows that

$$\begin{pmatrix} 1 & \overline{1} & 1 \\ 1 & 1 & 1 \\ \overline{2} & 0 & 1 \end{pmatrix} = \begin{pmatrix} J_{11} & J_{12} & J_{13} \\ J_{21} & J_{22} & J_{23} \\ J_{31} & J_{32} & J_{33} \end{pmatrix} \begin{pmatrix} 1 & \overline{1} & \overline{1} \\ 1 & 1 & \overline{1} \\ \overline{2} & 0 & \overline{1} \end{pmatrix}$$

Solving for (M J T), we get

$$(\mathrm{M\,J\,T}) = \frac{1}{6}\begin{pmatrix} 1 & \overline{1} & 1 \\ 1 & 1 & 1 \\ \overline{2} & 0 & 1 \end{pmatrix} \begin{pmatrix} 1 & 1 & \overline{2} \\ \overline{3} & 3 & 0 \\ \overline{2} & \overline{2} & \overline{2} \end{pmatrix} = \frac{1}{3}\begin{pmatrix} \overline{1} & \overline{2} & \overline{2} \\ \overline{2} & \overline{1} & \overline{2} \\ \overline{2} & \overline{2} & 1 \end{pmatrix}$$

11.2.1 Comments

(i) Equations like Equation 11.7 can be used to predict the likely ways in which different lattices might twin, especially when the determining factor is the magnitude of the twinning shear [13].

(ii) There actually are four different ways of generating the twin lattice from the parent crystal: (a) by reflection about the K_1 plane on which the twinning shear occurs, (b) by a rotation of π about η_1, the direction of the twinning shear, (c) by reflection about the plane normal to η_1 and (d) by a rotation of π about the normal to the K_1 plane.

Since most metals are centrosymmetric, operations (a) and (d) produce crystallographically equivalent results, as do (b) and (c). In the case of the fcc twin discussed above, the high symmetry of the cubic lattice means that all four operations are crystallographically equivalent. Twins which can be produced by the operations (a) and (d) are called type **I** twins; type II twins result form the other two twinning operations. The twin discussed in the above example is called a compound twin, since type **I** and type II twins cannot be crystallographically distinguished.

Figure 11.5b illustrates some additional features of twinning. The K_2 plane is the plane which (like K_1) is undistorted by the twinning shear, but unlike K_1, is rotated by the shear. The "plane of shear" is the plane containing η_1 and the perpendicular to K_1; its intersection with K_2 defines the undistorted but rotated direction η_2. In general, η_2 and K_1 are rational for type **I** twins, and η_1 and K_2 are rational for type II twins. The set of four twinning elements K_1, K_2, η_1 and η_2 are all rational for compound twins. From Figure 11.5b, η_2 makes an angle of $\arctan(s/2)$ with the normal to K_1 and simple geometry shows that $\eta_2 = [1\ 1\ 2]$ for the fcc twin of Example 11.2. The corresponding K_2 plane which contains η_2 and $\eta_1 \wedge \eta_2$ is therefore $(1\ 1\ \bar{1})$, giving the rational set of twinning elements

$$K_1 = (1\,1\,1) \qquad \eta_2 = [1\,1\,2] \qquad s = 2^{-\frac{1}{2}} \qquad \eta_1 = [1\,1\,\bar{2}] \qquad K_2 = (1\,1\,\bar{1})$$

In fact, it only is necessary to specify either K_1 and η_2 or K_2 and η_1 to completely describe the twin mode concerned.

The deformation matrix (M P M) describing the twinning shear can be deduced using Equation 11.3 and the information that $[\mathrm{M};\mathbf{d}] \parallel [1\ 1\ \bar{2}]$, $(\mathbf{p};\mathrm{M}^*) \parallel (1\ 1\ 1)$ and $s = 2^{-\frac{1}{2}}$ to give

$$(\mathrm{M\ P\ M}) = \frac{1}{6}\begin{pmatrix} 7 & 1 & 1 \\ 1 & 7 & 1 \\ \bar{2} & \bar{2} & 4 \end{pmatrix} \quad \text{and} \quad (\mathrm{M\ P\ M})^{-1} = \frac{1}{6}\begin{pmatrix} 5 & \bar{1} & \bar{1} \\ \bar{1} & 5 & \bar{1} \\ 2 & 2 & 8 \end{pmatrix} \tag{11.8}$$

and if a vector $\mathbf{u}$ is deformed into a new vector $\mathbf{v}$ by the twinning shear, then

$$(\mathrm{M\ P\ M})[\mathrm{M};\mathbf{u}] = [\mathrm{M};\mathbf{v}] \tag{11.9}$$

and if $\mathbf{h}$ is a plane normal which after deformation becomes $\mathbf{k}$, then

$$(\mathbf{h};\mathrm{M}^*)(\mathrm{M\ P\ M})^{-1} = (\mathbf{k};\mathrm{M}^*) \tag{11.10}$$

These laws can be used to verify that $\mathbf{p}$ and $\mathbf{d}$ are unaffected by the twinning shear, and that the magnitude of a vector originally along η_2 is not changed by the deformation; similarly, the spacing of the planes initially parallel to K_2 remains the same after deformation, although the planes are rotated.

11.3 Correspondence matrix

The property of the homogeneous deformations we have been considering is that points which are initially collinear remain so in spite of the deformation, and lines which are initially coplanar remain coplanar after the strain. Using the data from Equation 11.8, we see that the deformation (M P M) alters the vector $[\mathrm{M};\mathbf{u}] = [0\ 0\ 1]$ to a new vector $[\mathrm{M};\mathbf{v}] = \frac{1}{6}[1\ 1\ 4]$ i.e.,

$$(\mathrm{M\ P\ M})[0\ 0\ 1]_\mathrm{M} = \frac{1}{6}[1\ 1\ 4]_\mathrm{M} \tag{11.11a}$$

The indices of this new vector $\mathbf{v}$ relative to the twin basis T can be obtained using the coordinate transformation matrix (T J M), so that

$$(\mathrm{T\ J\ M})\frac{1}{6}[1\ 1\ 4]_\mathrm{M} = [\mathrm{T};\mathbf{v}] = \frac{1}{2}[\bar{1}\ \bar{1}\ 0]_\mathrm{T} \tag{11.11b}$$

Hence, the effect of the shear stress is to deform a vector $[0\ 0\ 1]_M$ of the parent lattice into a vector $\frac{1}{2}[\bar{1}\ \bar{1}\ 0]_T$ of the twin. Equations 11.11a,b could have been combined to obtain this result, as follows:

$$(\text{T J M})(\text{M P M})[\text{M};\mathbf{u}] = [\text{T}; \mathbf{v}]$$

or

$$(\text{T C M})[\text{M};\mathbf{u}] = [\text{T}; \mathbf{v}]$$

where

$$(\text{T J M})(\text{M P M}) = (\text{T C M}) \tag{11.12}$$

The matrix (T C M) is called the correspondence matrix; the initial vector **u** with indices [M;**u**] in the parent basis, due to deformation becomes a *corresponding* vector **v** with indices [T;**v**] in the twin basis. The correspondence matrix tells us that a certain vector in the twin is formed by deforming a particular corresponding vector of the parent. In the example considered above, the vector **u** has rational components in M (i.e., the components are small integers or fractions) and **v** has rational components in T. It follows that the elements of the correspondence matrix (T C M) must also be rational numbers or fractions. The correspondence matrix can usually be written from inspection since its columns are rational lattice vectors referred to the second basis produced by the deformation from the basis vectors of the first basis.

We can similarly relate planes in the twin to planes in the parent, the correspondence matrix being given by

$$(\text{M C T}) = (\text{M P M})^{-1}(\text{M J T})$$

where

$$(\mathbf{h}; \text{M}^*)(\text{M C T}) = (\mathbf{k}; \text{T}^*)$$

so that the plane $(\mathbf{k}; \text{T}^*)$ of the twin was formed by the deformation of the plane $(\mathbf{h}; \text{M}^*)$ of the parent.

11.4 An alternative to the Bain strain

The pure strain proposed originally by Bain [18] (page 106) is one way of homogeneously deforming the austenite lattice into that of a body-centered cubic or body-centered tetragonal unit cell of martensite. The deformation involves a uniform expansion along two of the cubic-austenite cell axes by the ratio $\eta_1 = \sqrt{2}a_\alpha/a_\gamma$, and a compression along the third axis by $\eta_3 = a_\alpha/a_\gamma$, where a_α and a_γ are the lattice parameters of the cubic unit cells of martensite and austenite, respectively. The Bain strain is not, however, a unique way of deforming the cubic-austenite unit cell into that of martensite. Figure 11.6 shows an alternative correspondence in terms of a triclinic body-centered cell of austenite [19, 20] that can be deformed into a body-centered cubic cell. In what follows, all vectors and operations are referred to the cubic austenite lattice.

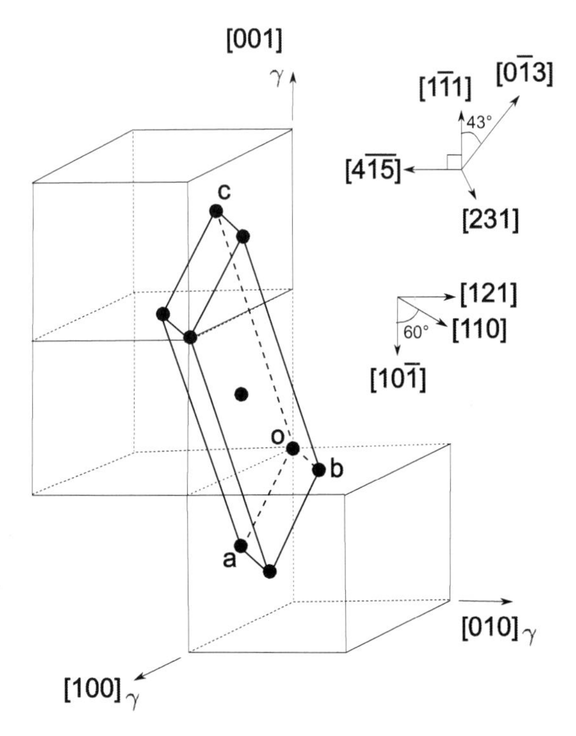

FIGURE 11.6
An alternative correspondence between the austenite, conventionally represented by the cubic unit cells illustrated, can also be represented by the body-centered triclinic cell. The main diagram is adapted from [19]. Not all the iron atoms in the cubic cells are illustrated for clarity. The directions oa, ob, and oc are, respectively, parallel to $[10\bar{1}]_\gamma$, $[110]_\gamma$, and $[0\bar{1}3]_\gamma$.

The basal plane of the triclinic cell consists of vectors $\mathrm{oa} = \frac{a_\gamma}{2}[10\bar{1}]$ and $\mathrm{ob} = \frac{a_\gamma}{2}[110]$, making an angle of 60° with each other. The normal to the basal plane is therefore parallel to the $[1\bar{1}1]$ direction but the vector $\mathrm{oc} = \frac{a_\gamma}{2}[0\bar{1}3]$. A shear $(\gamma\ \mathrm{S}_1\ \gamma)$ of $s_1 = \tan 43^\circ$ parallel to $[41\bar{5}]$ on $(1\bar{1}1)$ would make oc normal to the basal plane of the triclinic cell without affecting any vectors in the basal plane defined by the triclinic cell edges oa and ob. Defining $\mathbf{d} = \frac{1}{\sqrt{42}}[4\bar{1}\bar{5}]$ as the unit vector parallel to the shear direction, and $\mathbf{p} = \frac{1}{\sqrt{3}}[1\bar{1}1]$ as the unit normal to the shear plane, and defining $\phi = s_1 \frac{1}{\sqrt{3}} \frac{1}{\sqrt{42}}$, it follows that

$$
\begin{aligned}
(\gamma\ \mathrm{S}_1\ \gamma) &= \mathbf{I} + s_1[\gamma;\mathbf{d}](\mathbf{p};\gamma^*) \\
&= \begin{pmatrix} 1+4\phi & -4\phi & 4\phi \\ -\phi & 1+\phi & -\phi \\ -5\phi & 5\phi & 1-5\phi \end{pmatrix}
\end{aligned}
\tag{11.13}
$$

This deformation makes the original vector $[0\bar{1}3]$ rotated and distorted to $[1.33\ \overline{1.33}\ 1.33]\ \|\ [1\bar{1}1]$. It would need to be compressed by a ratio $\eta_3 = a_\alpha/2.3076a_\gamma$ so that its magnitude becomes the lattice parameter a_α of martensite.

To change the angle 60° between oa and ob to 90° without affecting any vectors normal to the basal plane requires a second shear $(\gamma\ S_2\ \gamma)$ of magnitude $s_2 = \tan 30°$ with $\mathbf{d} = \frac{1}{\sqrt{2}}[\bar{1}01]$ and $\mathbf{p} = \frac{1}{\sqrt{6}}[121]$. Taking $\psi = s_2 \frac{1}{\sqrt{2}} \frac{1}{\sqrt{6}}$:

$$\begin{aligned}(\gamma\ S_2\ \gamma) &= \mathbf{I} + s_2[\gamma;\mathbf{d}](\mathbf{p};\gamma^*) \\ &= \begin{pmatrix} 1-\psi & -2\psi & -\psi \\ 0 & 1 & 0 \\ \psi & 2\psi & 1+\psi \end{pmatrix}. \end{aligned} \tag{11.14}$$

This shear leaves $[\bar{1}01]$ unaffected, whereas a vector originally parallel to $[110]$ becomes $[\frac{1}{2}1\frac{1}{2}]$ which is at 90° to $[\bar{1}01]$. To match the lattice parameter a_α, the vector $\frac{1}{2}[\bar{1}01]$ needs to expanded by $\eta_1 = \sqrt{2}a_\alpha/a_\gamma$ whereas $\frac{1}{2}[\frac{1}{2}1\frac{1}{2}]$ would need to be expanded by the larger quantity $\eta 2 = 1.633a_\alpha/a_\gamma$ because the shear $(\gamma\ S_2\ \gamma)$ reduces the length of $\frac{1}{2}[110]$.

Therefore, the total deformation required to convert the triclinic austenite cell into the body-centered cubic cell of martensite is

$$(\gamma\ S\ \gamma) = \begin{pmatrix} 1+4\phi & -4\phi & 4\phi \\ -\phi & 1+\phi & -\phi \\ -5\phi & 5\phi & 1-5\phi \end{pmatrix} \begin{pmatrix} 1-\psi & -2\psi & -\psi \\ 0 & 1 & 0 \\ \psi & 2\psi & 1+\psi \end{pmatrix} \begin{pmatrix} \eta_1 & 0 & 0 \\ 0 & \eta_2 & 0 \\ 0 & 0 & \eta_3 \end{pmatrix}.$$

Since shears do not lead to any change in volume, the ratio of the volumes of the martensite and ferrite unit cells is given by the determinant of the distortion matrix, i.e., $\eta_1\eta_2\eta_3 = a_\alpha^3/a_\gamma^3$ as expected. These distortions, even though they are not measured along the principal axes, are large when compared with the ones associated with the Bain strain. As pointed out by Wayman, it might be concluded on this basis and and with the backing of other experimental evidence, that the Bain strain is the homogeneous deformation that leads to a change from fcc→bcc structure.

11.5 Stepped interfaces

A planar coherent twin boundary (unit normal **p**) can be generated from a single crystal by shearing on the twinning plane **p**, the unit shear direction and shear magnitude being **d** and m, respectively.

On the other hand, to generate a similar boundary but containing a step of height h requires additional virtual operations (Figure 11.7) [10, 21, 22]. The single crystal is first slit along a plane which is not parallel to **p** (Figure 11.7b), before applying the twinning shear. The shear which generates the twinned orientation also opens up the slit (Figure 11.7c), which then has to be rewelded (Figure 11.7d) along the original cut; this produces the required stepped interface. A Burgers circuit constructed around the stepped interface will, when compared with an equivalent circuit constructed around the unstepped interface exhibit a closure failure. This closure failure gives the Burgers vector $\mathbf{b}_{\mathrm{I}}$ associated with the step:

$$\mathbf{b}_{\mathrm{I}} = hm\mathbf{d}$$

The operations outlined above indicate one way of generating the required stepped interface. They are simply the virtual operations which allow us to produce the required defect – similar operations were first used by Volterra [23] in describing the elastic properties of cut and deformed cylinders, operations which were later recognised to generate the ordinary dislocations that metallurgists are so familiar with.

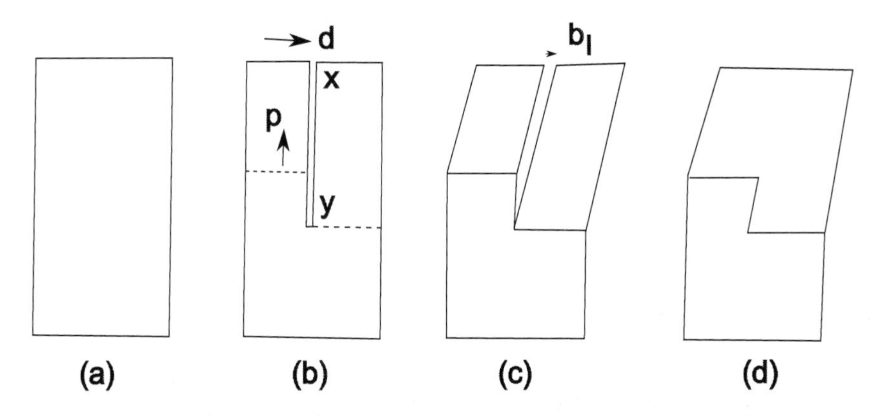

FIGURE 11.7
The virtual operations [24] used in determining $\mathbf{b}_{\mathrm{I}}$. (a) Single crystal. (b) Crystal cut along "xy". (c) Gap opens on shearing. The closure failure $\mathbf{b}_{\mathrm{I}}$ is the Burgers vector of the final step in the coherent twin boundary, that results on rewelding the gap as shown in (d).

Having defined $\mathbf{b}_{\mathrm{I}}$, we note that an initially planar coherent twin boundary can acquire a step if a dislocation of Burgers vector $\mathbf{b}_m$ crosses the interface. The height of the step is given by [10]:

$$h = \mathbf{b}_m.\mathbf{p}$$

so that

$$\mathbf{b}_{\mathrm{I}} = m(\mathbf{b}_m.\mathbf{p})\mathbf{d}. \tag{11.15}$$

From Equation 11.3, the invariant plane strain necessary to generate the twin from the parent lattice is given by $(\mathrm{M\ P\ M}) = \mathbf{I} + m[\mathrm{M};\mathbf{d}](\mathbf{p};\mathrm{M}^*)$ so that Equation 11.15 becomes

$$[\mathrm{M};\mathbf{b}_{\mathrm{I}}] = (\mathrm{M\ P\ M})[\mathrm{M};\mathbf{b}_m] - [\mathrm{M};\mathbf{b}_m] \tag{11.16}$$

Example 11.4: Interaction of dislocations with interfaces

Deduce the correspondence matrix for a fcc deformation twin and hence show that there are no geometrical restrictions to the passage of slip dislocations across coherent twin boundaries in fcc materials.

It was shown on page 161 that

$$(\mathrm{T\ J\ M}) = \frac{1}{3}\begin{pmatrix} 1 & \bar{2} & \bar{2} \\ \bar{2} & 1 & \bar{2} \\ \bar{2} & \bar{2} & 1 \end{pmatrix} \quad \text{and} \quad (\mathrm{M\ P\ M}) = \frac{1}{6}\begin{pmatrix} 7 & 1 & 1 \\ 1 & 7 & 1 \\ \bar{2} & \bar{2} & 4 \end{pmatrix}.$$

The correspondence matrix (T C M) which associates each vector of the parent with a corresponding vector in the twin is, from Equation 11.12, given by

$$(\mathrm{T\ C\ M}) = (\mathrm{T\ J\ M})(\mathrm{M\ P\ M})$$

$$(\mathrm{M\ C\ T}) = (\mathrm{M\ P\ M})^{-1}(\mathrm{M\ J\ T})$$

so that

$$(\mathrm{T\ C\ M}) = (\mathrm{M\ C\ T}) = \frac{1}{2}\begin{pmatrix} 1 & \bar{1} & \bar{1} \\ \bar{1} & 1 & \bar{1} \\ \bar{2} & \bar{2} & 0 \end{pmatrix}.$$

The character of a dislocation will in general be altered on crossing an interface. This is because the crossing process introduces a step in the interface, rather like the slip steps which arise at the free surfaces of deformed crystals. We consider the case where a dislocation crosses a coherent twin boundary. The interfacial step has dislocation character so that the original dislocation (Burgers vector $\mathbf{b}_m$) from the parent crystal is in effect converted into *two* dislocations, one being the step (Burgers vector $\mathbf{b}_\mathrm{I}$) and the other the dislocation (Burgers vector $\mathbf{b}_t$) which has penetrated the interface and entered into the twin lattice. If the total Burgers vector content of the system is to be preserved then it follows that in general, $\mathbf{b}_t \neq \mathbf{b}_m$, since $\mathbf{b}_m = \mathbf{b}_+\mathbf{b}_t$. Using this equation and Equation 11.16, we see that

$$[\mathrm{M};\mathbf{b}_t] = (\mathrm{M\ P\ M})[\mathrm{M};\mathbf{b}_m]$$

or

$$[\mathrm{T};\mathbf{b}_t] = (\mathrm{T\ J\ M})(\mathrm{M\ P\ M})[\mathrm{M};\mathbf{b}_m]$$

so that

$$[\mathrm{T};\mathbf{b}_t] = (\mathrm{T\ C\ M})[\mathrm{M};\mathbf{b}_m].$$

Clearly, dislocation glide across the coherent interface will not be hindered if $\mathbf{b}_t$ is a perfect lattice vector of the twin. If this is not the case and $\mathbf{b}_t$ is a partial dislocation in the twin, then glide across the interface will be hindered because the motion of $\mathbf{b}_t$ in the twin would leave a stacking fault trailing all the way from the interface to the position of the partial dislocation in the twin.

There is an additional condition to be fulfilled for easy glide across the interface; the *corresponding* glide planes $\mathbf{p}_m$ and $\mathbf{p}_t$ of dislocations $\mathbf{b}_m$ and $\mathbf{b}_t$ in the parent and twin lattices, respectively, must meet edge to edge in the interface. Now,

$$(\mathbf{p}_t; \mathrm{T}^*) = (\mathbf{p}_m; \mathrm{M}^*)(\mathrm{M\ C\ T}).$$

If the interface plane normal is $\mathbf{p}_i$, then the edge to edge condition is satisfied if $\mathbf{p}_m \wedge \mathbf{p}_i \parallel \mathbf{p}_t \wedge \mathbf{p}_i$.

Dislocations in fcc materials usually glide on close-packed $\{1\ 1\ 1\}$ planes and have Burgers vectors of type $\frac{a}{2}\langle 1\ \bar{1}\ 0\rangle$. Using the data of Table 11.1 it can easily be verified that all the close-packed planes of the parent lattice meet the corresponding glide planes in the twin edge to edge in the interface, which is taken to be the coherent $(1\ 1\ 1)_\mathrm{M}$ twinning plane. Furthermore, all the $\frac{a}{2}\langle 1\ \bar{1}\ 0\rangle$ Burgers vectors of glide dislocations in the parent correspond to perfect lattice dislocations in the twin. It must be concluded that the coherent twin boundary for $\{1\ 1\ 1\}$ twins in fcc metals does not offer any geometrical restrictions to the transfer of slip between the parent and product lattices.

TABLE 11.1
Corresponding glide planes and Burgers vectors

Parent	Twin
$\frac{a}{2}[1\ 1\ 0]$	$a[0\ 0\ \bar{1}]$
$\frac{a}{2}[1\ 0\ 1]$	$\frac{a}{2}[0\ \bar{1}\ \bar{1}]$
$\frac{a}{2}[0\ 1\ 1]$	$\frac{a}{2}[\bar{1}\ 0\ \bar{1}]$
$\frac{a}{2}[1\ \bar{1}\ 0]$	$\frac{a}{2}[1\ \bar{1}\ 0]$
$\frac{a}{2}[1\ 0\ \bar{1}]$	$\frac{a}{2}[1\ 0\ \bar{1}]$
$\frac{a}{2}[0\ 1\ \bar{1}]$	$\frac{a}{2}[0\ 1\ \bar{1}]$
$(1\ 1\ 1)$	$(\bar{1}\ \bar{1}\ \bar{1})$
$(1\ 1\ \bar{1})$	$(1\ 1\ \bar{1})$
$(1\ \bar{1}\ 1)$	$(0\ \bar{2}\ 0)$
$(\bar{1}\ 1\ 1)$	$(\bar{2}\ 0\ 0)$

These data (Table 11.1) show also that all dislocations with Burgers vectors in the $(1\ 1\ 1)_\mathrm{M}$ plane are unaffected, both in magnitude and direction, as a result of crossing into the twin. For example, $\frac{a}{2}[1\ \bar{1}\ 0]_\mathrm{M}$ becomes $\frac{a}{2}[1\ \bar{1}\ 0]_\mathrm{T}$ so that $|\mathbf{b}_m| = |\mathbf{b}_t|$, and using (T J M) it can be demonstrated that $[1\ \bar{1}\ 0]_\mathrm{M} \parallel [1\ \bar{1}\ 0]_\mathrm{T}$. This result is expected because these particular dislocations cannot generate a step in the $(1\ 1\ 1)_\mathrm{M}$ interface when they cross into the twin lattice (see Equation 11.15). Only dislocations with Burgers vectors not parallel to the interface cause the formation of steps.

The data further illustrate the fact that when $\mathbf{b}_m$ lies in the $(1\ 1\ 1)_\mathrm{M}$ plane, there is no increase in energy due to the reaction $\mathbf{b}_m \to \mathbf{b}_\mathrm{I} + \mathbf{b}_t$, which occurs when a dislocation crosses the interface. This is because $\mathbf{b}_\mathrm{I} = 0$ and $\mathbf{b}_t = \mathbf{b}_m$. For all other cases $\mathbf{b}_\mathrm{I}$ is not zero and since $|\mathbf{b}_t|$ is never less than $|\mathbf{b}_m|$, $\mathbf{b}_m \to \mathbf{b}_\mathrm{I} + \mathbf{b}_t$ is always energetically unfavorable. In fact, in the example being discussed, there can never be an energy reduction when an $\frac{a}{2}\langle 1\ \bar{1}\ 0\rangle$ dislocation penetrates the coherent twin boundary. The dislocations cannot therefore spontaneously cross the boundary. A trivial case where dislocations might spontaneously cross a boundary is when the latter is a free surface, assuming that the increase in surface area (and hence surface energy) due to the formation of a step is not prohibitive. Spontaneous penetration of the interface might also become favorable if the interface separates crystals with very different elastic properties.

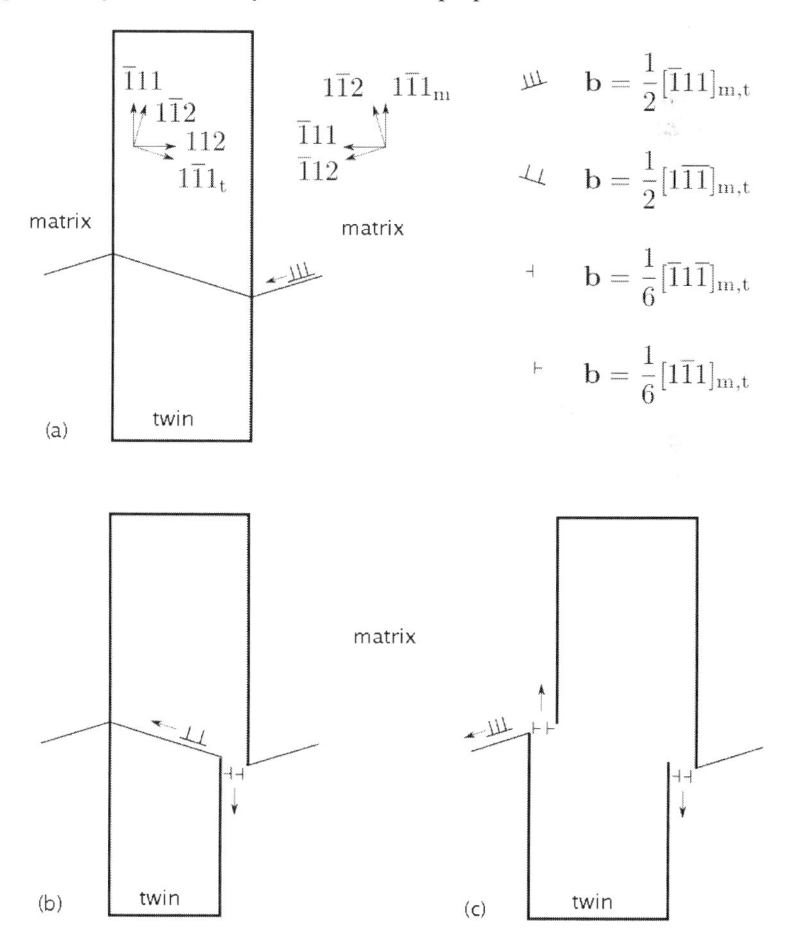

FIGURE 11.8
The passage of a slip dislocation across a coherent twin boundary in a bcc crystal. The twinning system is $\{\bar{1}\ 1\ 2\}\langle 1\ \bar{1}\ 1\rangle$, $s = 2^{-\frac{1}{2}}$. The subscripts m and t refer to

the twin and matrix, respectively; the arrows at the dislocation symbols indicate the sense of the Burgers vectors and the dislocation line vectors are all parallel to $[\bar{1}\ \bar{1}\ 0]_{m,t}$ where m refers to the matrix and t to the twin.

The results obtained show that single dislocations can glide into twins in fcc crystals without leaving a fault; there are no geometrical restrictions to the passage of slip dislocations across the coherent twin boundaries concerned. It can similarly be demonstrated that slip dislocations can comfortably traverse the coherent twin boundaries of $\{1\ 1\ 2\}$ twins in bcc or bct lattices and this has implications on the interpretation of the strength of martensite [22]. As will be discussed later, the substructure of martensite plates in steels (and in many non-ferrous alloys) often consists of very finely spaced $\{1\ 1\ 2\}$ transformation twins. It was at one time believed that the twins were mainly responsible for the high strength of ferrous martensites, because the numerous twin boundaries should hinder slip – the analysis above clearly suggests otherwise. Indeed, twinned martensites which do not contain carbon also do not exhibit exceptionally high strengths and it is now generally accepted that the strength of virgin ferrous martensites is largely due to interstitial solid solution hardening by carbon atoms, or in the case of lightly autotempered martensites due to carbon atom clustering or fine precipitation. Consistent with this, it is found that Fe-30Ni wt% twinned martensites are not particularly hard.

Finally, it should be mentioned that even when glide across coherent twin boundaries in martensites should be unhindered, the boundaries will cause a small amount of hardening, partly because the corresponding slip systems in the matrix and twin will in general be differently stressed [22, 25] (simply because they are not necessarily parallel) and partly due to the work necessary to create the steps in the interfaces. It is emphasised, however, that these should be relatively small contributions to the strength of martensite. Figure 11.8 illustrates the passage of a slip dislocation across a coherent $\{1\ 1\ 2\}$ twin interface in a bcc material.

Example 11.5: fcc to hcp transformation revisited

A Co-6.5Fe wt% alloy transforms from an fcc (γ) structure to a hcp martensite structure with zero change in density [26]. The invariant plane of the transformation is the close-packed $\{1\ 1\ 1\}_\gamma$ plane, the shear direction being $\langle 1\ 1\ \bar{2}\rangle_\gamma$. The magnitude of the shear is $8^{-\frac{1}{2}}$, which is half the normal twinning shear for fcc crystals. By factorizing the total transformation strain into a pure strain and a rigid body rotation, show that the maximum extension or contraction suffered by any vector of the parent lattice, as a result of the transformation, is less than 20%.

Representing the fcc parent lattice in an orthonormal basis Z, consisting of unit basis vectors parallel to [1 0 0], [0 1 0] and [0 0 1] fcc directions, respectively, and substituting $(\mathbf{p}; Z^*) = (3^{-\frac{1}{2}})(1\ 1\ 1)$, $[Z; \mathbf{d}] = (6^{-\frac{1}{2}})[1\ 1\ \bar{2}]$ and $m = 8^{-\frac{1}{2}}$ into

Equation 11.3, the total transformation strain (Z P Z) is found to be:

$$(\mathrm{Z\ P\ Z}) = \frac{1}{12}\begin{pmatrix} 13 & 1 & 1 \\ 1 & 13 & 1 \\ -2 & -2 & 10 \end{pmatrix}$$

This can be factorized into a pure strain (Z Q Z) and a rigid body rotation (Z J Z) (Figure 10.5). Writing $(\mathrm{Z\ T\ Z}) = (\mathrm{Z\ P'\ Z})(\mathrm{Z\ P\ Z})$, we obtain:

$$(\mathrm{Z\ T\ Z}) = \frac{1}{144}\begin{pmatrix} 174 & 30 & -6 \\ 30 & 174 & -6 \\ -6 & -6 & 102 \end{pmatrix}.$$

The eigenvalues and eigenvectors of (Z T Z) are

$$\begin{array}{lll} \lambda_1 = 1.421535 & [\mathrm{Z}; \mathbf{u}] = & [0.704706 \quad 0.704706 \quad -0.082341] \\ \lambda_2 = 1.000000 & [\mathrm{Z}; \mathbf{v}] = & [0.707107 \quad -0.707107 \quad 0.000000] \cdot \\ \lambda_3 = 0.703465 & [\mathrm{Z}; \mathbf{x}] = & [0.058224 \quad 0.058224 \quad 0.996604] \end{array}$$

Notice that the eigenvectors form an orthogonal set and that consistent with the fact that **v** lies in the invariant plane, λ_2 has a value of unity. **u**, **v**, and **x** are also the eigenvectors of (Z Q Z). The eigenvalues of (Z Q Z) are given by the square roots of the eigenvalues of (Z T Z); they are 1.192282, 1.0 and 0.838728. Hence, the maximum extensions and contractions are less than 20% since each eigenvalue is the ratio of the final to initial length of a vector parallel to an eigenvector. The maximum extension occurs along **u** and the maximum contraction along **x**.

The matrix (Z Q Z) is given by Equation 10.12 as:

$$\begin{aligned} (\mathrm{Z\ Q\ Z}) &= \begin{pmatrix} 0.70471 & 0.70711 & 0.05822 \\ 0.70471 & -0.70711 & 0.05822 \\ -0.08234 & 0.00000 & 0.99660 \end{pmatrix}\begin{pmatrix} 1.19228 & 0.0 & 0.0 \\ 0.0 & 1.00000 & 0.0 \\ 0.0 & 0.0 & 0.83873 \end{pmatrix} \\ &\quad\times \begin{pmatrix} 0.70471 & 0.70471 & -0.08234 \\ 0.70711 & -0.70711 & 0.00000 \\ 0.05822 & 0.05822 & 0.99660 \end{pmatrix} \\ &= \begin{pmatrix} 1.094944 & 0.094943 & -0.020515 \\ 0.094943 & 1.094944 & -0.020515 \\ -0.020515 & -0.020515 & 0.841125 \end{pmatrix} \end{aligned} \tag{11.17}$$

and

$$(\mathrm{Z\ Q\ Z})^{-1} = \begin{pmatrix} 0.920562 & -0.079438 & 0.020515 \\ -0.079438 & 0.920562 & 0.020515 \\ 0.020515 & 0.020515 & 1.189885 \end{pmatrix}$$

From Equation 10.13, $(\mathrm{Z\ J\ Z}) = (\mathrm{Z\ P\ Z})(\mathrm{Z\ Q\ Z})^{-1}$

$$(\mathrm{Z\ J\ Z}) = \begin{pmatrix} 0.992365 & -0.007635 & 0.123091 \\ -0.007635 & 0.992365 & 0.123091 \\ -0.123092 & -0.123092 & 0.984732 \end{pmatrix}. \tag{11.18}$$

The matrix (Z J Z) represents a right-handed rotation of 10.03° about $[1\ \overline{1}\ 0]_{\mathrm{Z}}$ axis.

It is interesting to examine what happens to the vector $[1\ 1\ \overline{2}]_{\mathrm{Z}}$ due to the operations (Z Q Z) and (Z J Z):

$$(\mathrm{Z\ Q\ Z})[1\ 1\ \overline{2}]_{\mathrm{Z}} = [1.230916\ 1.230916\ -1.723280]_{\mathrm{Z}}$$

where the new vector can be shown to have the same magnitude as $[1\ 1\ \overline{2}]$ but points in a different direction. The effect of the pure rotation is

$$(\mathrm{Z\ J\ Z})[1.230916\ 1.230916\ -1.723280]_{\mathrm{Z}} = [1\ 1\ \overline{2}]_{\mathrm{Z}}.$$

Thus, the pure strain deforms $[1\ 1\ \overline{2}]_{\mathrm{Z}}$ into another vector of identical magnitude and the pure rotation brings this new vector back into the $[1\ 1\ \overline{2}]_{\mathrm{Z}}$ direction, the net operation leaving it invariant, as expected, since $[1\ 1\ \overline{2}]_{\mathrm{Z}}$ is the shear direction which lies in the invariant plane. Referring to Figure 10.5, the direction $fe = [1\ \overline{1}\ 0]_{\mathrm{Z}}$, $yz = [1\ 1\ \overline{2}]_{\mathrm{Z}}$ and $y'z' = [1.230916\ 1.230916\ -1.723280]_{\mathrm{Z}}$. $y'z'$ is brought into coincidence with yz by the rigid body rotation (Z J Z) to generate the invariant plane containing fe and yz.

Physically, the fcc to hcp transformation occurs by the movement of a single set of Shockley partial dislocations, Burgers vector $\mathbf{b} = \frac{a}{6}\langle 1\ 1\ \overline{2}\rangle_{\gamma}$ on *alternate* close-packed $\{1\ 1\ 1\}_{\gamma}$ planes. To produce a fair thickness of hcp martensite, a mechanism has to be sought which allows Shockley partials to be generated on every other slip plane. Some kind of a pole mechanism (see, for example, p. 310 of [27]) would allow this to happen, but there is as yet no experimental evidence confirming this. Motion of the partials would cause a shearing of the γ lattice, on the system $\{1\,1\,1\}_{\gamma}\langle 1\ 1\ \overline{2}\rangle_{\gamma}$, the average magnitude $\overline{s}$ of the shear being $\overline{s} = |\mathbf{b}|/2d$, where d is the spacing of the close-packed planes. Hence, $\overline{s} = 6^{-\frac{1}{2}}a/2(3^{-\frac{1}{2}}a) = 8^{-\frac{1}{2}}$. This is exactly the shear system we used in generating the matrix (Z P Z) and the physical effect of the shear on the shape of an originally flat surface is, in general, to tilt the surface (about a line given by its intersection with the hcp habit plane) through some angle dependant on the indices of the free surface. By measuring such tilts it is possible to deduce $\overline{s}$, which has been experimentally confirmed to equal half the twinning shear.

In fcc crystals, the close-packed planes stack in the sequence $\ldots ABCABC\ldots$; the passage of a single Shockley partial causes the sequence to change to $\ldots ABA\ldots$

creating a three layer thick region of hcp phase since the stacking sequence of close-packed planes in the hcp lattice has a periodicity of 2. This then is the physical manner in which the transformation occurs, the martensite having a $\{1\ 1\ 1\}_\gamma$ habit plane – if the parent product interface deviates slightly from $\{1\ 1\ 1\}_\gamma$, then it will consist of stepped sections of close-packed plane, the steps representing the Shockley partial transformation dislocations. The spacing of the partials along $\langle 1\ 1\ 1\rangle_\gamma$ would be $2d$. In other words, in the stacking sequence ABC, the motion of a partial on B would leave A and B unaffected though C would be displaced by $6^{-\frac{1}{2}}a\langle 1\ 1\ 2\rangle_\gamma$ to a new position A, giving ABA stacking. Partials could thus be located on every alternate plane of the fcc crystal.

Hence, we see that the matrix (Z P Z) is quite compatible with the microscopic dislocation based mechanism of transformation. (Z P Z) predicts the correct macroscopic surface relief effect and its invariant plane is the habit plane of the martensite. However, if (Z P Z) is considered to act homogeneously over the entire crystal, then it would carry half the atoms into the wrong positions. For instance, if the habit plane is designated A in the sequence ABC of close packed planes, then the effect of (Z P Z) is to leave A unchanged, shift the atoms on plane C by $2sd$ and those on plane B by sd along $\langle 1\ 1\ \overline{2}\rangle_\gamma$. Of course, this puts the atoms originally in C sites into A sites, as required for hcp stacking. However, the B atoms are located at positions half way between B and C sites, through a distance $\frac{a}{12}\langle 1\ 1\ \overline{2}\rangle_\gamma$. *Shuffles* are thus necessary to bring these atoms back into the original B positions and to restore the $\ldots ABA \ldots$ hcp sequence. These atomic movements in the middle layer are called shuffles because they occur through very small distances (always less than the interatomic spacing) and do not affect the macroscopic shape change [21]. The shuffle here is a purely formal concept; consistent with the fact that the Shockley partials glide over alternate close-packed planes, the deformation (Z P Z) must in fact be considered homogeneous only on a scale of every two planes. By locking the close-packed planes together in pairs, we avoid displacing the B-site atoms to the wrong positions and thus automatically avoid the reverse shuffle displacement.

In the particular example discussed above, the dislocation mechanism is established experimentally and physically reasonable shear systems were used in determining (Z P Z). However, in general it is possible to find an infinite number of deformations [8, 21] which may accomplish the same lattice change and slightly empirical criteria have to be used in selecting the correct deformation. One such criterion could involve the selection of deformations which involve the minimum principal strains and the minimum degree of shuffling, but intuition and experimental evidence is almost always necessary to reach a decision. The Bain strain which transforms the fcc lattice to the bcc lattice is believed to be the correct choice because it seems to involve the least atomic displacements and zero shuffling of atoms [20]. The absence of shuffles can be deduced from the Bain correspondence matrix $(\alpha\ \mathrm{C}\ \gamma)$ which can be deduced from inspection since its columns are rational lattice vectors referred to the α basis, produced by the deformation of the basis vectors of the γ basis; since $[1\ 0\ 0]_\gamma$ is deformed to $[1\ 1\ 0]_\alpha$, $[0\ 1\ 0]_\gamma$ to $[\overline{1}\ 1\ 0]_\alpha$ and $[0\ 0\ 1]_\gamma$ to $[0\ 0\ 1]_\alpha$, by the

Bain strain (Figure 8.2), the correspondence matrix is simply:

$$(\alpha \text{ C } \gamma) = \begin{pmatrix} 1 & \bar{1} & 0 \\ 1 & 1 & 0 \\ 0 & 0 & 1 \end{pmatrix} \tag{11.19}$$

If $\mathbf{u}$ is a vector defining the position of an atom in the γ unit cell, then it can be verified that $(\alpha \text{ C } \gamma)[\gamma; \mathbf{u}]$ always gives a corresponding vector in the α lattice which terminates at a lattice point. For example, $\frac{1}{2}[1\ 0\ 1]_\gamma$ corresponds to $\frac{1}{2}[1\ 1\ 1]_\alpha$; both these vectors connect the origins of their respective unit cells to an atomic position. The Bain correspondence thus defines the position of each and every atom in the α lattice relative to the γ lattice. It only is possible to obtain a correspondence matrix like this when the primitive cells of each of the lattices concerned contain just one atom [8].

The primitive cell of the hcp lattice contains two atoms and any lattice correspondence will only define the final positions of an integral fraction of the atoms, the remainder having to shuffle into their correct positions in the product lattice. This can be demonstrated with the correspondence matrix for the example presented above. It is convenient to represent the conventional hcp lattice (basis H) in an alternative orthorhombic basis (symbol O), with basis vectors:

$$[1\,0\,0]_{\text{O}} = \frac{1}{2}[0\,\bar{1}\,1]_\gamma = \frac{1}{3}[\bar{1}\,\bar{1}\,2\,0]_{\text{H}}$$

$$[0\,1\,0]_{\text{O}} = \frac{1}{2}[2\,\bar{1}\,\bar{1}]_\gamma = [\bar{1}\,1\,0\,0]_{\text{H}}$$

$$[0\,0\,1]_{\text{O}} = \frac{2}{3}[1\,1\,1]_\gamma = [0\,0\,0\,1]_{\text{H}}$$

The orthorhombic unit cell thus contains three close-packed layers of atoms parallel to its (0 0 1) faces. The middle layer has atoms located at $[0\ \frac{1}{3}\ \frac{1}{2}]_{\text{O}}$, $[1\ \frac{1}{3}\ \frac{1}{2}]_{\text{O}}$ and $[\frac{1}{2}\ \frac{5}{6}\ \frac{1}{2}]_{\text{O}}$. The other two layers have atoms located at each corner of the unit cell and in the middle of each (0 0 1) face, as illustrated in Figure 11.9.

From our earlier definition of a correspondence matrix, $(\text{O C } \gamma)$ can be written directly from the relations (between basis vectors) stated earlier:

$$(\gamma \text{ C O}) = \frac{1}{2}\begin{pmatrix} 0 & 2 & 1 \\ -1 & -1 & 1 \\ 1 & -1 & 2 \end{pmatrix}$$

Alternatively, the correspondence matrix $(\text{O C } \gamma)$ may be derived (page 164) as follows:

$$(\text{O C } \gamma) = (\text{O J } \gamma)(\gamma \text{ P } \gamma)$$

The matrix $(\gamma \text{ P } \gamma)$ is the total strain, which transforms the fcc lattice into the hcp lattice; it is equal to the matrix (Z P Z) derived previously (page 172), since the basis

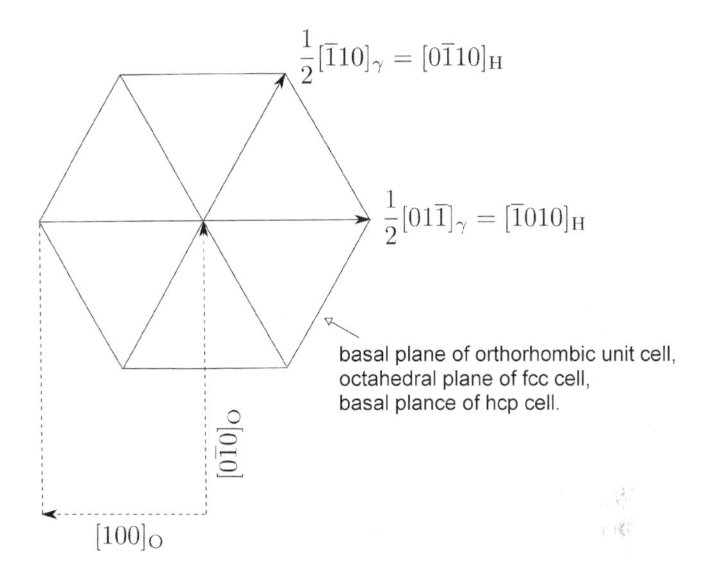

FIGURE 11.9
Representation of bases O, H, and γ. The directions in the hexagonal cell are expressed in the Weber notation.

vectors of the orthonormal basis Z are parallel to the corresponding basis vectors of the orthogonal basis γ. It follows that:

$$(\mathrm{O}\ \mathrm{C}\ \gamma) = \begin{pmatrix} 0 & -1 & 1 \\ 2/3 & -1/3 & -1/3 \\ 1/2 & 1/2 & 1/2 \end{pmatrix} \begin{pmatrix} 13/12 & 1/12 & 1/12 \\ 1/12 & 13/12 & 1/12 \\ -2/12 & -2/12 & 10/12 \end{pmatrix}$$

$$= \begin{pmatrix} -1/4 & -5/4 & 3/4 \\ 3/4 & -1/4 & -1/4 \\ 1/2 & 1/2 & 1/2 \end{pmatrix}$$

and

$$(\gamma\ \mathrm{C}\ \mathrm{O}) = \frac{1}{2} \begin{pmatrix} 0 & 2 & 1 \\ -1 & -1 & 1 \\ 1 & -1 & 2 \end{pmatrix}.$$

Using this correspondence matrix, we can show that all the atoms, except those in the middle close-packed layer in the unit cell, have their positions relative to the parent lattice defined by the correspondence matrix. For example, the atom at the position $[1\ 0\ 0]_\mathrm{O}$ corresponds directly to that at $[0\ \bar{1}\ 1]_\gamma$ in the fcc lattice. However, $[0\ \frac{1}{3}\ \frac{1}{2}]_\mathrm{O}$ corresponds to $\frac{1}{2}[7\ 1\ 4]_\gamma$ and there is no atom located at these coordinates in the γ

lattice. The generation of the middle layer thus involves shuffles of $\frac{1}{12}[1\ 1\ \overline{2}]_\gamma$, as discussed earlier; we note that $\frac{1}{12}[7\ 1\ 4]_\gamma - \frac{1}{12}[1\ 1\ \overline{2}]_\gamma = \frac{1}{2}[1\ 0\ 1]_\gamma$. Thus, the atom at $[0\ \frac{1}{3}\ \frac{1}{2}]_O$ is derived from that at $\frac{1}{2}[1\ 0\ 1]_\gamma$ in addition to a shuffle displacement through $\frac{a}{12}[1\ 1\ \overline{2}]_\gamma$.

11.6 Conjugate of an invariant-plane strain

We have seen already that an fcc lattice can be transformed to an hcp lattice by shearing the former on the system $\{1\ 1\ 1\}\langle 1\ 1\ \overline{2}\rangle$, $s = 8^{-\frac{1}{2}}$. This shear represents an invariant-plane strain (Z P Z) which can be factorised into a pure strain (Z Q Z) and a rigid body rotation (Z J Z), as in section 10.4. The pure deformation (Z Q Z) accomplishes the required lattice change from fcc to hcp, but is not an invariant-plane strain. As illustrated in Figure 10.5 and from Equation 11.18, it is the rigid body rotation of 10.03° about $\langle 1\ \overline{1}\ 0\rangle$ that makes the $\{1\ 1\ 1\}$ plane invariant and in combination with (Z Q Z) produces the final orientation relation implied by (Z P Z).

Referring to Figure 10.5a,b, we see that there are in fact two ways [21] in which (Z Q Z) can be converted into an invariant-plane strain which transforms the fcc lattice to the hcp lattice. The first involves the rigid body rotation (Z J Z) in which $y'z'$ is brought into coincidence with yz, as shown in Figure 10.12c. The alternative would be to employ a rigid body rotation (Z J_2 Z), involving a rotation of 10.03° about $\langle \overline{1}\ 1\ 0\rangle$, which would bring $x'w'$ into coincidence with xw, making xw the trace of the invariant-plane. Hence, (Z Q Z) when combined with (Z J_2 Z) would result in a different invariant-plane strain (Z P_2 Z) which also shears the fcc lattice to the hcp lattice. From Equation 7.9, (Z J_2 Z) is given by:

$$(\text{Z J}_2\text{ Z}) = \begin{pmatrix} 0.992365 & -0.007635 & -0.123091 \\ -0.007635 & 0.992365 & -0.123091 \\ 0.123092 & 0.123092 & 0.984732 \end{pmatrix}$$

From Example 16, (Z Q Z) is given by:

$$(\text{Z Q Z}) = \begin{pmatrix} 1.094944 & 0.094943 & -0.020515 \\ 0.094943 & 1.094944 & -0.020515 \\ -0.020515 & -0.020515 & 0.841125 \end{pmatrix}$$

From Equation 10.10, (Z P_2 Z) = (Z J_2 Z)(Z Q Z)

$$(\text{Z P}_2\text{ Z}) = \begin{pmatrix} 1.0883834 & 0.088384 & -0.123737 \\ 0.088384 & 1.088384 & -0.123737 \\ 0.126263 & 0.126263 & 0.823232 \end{pmatrix}.$$

On comparing this with Equation 11.3, we see that (Z P_2 Z) involves a shear of magnitude $s = 8^{-\frac{1}{2}}$ on $\{5\ 5\ \overline{7}\}_Z\langle 7\ 7\ 10\rangle_Z$. It follows that there are two ways of accomplishing the fcc to hcp change:

$$\text{Mode 1: Shear on} \quad \{5\,5\,\overline{7}\}_Z\langle 7\,7\,10\rangle_Z \quad s = 8^{-\frac{1}{2}}$$

$$\text{Mode 2: Shear on} \quad \{1\,1\,1\}_Z\langle 1\,1\,\overline{2}\rangle_Z \quad s = 8^{-\frac{1}{2}}$$

Both shears can generate a fully coherent interface between the fcc and hcp lattices (the coherent interface plane being coincident with the invariant-plane). Of course, while the $\{1\ 1\ 1\}$ interface of mode 2 would be atomically flat, the $\{5\ 5\ 7\}$ interface of mode 1 must probably be stepped on an atomic scale. The orientation relations between the fcc and hcp lattices would be different for the two mechanisms. In fact, (Z J_2 Z) is

$$(\mathrm{Z}\ \mathrm{J}_2\ \mathrm{Z}) = \begin{pmatrix} -0.2121216 & -1.2121216 & 0.6969703 \\ 0.7373739 & -0.2626261 & -0.2323234 \\ 0.3484848 & 0.3484848 & 0.7121212 \end{pmatrix}.$$

It is intriguing that only the second mode has been observed experimentally, even though both involve identical shear magnitudes.

A general conclusion to be drawn from the above analysis is that whenever two lattices can be related by an IPS (i.e., whenever they can be joined by a fully coherent interface), it is always possible to find a *conjugate* IPS which in general allows the two lattices to be differently orientated but still connected by a fully coherent interface. This is clear from Figure 10.5 where we see that there are two ways of carrying out the rigid body rotation in order to obtain an IPS which transforms the fcc lattice to the hcp lattice. The deformation involved in twinning is also an IPS so that for a given twin mode it ought to be possible to find a conjugate twin mode. In Figure 11.5c, a rigid body rotation about $[\overline{1}\ 1\ 0]$, which brings K_2 into coincidence with K_2' would give the conjugate twin mode on $(1\ 1\ \overline{1})[1\ 1\ 2]$.

We have used the pure strain (Z Q Z) to transform the fcc crystal into a hcp crystal. However, before this transformation, we could use any of an infinite number of operations (e.g., a symmetry operation) to bring the fcc lattice into self-coincidence. Combining any one of these operations with (Z Q Z) then gives us an alternative deformation which can accomplish the fcc→hcp lattice change without altering the orientation relationship. It follows that two lattices can be deformed into one another in an infinite number of ways. Hence, *prediction* of the transformation strain is not possible in the sense that intuition or experimental evidence has to be used to choose the "best" or "physically most meaningful" transformation strain.

Example 11.6: Combined effect of two invariant-plane strains

Show that the combined effect of the operation of two arbitrary invariant-plane strains is equivalent to an invariant-line strain (ILS). Hence prove that if the two invariant-plane strains have the same invariant-plane, or the same displacement direction, then their combined effect is simply another IPS [2].

The two invariant-plane strains are referred to an orthonormal basis X and are designated (X P X) and (X Q X), such that m and n are their respective magnitudes, **d** and **e** their respective unit displacement directions and **p** and **q** their respective unit invariant-plane normals. If (X Q X) operates first, then the combined effect of the two strains is

$$\begin{aligned}(\mathrm{X\,P\,X})(\mathrm{X\,Q\,X}) &= \{\mathbf{I}+m[\mathrm{X};\mathbf{d}](\mathbf{p};\mathrm{X}^*)\}\{\mathbf{I}+n[\mathrm{X};\mathbf{e}](\mathbf{q};\mathrm{X}^*)\}\\ &= \mathbf{I}+m[\mathrm{X};\mathbf{d}](\mathbf{p};\mathrm{X}^*)+n[\mathrm{X};\mathbf{e}](\mathbf{q};\mathrm{X}^*)\\ &\quad +mn[\mathrm{X};\mathbf{d}](\mathbf{p};\mathrm{X}^*)[\mathrm{X};\mathbf{e}](\mathbf{q};\mathrm{X}^*)\\ &= \mathbf{I}+m[\mathrm{X};\mathbf{d}](\mathbf{p};\mathrm{X}^*)+n[\mathrm{X};\mathbf{e}](\mathbf{q};\mathrm{X}^*)\\ &\quad +g[\mathrm{X};\mathbf{d}](\mathbf{q};\mathrm{X}^*)\end{aligned} \tag{11.20}$$

where g is the scalar quantity $g = mn(\mathbf{p};\mathrm{X}^*)[\mathrm{X};\mathbf{e}]$.

If **u** is a vector which lies in both the planes represented by **p** and **q**, i.e., it is parallel to $\mathbf{p}\wedge\mathbf{q}$, then it is obvious (Equation 11.20) that $(\mathrm{X\,P\,X})(\mathrm{X\,Q\,X})[\mathrm{X};\mathbf{u}] = [\mathrm{X};\mathbf{u}]$, since $(\mathbf{p};\mathrm{X}^*)[\mathrm{X};\mathbf{u}] = 0$ and $(\mathbf{q};\mathrm{X}^*)[\mathrm{X};\mathbf{u}] = 0$. It follows that **u** is parallel to the invariant line of the total deformation $(\mathrm{X\,S\,X}) = (\mathrm{X\,P\,X})(\mathrm{X\,Q\,X})$. This is logical since (X P X) should leave every line on **p** invariant and (X Q X) should leave all lines on **q** invariant. The line that is common to both **p** and **q** should therefore be unaffected by (X P X)(X Q X). Hence, the combination of two arbitrary invariant-plane strains (X P X)(X Q X) gives and *invariant-line strain* (X S X).

If **d** = **e**, then from Equation 11.20

$$\begin{aligned}(\mathrm{X\,P\,X})(\mathrm{X\,Q\,X}) &= \mathbf{I}+[\mathrm{X};\mathbf{d}](\mathbf{r};\mathrm{X}^*)\\ &\quad \text{where } (\mathbf{r};\mathrm{X}^*) = m(\mathbf{p};\mathrm{X}^*)+n(\mathbf{q};\mathrm{X}^*)+g(\mathbf{q};\mathrm{X}^*)\end{aligned}$$

which is simply another IPS on a plane whose normal is parallel to **r**. If $\mathbf{p} = \mathbf{q}$, then it follows that

$$\begin{aligned}(\mathrm{X\,P\,X})(\mathrm{X\,Q\,X}) &= \mathbf{I}+[\mathrm{X};\mathbf{f}](\mathbf{p};\mathrm{X}^*)\\ &\quad \text{where } [\mathrm{X};\mathbf{f}] = m[\mathrm{X};\mathbf{d}]+n[\mathrm{X};\mathbf{e}]+g[\mathrm{X};\mathbf{d}]\end{aligned}$$

which is an IPS with a displacement direction parallel to [X;**f**].

Hence, in the special case where the two IPSs have their displacement directions parallel, or have their invariant-plane normals parallel, their combined effect is simply another IPS.

It is interesting to examine how plane normals are affected by invariant-line strains. Taking the inverse of (X S X), we see that

$$(\mathrm{X\,S\,X})^{-1} = (\mathrm{X\,Q\,X})^{-1}(\mathrm{X\,P\,X})^{-1}$$

or from Equation 11.6,

$$\begin{aligned}(\mathrm{X\,S\,X})^{-1} &= \mathbf{I}-an[\mathrm{X};\mathbf{e}](\mathbf{q};\mathrm{X}^*)\mathbf{I}-bm[\mathrm{X};\mathbf{d}](\mathbf{p};\mathrm{X}^*)\\ &= \mathbf{I}-an[\mathrm{X};\mathbf{e}](\mathbf{q};\mathrm{X}^*)-bm[\mathrm{X};\mathbf{d}](\mathbf{p};\mathrm{X}^*)\\ &\quad +cnm[\mathrm{X};\mathbf{e}](\mathbf{p};\mathrm{X}^*)\end{aligned} \tag{11.21}$$

where a, b and c are scalar constants given by $1/a = \det(\mathrm{X\,Q\,X})$, $1/b = \det(\mathrm{X\,P\,X})$ and $c = ab(\mathbf{q};\mathrm{X}^*)[\mathrm{X};\mathbf{d}]$.

If $\mathbf{h} = \mathbf{e} \wedge \mathbf{d}$, then $\mathbf{h}$ is a reciprocal lattice vector representing the plane which contains both $\mathbf{e}$ and $\mathbf{d}$. It is evident from Equation 11.21 that $(\mathbf{h};\mathrm{X}^*)(\mathrm{X\,S\,X})^{-1} = (\mathbf{h};\mathrm{X}^*)$, since $(\mathbf{h};\mathrm{X}^*)[\mathrm{X};\mathbf{e}] = 0$ and $(\mathbf{h};\mathrm{X}^*)[\mathrm{X};\mathbf{d}] = 0$. In other words, the plane normal $\mathbf{h}$ is an invariant normal of the invariant-line strain $(\mathrm{X\,S\,X})^{-1}$.

We have found that an ILS has two important characteristics: it leaves a line **u** invariant and also leaves a plane normal **h** invariant. If the ILS is factorized into two IPS's, then **u** lies at the intersection of the invariant-planes of these component IPS's, and **h** defines the plane containing the two displacement vectors of these IPS's. These results will be useful in gaining a deeper understanding of martensitic transformations.

11.7 Summary

In mathematics, the term *invariance* generally refers to a property that is unchanged by a transformation, which in the present context can be a deformation. The deformations considered here leave planes or lines invariant *and* have physical consequences that can be detected experimentally. Ordinary slip in solids leaves traces at free surfaces, that for crystalline materials reflect the anisotropy of atomic arrangements but there is no change in volume. In contrast, solid-state phase transformations are usually associated with changes in density, but the displacements associated with the volume change can be directed normal to the broad interface between the parent and product crystal, so that an invariant plane is preserved. An important result from the methods described in this chapter is that there is unique connection between the orientation relationship, invariant plane (habit plane) and the shape deformation.

References

1. L. Gold: Evaluation of the stiffness coeffcients for beryllium from ultrasonic measurements in polycrystalline and single crystal specimens, *Physical Review*, 1949, **77**, 390–395.

2. J. S. Bowles, and J. K. Mackenzie: The crystallography of martensite transformations, part I, *Acta Metallurgica*, 1954, **2**, 129–137.

3. J. W. Christian: The origin of surface relief effects in phase transformations, In: V. F. Zackay, and H. I. Aaronson, eds. *Decomposition of austenite by diffusional processes*. New York: Interscience, 1962:371–386.

4. J. W. Christian, and D. V. Edmonds: The bainite transformation, In: A. R. Marder, and J. I. Goldstein, eds. *Phase transformations in ferrous alloys*. Warrendale, Pennsylvania: TMS-AIME, 1984:293–327.

5. F. P. Bundy: Pressure-temperature phase diagram of iron to 200 kbar, 900°C, *Journal of Applied Physics*, 1965, **36**, 616–620.

6. J. W. Brooks, M. H. Loretto, and R. E. Smallman: Direct observations of martensite nuclei in stainless steel, *Acta Metallurgica*, 1979, **27**, 1839–1847.

7. C. M. Wayman: *Introduction to the crystallography of martensitic transformations*: New York: Macmillan, 1964.

8. J. W. Christian: *Theory of transformations in metals and alloys*, Part I: 2nd ed., Oxford, U. K.: Pergamon Press, 1975.

9. E. Schmid, and W. Boas: *Plasticity of crystals* (translated from the 1935 edition of Kristalplastizitaet): London, U.K.: F. A. Hughes and Co., 1950.

10. J. W. Christian: Deformation by moving interfaces, *Metallurgical Transactions A*, 1982, **13**, 509–538.

11. A. Kelly, and K. M. Knowles: *Crystallography and crystal defects*: 2nd ed., New York: John Wiley & Sons, Inc., 2012.

12. H. K. D. H. Bhadeshia: TRIP-assisted steels?, *ISIJ International*, 2002, **42**, 1059–1060.

13. J. W. Christian: *Theory of transformations in metals and alloys*: Oxford, U. K.: Pergamon Press, 1965.

14. J. W. Christian, and S. Mahajan: Deformation twinning, *Progress in Materials Science*, 1995, **39**, 1–157.

15. J. D. Eshelby: The determination of the elastic field of an ellipsoidal inclusion and related problem, *Proceedings fo the Royal Society A*, 1957, **241**, 376–396.

16. J. W. Christian: Accommodation strains in martensite formation, the use of the dilatation parameter, *Acta Metallurgica*, 1958, **6**, 377–379.

17. J. W. Christian: Thermodynamics and kinetics of martensite, In: G. B. Olson, and M. Cohen, eds. *International Conference on Martensitic Transformations ICOMAT '79*. Massachusetts: Alpine Press, 1979:220–234.

18. E. C. Bain: The nature of martensite, *Trans. AIME*, 1924, **70**, 25–46.

19. C. M. Wayman: The crystallography of martensitic transformations in alloys of iron': In: H. Hermans, ed. *Advances in materials research*, vol. 3. John Wiley & Sons, Inc., 1968:147–304.

20. J. S. Bowles, and C. M. Wayman: The Bain strain, lattice correspondences, and deformations related to martensitic transformations, *Metallurgical Transactions*, 1972, **3**, 1113–1121.

21. J. W. Christian, and A. G. Crocker: *Dislocations in solids*, ed. F. R. N. Nabarro: Amsterdam, Holland: North Holland, 1980.

22. J. W. Christian: The strength of martensite, In: A. Kelly, and R. Nicholson, eds. *Strengthening methods in crystals*. Netherlands: Elsevier, 1971:261–329.

23. V. Volterra: Sur l'équilibre des corps élastiques multiplement connexes, *Annales scientifiques de l'École normale supérieure*, 1907, **24**, 401–517.

24. J. W. Christian, and A. G. Crocker: *Dislocations in Solids*, ed. F. R. N. Nabarro, vol. 3, chap. 11: Amsterdam, Holland: North Holland, 1980:165–252.

25. P. M. Kelly, and G. Pollard: The movement of slip dislocations in internally twinned martensite, *Acta Metallurgica*, 1969, **17**, 1005–1008.

26. J. W. Brooks, M. H. Loretto, and R. E. Smallman: In situ observations of martensite formation in stainless steel, *Acta Metallurgica*, 1979, **27**, 1829–1838.

27. A. Kelly, and G. W. Groves: *Crystallography and crystal defects*: London, U.K.: Longmans, 1970.

12

Martensite

Abstract

Martensitic transformations can be quite simple to understand in systems where the transformation strain is equal to the observed shape deformation. This is the case when hcp martensite grows in fcc austenite. The odd observation that requires explanation is that during the fcc→bcc or bct transformation, the shape deformation is inconsistent with the strain needed to change the change the lattice. The crystallographic theory described in this chapter deals with this anomaly using a most elegant formalism that is quantitative.

12.1 Introduction

The basic principles of the characteristics of martensitic transformation have already been covered in Chapter 8, albeit qualitatively. The purpose here is to explain the mathematical framework developed by Bowles, MacKenzie, Wechsler, Lieberman and Read [1–4], a framework that represents one of the most elegant and complete theories that helped explain an accumulation of findings that could not be reconciled. And there were predictions made which were only to be verified when advanced instrumentation such as transmission electron microscopy became available. The history of this wonderful theory is described by Wayman [5]; a classic example of a subject receptive to a new way of thinking when a discipline is in a state of crisis [6, 7]. The theory has had many consequences, implicitly or directly, in unrelated subjects such as in the estimation of interfacial structure [8–11], the calculation of stress-induced crystallographic texture [12–14] and the general understanding of the relative orientations of crystals that precipitate in the solid state.

12.2 Shape deformation

All martensitic transformations involve coordinated movements of atoms and are diffusionless. Since the shape of the pattern in which the atoms in the parent crystal are arranged nevertheless changes to generate the product phase, it follows that there must be a visible change in the macroscopic shape of the parent crystal during transformation [15]. The shape deformation and its significance can best be illustrated by reference to Figure 12.1, where a comparison is made between reconstructive and diffusionless transformations. For simplicity, the diagram refers to a case where the transformation strain is an invariant-plane strain and a fully coherent interface exists between the parent and product lattices, irrespective of the mechanism of transformation.

Considering the shear transformation first, we note that since the pattern of atomic arrangement is changed on transformation, and since the transformation is diffusionless, the macroscopic shape of the crystal changes. The shape deformation has the exact characteristics of an IPS. The initially flat surface normal to da becomes tilted about the line formed by the intersection of the interface plane with the surface normal to da. The straight line ab is bent into two connected and straight segments ae and eb. Hence, an observer looking at a scratch that is initially along ab and in the surface abcd would note that on martensite formation, the scratch becomes homogeneously deflected about the point e where it intersects the trace of the interface plane. Furthermore, the scratches ae and eb would be seen to remain connected at the point e. This amounts to proof that the shape deformation has, on a macroscopic scale, the characteristics of an IPS and that the interface between the parent and product lattices does not contain any distortions (i.e., it is an invariant-plane). Observing the deflection of scratches is one way of deducing the nature of shape deformations accompanying transformations.

In Figure 12.1 it also is implied that martensitic transformation is diffusionless; labelled rows of atoms in the parent crystal are expected to remain in the correct sequence in the martensite lattice. It is possible therefore to suggest that a particular atom in the martensite must have originated from a corresponding particular atom in the parent crystal. A formal way of expressing this property is to say that there exists an *atomic correspondence* between the parent and product lattices.

In the case of the reconstructive transformation illustrated in Figure 12.1, it is evident that the product phase can be of a different composition from the parent. In addition, there has been much mixing up of atoms during transformation and the order of arrangement of atoms in the product lattice is different from that in the parent lattice – the atomic correspondence has been destroyed. Because the transformation involves a reconstruction of the parent lattice, atoms are able to diffuse around in such a way that the IPS shape deformation and its accompanying strain energy, do not arise. The scratch ab remains straight across the interface and is unaffected by the transformation.

In summary, martensitic transformations always are accompanied by a change in the shape of the parent crystal, characterised as an invariant-plane strain when examined on a macroscopic scale. The occurrence of such a shape deformation is taken to imply the existence of an atomic correspondence between the parent and product lattices. It is possible to state that a particular atom in the product occupied a particular corresponding site in the parent lattice.

These results have some interesting consequences. The formation of martensite in a constrained environment causes distortions in its surroundings. The strain energy

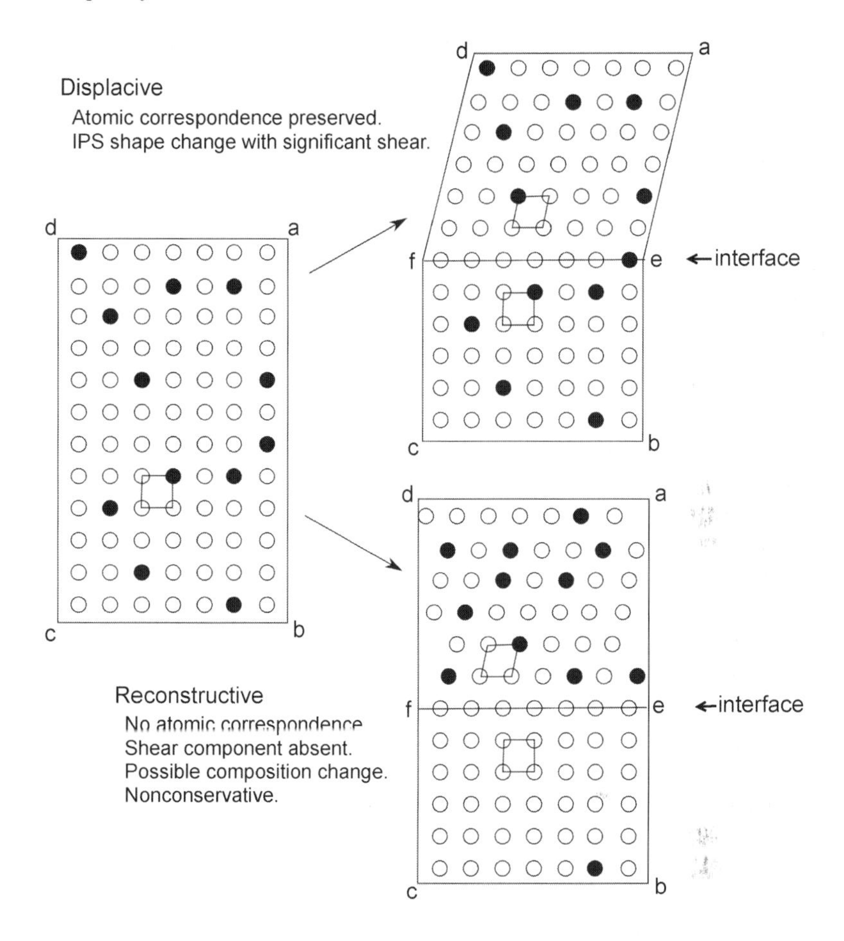

FIGURE 12.1
Schematic illustration of the mechanisms of diffusional and shear transformations.

due to this distortion, per unit volume of martensite, is given by [16–18]

$$E \approx \frac{c}{r}\mu(s^2 + \delta^2) \tag{12.1}$$

where μ is the shear modulus of the parent lattice, c/r is the thickness to length ratio of the martensite plate and s and δ are the shear and dilatational components of the shape deformation strain. It follows that martensite must always have a thin plate morphology consistent with the minimising of strain energy but achieving transformation at the same time. E usually amounts to about 600 J mol^{-1} for martensite in steels [18], when the shape deformation is entirely elastically accommodated. If the austenite is soft, then some plastic accommodation driven by the shape deformation may relieve some of the strains, but the E value calculated on the basis of purely elastic accommodation should be taken to be the upper limit of the stored energy due to the shape change accompanying martensitic transformation. This is because the plastic accommodation is driven by the shape deformation [18]. In the event that plastic accommodation occurs, dislocations and other defects may be generated both in the parent and product lattices, which would make it impossible to reverse the motion of the interface, thus eliminating the possibility of a shape memory effect.

Martensitic transformation does not require diffusion, can occur at very low temperatures and the interface can move rapidly. The interface between martensite and the parent phase must therefore be glissile, so we begin with a discussion of what the appropriate structure of such an interface might be.

12.3 Interfacial structure of martensite

All the mathematical operations that we describe in the context of martensite occur at the transformation interface, which separates the perfect austenite from the perfect martensite. The transformation is thermodynamically of first order [19] which implies a sharp interface between the parent and product crystals which can therefore coexist. This is obvious when we observe routinely a mixture of martensite plates embedded in austenite.

That martensite can form at cryogenic temperatures [20] indicates that the process cannot rely on thermal activation. In particular, the (sharp) interface connecting the martensite with the parent phase must be able to move in the manner of dislocation glide rather than dislocation climb. It therefore cannot be incoherent but it is possible for certain semi-coherent or fully coherent boundaries to be glissile [21]. However, stress-free coherence is only possible when the parent and product lattices can be related by a strain which is an invariant-plane strain [22] so such interfaces are rare for particles of appreciable size. The fcc→hcp transformation is one example where a fully coherent interface is possible. Martensitic transformation in ordered Fe_3Be occurs by a simple shearing of the lattice (an IPS) [23], so a fully coherent interface is again possible. More generally, the interfaces tend to be semi-coherent. It was demonstrated in Chapter 8 that a fcc austenite lattice cannot be transformed into a bcc martensite lattice by a strain which is an IPS, so that the as semi-coherent interface is expected.

The semi-coherent interface should consist of coherent regions separated periodically by discontinuities which prevent the misfit in the interface plane from accumulating over large distances, in order to minimize the elastic strains associated with the interface. There are two kinds of semi-coherency [22, 24]; if these discontinuities are intrinsic dislocations with Burgers vectors in the interface plane, not parallel to the dislocation line, then the interface is said to be epitaxially semi-coherent. The term 'intrinsic' means that the dislocations are a necessary part of the interface structure and have not simply strayed into the boundary – they do not have a long-range strain field. The normal displacement of such an interface requires the thermally activated climb of intrinsic dislocations, so that the interface can only move in a non-conservative manner, with relatively restricted or zero mobility at low temperatures. A martensite interface cannot therefore be epitaxially semi-coherent.

In the second type of semi-coherency, the discontinuities discussed above are screw dislocations, or dislocations whose Burgers vectors do not lie in the interface plane. This kind of semi-coherency is of the type associated with glissile martensite interfaces, whose motion is conservative (i.e., the motion does not lead to the creation or destruction of lattice sites). Such an interface should have a high mobility since the migration of atoms is not necessary for its movement. Actually, two further conditions must be satisfied before even this interface can be said to be glissile:

(i) A glissile interface requires that the glide planes of the intrinsic dislocations associated with the product lattice must meet the corresponding glide planes of the parent lattice edge to edge in the interface [24], along the dislocation lines.

(ii) If more than one set of intrinsic dislocations exist, then these should either have the same line vector in the interface, or their respective Burgers vectors must be parallel [24]. This condition ensures that the interface can move as an integral unit. It implies in addition, that the deformation caused by the intrinsic dislocations, when the interface moves, can always be described as a simple shear (caused by a resultant intrinsic dislocation which is a combination of all the intrinsic dislocations) on some plane which makes a finite angle with the interface plane, and intersects the latter along the line vector of the resultant intrinsic dislocation.

Obviously, if the intrinsic dislocation structure consists of just a single set of parallel dislocations, or of a set of different dislocations which can be summed to give a single glissile intrinsic dislocation, then it follows that there must exist in the interface, a line which is parallel to the resultant intrinsic dislocation line vector, along which there is zero distortion. Because this line exists in the interface, it is also unrotated. It is an *invariant-line* in the interface between the parent and product lattices. When full coherency between the parent and and martensite lattices is not possible, then for the interface to be glissile, the transformation strain relating the two lattices must be an invariant-line strain, with the invariant-line being in the interface plane.

The interface between the martensite and the parent phase usually is called the "habit plane"; when the transformation occurs without any constraint, the habit plane is macroscopically flat, as illustrated in Figure 12.2. When the martensite forms in a constrained environment, it grows in the shape of a thin lenticular plate or lath and the habit plane is a little less clear in the sense that the interface is curved on a macroscopic scale. However, it is found that the average plane of the plate (the plane containing the major circumference of the lens) corresponds closely to that expected from crystallographic theory, and to that determined under conditions of unconstrained transformation. The aspect ratio (maximum thickness to length ratio) of lenticular plates is usually less than 0.05, so that the interface plane does not depart very much from the average plane of the plate. Some examples of habit plane indices (relative to the austenite lattice) are given in Table 8.1.

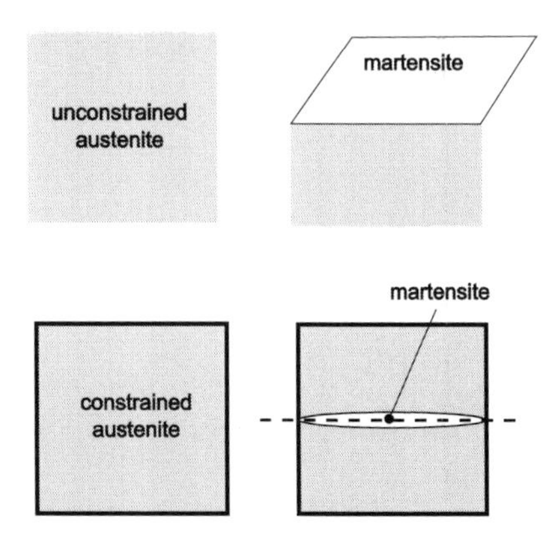

FIGURE 12.2
The habit plane of martensite (α') under conditions of unconstrained and constrained transformation, respectively. In the latter case, the dashed line indicates the trace of the habit plane.

12.4 Phenomenological theory of martensite crystallography

We have emphasised that an essential feature of martensitic transformation is its shape deformation, which on a macroscopic scale has the characteristics of an invariant-plane strain. The magnitude m of the shape deformation can be determined experimentally as can its unit displacement vector **d**. The habit plane of the martensite (unit normal **p**) is the invariant-plane of the shape deformation. The shape deformation can be represented by means of a shape deformation matrix (F P F) such that:

$$(\mathrm{F\ P\ F}) = \mathbf{I} + m[\mathrm{F};\mathbf{d}](\mathbf{p};\mathrm{F}^*)$$

where the basis F is for convenience chosen to be orthonormal, although the equation is valid for any basis.

For the shear transformation illustrated in Figure 12.1 and for the fcc→hcp martensite reaction, the lattice transformation strain is itself an IPS and there is no difficulty in reconciling the transformation strain and the observed shape deformation. In other words, if the parent lattice is operated on by the shape deformation matrix, then the correct product lattice is generated if shuffles are allowed; the transformation strain is the same as the shape deformation.

This is not the case [25] for the fcc→ bcc martensite reaction and for many other martensite transformations where the lattice transformation strain (F S F) does not equal the observed shape deformation (F P F). Figure 8.4 illustrated the fact that the Bain strain (F B F) when combined with an appropriate rigid body rotation (F J F) gives an invariant-line strain which when applied to the fcc lattice generates the bcc martensite lattice. However, the shape deformation that accompanies the formation of bcc martensite from austenite is nevertheless found experimentally to be an invariant-plane strain. This is the major anomaly that the theory of martensite crystallography attempts to resolve: the experimentally observed shape deformation is inconsistent with the lattice transformation strain. If the observed shape deformation is applied to the parent lattice then the austenite lattice is deformed into an intermediate lattice (not experimentally observed) but not into the required bcc lattice.

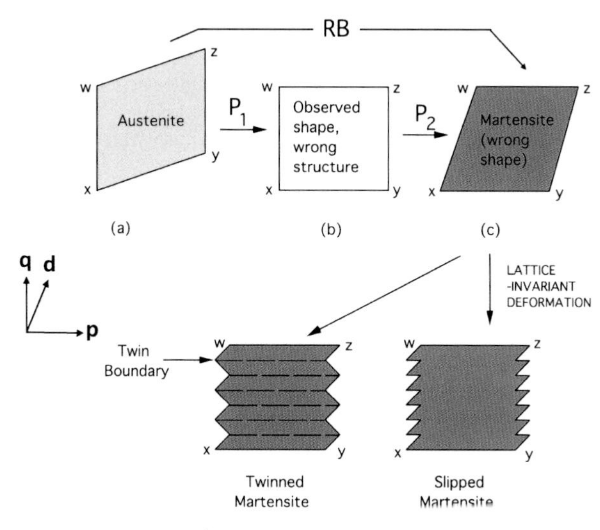

FIGURE 12.3
Schematic illustration of the phenomenological theory of martensite. (a) represents the austenite crystal and (c) has a bcc structure. (b) has a structure between fcc and bcc, **p** is the habit plane unit normal and **q** is the unit normal to the plane on which the lattice-invariant shear occurs. The heavy horizontal lines in (e) are coherent twin boundaries. Note that the vector **e** is normal to **q** but does not lie in the plane of the diagram.

This problem is illustrated in Figures 12.3. The shape of the starting austenite crystal with the fcc structure is shown in Figure 12.3a. On martensitic transformation its shape alters to that illustrated in Figure 12.3b and the shape deformation on going from (a) to (b) is clearly an IPS on the plane with unit normal **p** and in the unit displacement direction **d**. However, the structure of the crystal in Figure 12.3b is some intermediate lattice which is not bcc, since an IPS cannot on its own change the fcc structure to the bcc structure. An invariant-line strain can however transform fcc to bcc, and since an ILS can be factorised into two invariant-plane strains, it follows that the further deformation (F Q F) needed to change the intermediate structure of Figure 12.3b to the bcc structure (Figure 12.3c) is another IPS. If the deformation (F Q F) is of magnitude n on a plane with unit normal **q** and in a unit direction **e**, then:

$$(\mathrm{F\ Q\ F}) = \mathbf{I} + n[\mathrm{F};\mathbf{e}](\mathbf{q};\mathrm{F}^*)$$

(F Q F) has to be chosen in such a way that (F P F)(F Q F) = (F S F), where (F S F) is an invariant-line strain which transforms the fcc lattice to the bcc lattice. Hence, a combination of two invariant-plane strains can accomplish the necessary lattice change but this then gives the wrong shape change as the extra shape change due to (F Q F), in changing (b) to (c), is not observed.

Numerous experiments have demonstrated that the shape deformation due to the fcc→bcc martensite transformation is an IPS [11], and it seems that the effect of (F Q F) on the macroscopic shape is invisible. If a way can be discovered of making the shape altering effect of (F Q F) invisible, then the problem is essentially determined.

(F Q F) can be made invisible by applying another deformation to (c) such that the shape of (c) is brought back to that of (b), without altering the bcc structure of (c). Such a deformation must therefore be *lattice-invariant* because it must not alter the symmetry or unit cell dimensions of the parent crystal structure. Ordinary slip does not change the nature of the lattice and is one form of a lattice-invariant deformation. Hence, slip deformation on the planes **q** and in the direction -**e** would make the shape change due to (F Q F) invisible on a macroscopic scale, as illustrated in Figure 12.3d. The magnitude of this lattice-invariant slip shear is of course determined by that of (F Q F) and we know that it is not possible to continuously vary the magnitude of slip shear, since the Burgers vectors of slip dislocations are discrete. (F Q F) on the other hand can have any arbitrary magnitude. This difficulty can be overcome by applying the slip shear *inhomogeneously*, by the passage of a discrete slip dislocation on say every nth plane, which has the effect of allowing the magnitude of the lattice-invariant shear to vary as a function of n. In applying the lattice-invariant shear to (c) in order to obtain (d), the bcc structure of (c) is completely unaffected, while is shape is deformed inhomogeneously to correspond to that of (b), as illustrated in Figure 12.3d.

This then is the essence of the theory of martensite crystallography [1–4], which explains the contradiction that the lattice transformation strain is an ILS but the macroscopic shape deformation is an IPS. The lattice transformation strain when combined with an inhomogeneous lattice-invariant shear produces a macroscopic shape change which is an IPS.

Twinning is another deformation which does not change the crystal structure, simply reorientates it; the shape (c) of Figure 12.3 could be deformed to correspond macroscopically to that of (b), without changing its bcc nature, by twinning, as illustrated in Figure 12.3e. The magnitude of the lattice-invariant deformation can be adjusted by varying the volume fraction of the twin. This explains the twin substructure found in many ferrous martensites, and such twins are called transformation twins. The irrationality of the habit planes arises because the indices of the habit plane depend on the amount of lattice-invariant deformation, a quantity which does not necessarily correlate with displacements equal to discrete lattice vectors.

Since the effect of the homogeneous strain (F Q F) on the shape of the parent crystal has to be cancelled by another opposite but inhomogeneously applied lattice-invariant deformation, it follows that (F Q F) is restricted to being a simple shear with the displacement vector **e** being confined to the invariant-plane of (F Q F). In other words, (F Q F) must have a zero dilatational component, since it cannot otherwise be cancelled by another IPS which preserves the lattice. As noted earlier, lattice-invariant deformations cannot alter the volume or symmetry of the lattice. The determinant of a deformation matrix gives the ratio of the volume after deformation to that prior to deformation, so that $\det(\mathrm{F\ Q\ F}) = 1$. This means that the total volume change of transformation is given by $\det(\mathrm{F\ P\ F}) = \det(\mathrm{F\ S\ F})$.

In summary, the martensite transformation in iron requires an invariant-line strain (F S F) to change the fcc lattice to the bcc martensite lattice and to obtain the experimentally observed orientation relation. This can be imagined to consist of two homogeneous invariant-plane strains (F P F) and (F Q F), such that $(\mathrm{F\ S\ F}) = (\mathrm{F\ P\ F})(\mathrm{F\ Q\ F})$. However, the shape change due to the simple shear (F Q F) is rendered invisible on a macroscopic scale since there is also an inhomogeneous lattice-invariant deformation (which can be slip or twinning) which cancels out the shape change due to (F Q F), without altering the lattice structure. It follows that the macroscopic shape change observed is solely due to (F P F) and therefore has the characteristics of an invariant-plane strain, as experimentally observed. We have seen already that the transformation strain (F S F) can be factorised into a Bain strain (F B F) combined with an appropriate rigid body rotation (F J F), such that (F S F)=(F J F)(F B F) and is an invariant-line strain, with the invariant-line lying in the planes **p** and **q**, and the invariant-normal of (F S F) defining a plane containing **d** and **e**. Hence, the theory of martensite can be summarized in terms of the equation

$$(\mathrm{F\ S\ F}) = (\mathrm{F\ J\ F})(\mathrm{F\ B\ F}) = (\mathrm{F\ P\ F})(\mathrm{F\ Q\ F}). \tag{12.2}$$

12.5 Stage 1: Calculation of lattice transformation strain

Two arbitrary lattices can be transformed into one another by an infinite number of different transformation strains, but only some of these may have reasonably small principal deformations. The choice available can be further reduced by considering only those strains which involve the minimum degree of shuffling of atoms and by considering the physical implications of such strains. In the case of martensitic transformations, a further condition has to be satisfied; the lattice transformation strain must also be an invariant-line strain if the interface is to be glissile [26].

For the fcc→bcc martensitic transformation, the Bain strain, which is a pure deformation, involves the smallest atomic displacements during transformation. When it is combined with an appropriate rigid body rotation, the total strain amounts to an invariant-line strain. For martensitic transformations, the rigid body rotation has to be chosen in such a way that the invariant-line lies in the plane of the lattice-invariant shear and also in the habit plane of the martensite; these planes are the invariant planes of (F Q F) and (F P F), respectively, so that the line common to these planes is not affected by these deformations. Furthermore, the invariant normal of the ILS must define a plane which contains the displacement directions of the lattice invariant shear and of the shape deformation. This ensures that the spacing of this plane is not affected by (F Q F) or (F P F).

The next example illustrates how the transformation strain can be determined once the pure deformation which accomplishes the lattice change is deduced. To ensure that the invariant-line and invariant-normal of the transformation strain are compatible with the mode of lattice-invariant shear, we first need to specify the latter. It is assumed here that the plane and direction of the lattice-invariant shear are $(1\ 0\ 1)_{\mathrm{F}}$ and $[1\ 0\ \bar{1}]_{\mathrm{F}}$, respectively. One variant of the Bain strain is illustrated in Figure 8.2, where we see that $[1\ 0\ 0]_{\gamma}$ is deformed into $[1\ 1\ 0]_{\alpha}$, $[0\ 1\ 0]_{\gamma}$ to $[\bar{1}\ 1\ 0]_{\alpha}$ and $[0\ 0\ 1]_{\gamma}$ to $[0\ 0\ 1]_{\alpha}$, so that the variant of the Bain correspondence matrix is given by Equation 11.19. We will use this variant of the Bain correspondence matrix throughout the text, but we note that there are two other possibilities, where $[0\ 0\ 1]_{\alpha}$ can be derived from either $[1\ 0\ 0]_{\gamma}$ or $[0\ 1\ 0]_{\gamma}$, respectively.

Example 12.1: Determination of lattice transformation strain

The deformation matrix representing the Bain strain, which carries the fcc austenite lattice (Figure 8.2) to the bcc martensite lattice is given by

$$(\mathrm{F\ B\ F}) = \begin{pmatrix} \eta_1 & 0 & 0 \\ 0 & \eta_2 & 0 \\ 0 & 0 & \eta_3 \end{pmatrix}$$

where F is an orthonormal basis consisting of unit basis vectors $\mathbf{f}_i$ parallel to the crystallographic axes of the conventional fcc austenite unit cell (Figure 8.2, $\mathbf{f}_1 \parallel \mathbf{a}_1$, $\mathbf{f}_2 \parallel \mathbf{a}_2$ & $\mathbf{f}_3 \parallel \mathbf{a}_3$). η_i are the principal deformations of the Bain strain, given by $\eta_1 = \eta_2 = 1.136071$ and $\eta_3 = 0.803324$.

Find the rigid body rotation (F J F) which when combined with the Bain strain gives an invariant-line strain $(\mathrm{F\ S\ F}) = (\mathrm{F\ J\ F})(\mathrm{F\ B\ F})$, subject to the condition that the invariant-line of (F S F) must lie in $(1\ 0\ 1)_{\mathrm{F}}$ and that the plane defined by the invariant-normal of (F S F) contains $[1\ 0\ \bar{1}]_{\mathrm{F}}$.

Writing the invariant-line as $[\mathrm{F};\mathbf{u}] = [u_1 u_2 u_3]$, we note that for $\mathbf{u}$ to lie in $(1\ 0\ 1)_{\mathrm{F}}$, its components must satisfy the equation

$$u_1 = -u_3 \tag{12.3}$$

Prior to deformation,

$$|\mathbf{u}|^2 = (\mathbf{u};\mathrm{F})[\mathrm{F};\mathbf{u}] = 1 \tag{23b}$$

$\mathbf{u}$, as a result of deformation becomes a new vector $\mathbf{x}$ with

$$\begin{aligned} |\mathbf{x}|^2 &= (\mathbf{x};\mathrm{F})[\mathrm{F};\mathbf{x}] \\ &= (\mathbf{u};\mathrm{F})(\mathrm{F\ B'\ F})(\mathrm{F\ B\ F})[\mathrm{F};\mathbf{u}] \\ &= (\mathbf{u};\mathrm{F})(\mathrm{F\ B\ F})^2[\mathrm{F};\mathbf{u}]. \end{aligned}$$

If the magnitude of **u** is not to change on deformation then $|\mathbf{u}| = |\mathbf{x}|$ or

$$u_1^2 + u_2^2 + u_3^2 = \eta_1^2 u_1^2 + \eta_2^2 u_2^2 + \eta_3^2 u_3^2. \tag{12.4}$$

Equations 12.3 and 12.4 can be solved simultaneously to give *two* solutions for undistorted lines:

$$\begin{aligned} [\mathrm{F};\mathbf{u}] &= [-0.671120 \ -0.314952 \ 0.671120], \\ [\mathrm{F};\mathbf{v}] &= [-0.671120 \ 0.314952 \ 0.671120]. \end{aligned}$$

To solve for the invariant normal of the ILS, we proceed as follows. Writing $(\mathbf{h};\mathrm{F}^*) = (h_1 h_2 h_3)$, we note that for **h** to contain $[1\ 0\ \overline{1}]_{\mathrm{F}}$, its components must satisfy the equation

$$h_1 = h_3 \tag{12.5}$$

and furthermore,

$$(\mathbf{h};\mathrm{F}^*)[\mathrm{F}^*;\mathbf{h}] = 1.$$

h, on deformation becomes a new plane normal **l** and if $|\mathbf{h}| = |\mathbf{l}|$ then

$$\begin{aligned} |\mathbf{l}|^2 &= (\mathbf{l};\mathrm{F}^*)[F^*;\mathbf{l}] \\ &= (\mathbf{h};\mathrm{F}^*)(\mathrm{F\ B\ F})^{-1}(\mathrm{F\ B'\ F})^{-1}[F^*;\mathbf{h}] \end{aligned}$$

so that

$$h_1^2 + h_2^2 + h_3^2 = (l_1/\eta_1)^2 + (l_2/\eta_2)^2 + (l_3/\eta_3)^2 \tag{12.6}$$

Solving Equations 12.5 and 12.6 simultaneously, we obtain the two possible solutions for the undistorted-normals as

$$\begin{aligned} (\mathbf{h};\mathrm{F}^*) &= (0.539127 \ 0.647058 \ 0.539127), \\ (\mathbf{k};\mathrm{F}^*) &= (0.539127 \ -0.647058 \ 0.539127). \end{aligned}$$

To convert (F B F) into an invariant-line strain (F S F) we have to employ a rigid body rotation (F J F) which simultaneously brings an undistorted line (such as **x**) and an undistorted normal (such as **l**) back into their original directions along **u** and **h** respectively. This is possible because the angle between **x** and **l** is the same as that between **u** and **h**, as shown below:

$$\begin{aligned} \mathbf{l}.\mathbf{x} &= (\mathbf{l};\mathrm{F}^*)[\mathrm{F};\mathbf{x}] \\ &= (\mathbf{h};\mathrm{F}^*)(\mathrm{F\ B\ F})^{-1}(\mathrm{F\ B\ F})[\mathrm{F};\mathbf{u}] \\ &= (\mathbf{h};\mathrm{F}^*)[\mathrm{F};\mathbf{u}] \\ &= \mathbf{h}.\mathbf{u} \end{aligned}$$

Hence, one way of converting (F B F) into an ILS is to employ a rigid body rotation which simultaneously rotates **l** into **h** and **x** into **u**. We have found that there are two undistorted lines and two undistorted normals which satisfy the conditions of the original question, so there are four ways of choosing pairs of undistorted lines and undistorted normals, but in this case, the four solutions are crystallographically equivalent. There are therefore four solutions (different in general) to the problem of converting (F B F) to (F S F), subject to the condition that the invariant-line should be in (1 0 1) and that the invariant normal defines a plane containing $[1\ 0\ \overline{1}]$. We will concentrate on the solution obtained using the pair **u** and **h**:

$$\mathbf{l} = (\mathbf{h}; \mathrm{F}^*)(\mathrm{F\ B\ F})^{-1} = (0.474554\ 0.569558\ 0.671120)$$

$$\mathbf{x} = (\mathrm{F\ B\ F})[\mathrm{F};\mathbf{u}] = [-0.762440\ -0.357809\ 0.539127]$$

$$\mathbf{a} = \mathbf{u} \wedge \mathbf{h} = (-0.604053\ 0.723638\ -0.264454)$$

$$\mathbf{b} = \mathbf{x} \wedge \mathbf{l} = (-0.547197\ 0.767534\ -0.264454)$$

The required rigid body rotation should rotate **x** back to **u**, **l** back to **h** and **b** to **a**, giving the three equations:

$$[\mathrm{F};\mathbf{u}] = (\mathrm{F\ J\ F})[\mathrm{F};\mathbf{x}]$$

$$[\mathrm{F};\mathbf{h}] = (\mathrm{F\ J\ F})[\mathrm{F};\mathbf{l}]$$

$$[\mathrm{F};\mathbf{a}] = (\mathrm{F\ J\ F})[\mathrm{F};\mathbf{b}]$$

which can be expressed as a 3×3 matrix equation

$$\begin{pmatrix} u_1 & h_1 & a_1 \\ u_2 & h_2 & a_2 \\ u_3 & h_3 & a_3 \end{pmatrix} = \begin{pmatrix} J_{11} & J_{12} & J_{13} \\ J_{21} & J_{22} & J_{23} \\ J_{31} & J_{32} & J_{33} \end{pmatrix} \begin{pmatrix} w_1 & l_1 & b_1 \\ w_2 & l_2 & b_2 \\ w_3 & l_3 & b_3 \end{pmatrix}$$

it follows that

$$\begin{pmatrix} -0.671120 & 0.539127 & -0.604053 \\ -0.314952 & 0.647058 & 0.723638 \\ 0.671120 & 0.539127 & -0.264454 \end{pmatrix}$$

$$= (\mathrm{F\ J\ F}) \begin{pmatrix} -0.762440 & 0.474554 & -0.547197 \\ -0.357808 & 0.569558 & 0.767534 \\ 0.539127 & 0.671120 & -0.264454 \end{pmatrix}$$

which on solving gives

$$(\mathrm{F\ J\ F}) = \begin{pmatrix} 0.990534 & -0.035103 & 0.132700 \\ 0.021102 & 0.994197 & 0.105482 \\ -0.135633 & -0.101683 & 0.985527 \end{pmatrix}$$

which is a rotation of 9.89° about $[0.602879 \; -0.780887 \; 0.163563]_{\mathrm{F}}$. The invariant-line strain (F S F) = (F J F)(F B F) is thus

$$(\mathrm{F\ S\ F}) = \begin{pmatrix} 1.125317 & -0.039880 & 0.106601 \\ 0.023973 & 1.129478 & 0.084736 \\ -0.154089 & -0.115519 & 0.791698 \end{pmatrix} \qquad (12.7a)$$

and we note that $(\mathrm{F\ S\ F})^{-1} = (\mathrm{F\ B\ F})^{-1}(\mathrm{F\ J\ F})^{-1}$ is given by

$$(\mathrm{F\ S\ F})^{-1} = \begin{pmatrix} 0.871896 & 0.018574 & -0.119388 \\ -0.030899 & 0.875120 & -0.089504 \\ 0.165189 & 0.131307 & 1.226811 \end{pmatrix}. \qquad (12.7b)$$

12.6 Stage 2: Determination of the orientation relationship

The orientation relationship between the austenite and martensite is best expressed in terms of a coordinate transformation matrix $(\alpha \ \mathrm{J} \ \gamma)$. Any vector **u** or any plane normal **h** can then be expressed in either crystal basis by using the equations

$$[\alpha; \mathbf{u}] = (\alpha \ \mathrm{J} \ \gamma)[\gamma; \mathbf{u}]$$
$$[\gamma; \mathbf{u}] = (\gamma \ \mathrm{J} \ \alpha)[\alpha; \mathbf{u}]$$
$$(\mathbf{h}; \alpha^*) = (\mathbf{h}; \gamma^*)(\gamma \ \mathrm{J} \ \alpha)$$
$$(\mathbf{h}; \gamma^*) = (\mathbf{h}; \alpha^*)(\alpha \ \mathrm{J} \ \gamma).$$

Example 12.2: Martensite-austenite orientation relationship

For the martensite reaction considered in example on page 194, determine the orientation relationship between the parent and product lattices.

The orientation relationship can be expressed in terms of a co-ordinate transformation matrix $(\alpha \ \mathrm{J} \ \gamma)$, which is related to the transformation strain $(\gamma \ \mathrm{S} \ \gamma)$ via Equation 11.12, so that

$$(\alpha \ \mathrm{J} \ \gamma)(\gamma \ \mathrm{S} \ \gamma) = (\alpha \ \mathrm{C} \ \gamma)$$

where $(\alpha \ \mathrm{C} \ \gamma)$ is the Bain correspondence matrix (Equation 11.19), and it follows that

$$(\alpha \ \mathrm{J} \ \gamma) = (\alpha \ \mathrm{C} \ \gamma)(\gamma \ \mathrm{S} \ \gamma)^{-1}$$

In Equation 12.7 the matrices representing the transformation strain and its inverse were determined in the basis F, and can be converted into the basis γ by means of a similarity transformation. However, because $\mathbf{f}_i$ are parallel to $\mathbf{a}_i$, it can easily be demonstrated that (F S F) = (γ S γ) and $(\text{F S F})^{-1} = (\gamma \text{ S } \gamma)^{-1}$. Hence, using the data from Equation 12.7 and the Bain correspondence matrix from Equation 12.1a, we see that

$$(\alpha \text{ J } \gamma) = \begin{pmatrix} 1 & \bar{1} & 0 \\ 1 & 1 & 0 \\ 0 & 0 & 1 \end{pmatrix} \begin{pmatrix} 0.871896 & 0.018574 & -0.119388 \\ -0.030899 & 0.875120 & -0.089504 \\ 0.165189 & 0.131307 & 1.226811 \end{pmatrix} \quad (12.8)$$

$$(\alpha \text{ J } \gamma) = \begin{pmatrix} 0.902795 & -0.856546 & -0.029884 \\ 0.840997 & 0.893694 & -0.208892 \\ 0.165189 & 0.131307 & 1.226811 \end{pmatrix}$$

$$(\gamma \text{ J } \alpha) = \begin{pmatrix} 0.582598 & 0.542718 & 0.106602 \\ -0.552752 & 0.576725 & 0.084736 \\ -0.019285 & -0.134804 & 0.791698 \end{pmatrix}.$$

These coordinate transformation matrices can be used to show that

$$(1\ 1\ 1)_\gamma = (0.016365\ 1.525799\ 1.523307)_\alpha$$

and

$$[\bar{1}\ 0\ 1]_\gamma = [-0.932679\ -1.049889\ 1.061622]_\alpha$$

which means that $(1\ 1\ 1)_\gamma$ is very nearly parallel to $(0\ 1\ 1)_\alpha$ and $[\bar{1}\ 0\ 1]_\gamma$ is about 3° from $[\bar{1}\ \bar{1}\ 1]_\alpha$. The orientation relationship is illustrated in Figure 12.4.

12.7 Stage 3: Nature of the shape deformation

The shape deformation (F P F) can be obtained by factorising the transformation strain (F S F). The deformation (F P F) is physically significant because it can be experimentally determined and describes the macroscopic change in shape of the parent crystal. In practise, all the changes necessary for transformation occur simultaneously at the moving interface [27]. Transformation dislocations (atomic height steps) in the interface cause the fcc lattice to change to the bcc lattice as the interface moves and the deformation that this produces is described by (F S F). The intrinsic dislocations which lie along the invariant-line in the interface have Burgers vectors which are perfect lattice vectors of the parent lattice. They cannot therefore take part in the actual transformation of the lattice, but as the interface moves, they inhomogeneously shear the volume of the material swept by the interface [24]. This of course is the lattice-invariant shear discussed above, which in combination with the shape de-

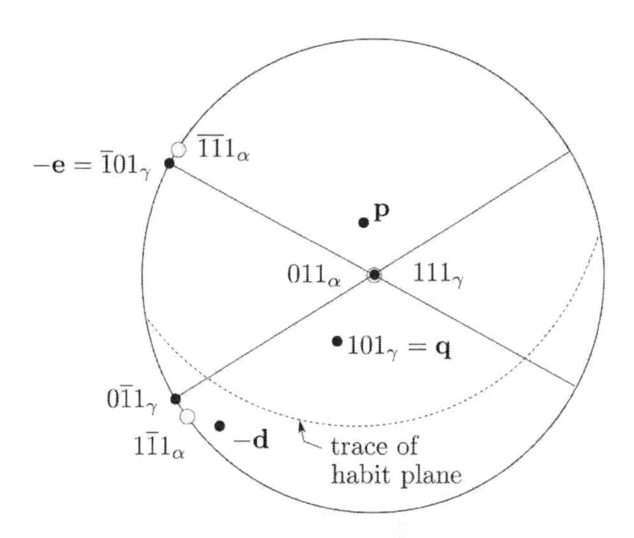

FIGURE 12.4
Stereographic representation of the orientation relationship between martensite and austenite, as deduced in Equation 12.8. The lattice-invariant shear plane (**q**) and direction (-**e**), and the habit plane (**p**) and unit displacement vector (**d**) also are illustrated.

formation due to (F S F) gives the experimentally observed IPS surface relief (F P F). This then is the physical interpretation of the transformation process.

The crystallographic theory of martensite is on the other hand called phenomenological; the steps into which the transformation is factorized (e.g., the two "shears" (F P F) and (F Q F)) are not unique and do not necessarily describe the actual path by which the atoms move from one lattice to the other. The theory simply provides a definite link between the initial and final states without being certain of the path in between.

Example 12.3: Habit plane and the shape deformation

For the martensite reaction considered on page 194, determine the habit plane of the martensite plate, assuming that the lattice-invariant shear occurs on the system $(1\ 0\ 1)_\gamma[\bar{1}\ 0\ 1]_\gamma$. Comment on the choice of this shear system and determine the nature of the shape deformation.

The lattice invariant shear is on $(1\ 0\ 1)[\bar{1}\ 0\ 1]$ and since its effect is to cancel the shape change due to (F Q F), the latter must be a shear on $(1\ 0\ 1)[1\ 0\ \bar{1}]$. To solve for the habit plane (unit normal **p**) it is necessary to factorise (F S F) into the two invariant-plane strains $(\text{F P F}) = \mathbf{I} + m[\text{F};\mathbf{d}](\mathbf{p};\text{F}^*)$ and $(\text{F Q F}) = \mathbf{I} + n[\text{F};\mathbf{e}](\mathbf{q};\text{F}^*)$.

The transformation strain (F S F) of Equation 12.7 was calculated by phenomenologically combining the Bain strain with a rigid body rotation, with the latter chosen to make (F S F) an invariant line strain, subject to the condition that the invariant line **u** of (F S F) must lie in $(1\ 0\ 1)_\gamma$ and that the invariant-normal **h** of (F S F) must define a plane containing $[1\ 0\ \bar{1}]_\gamma$. This is of course compatible with the lattice-invariant shear system chosen in the present example since $\mathbf{u.q} = \mathbf{e.h} = 0$. From Equation 12.2,

$$(\mathrm{F\ S\ F}) = (\mathrm{F\ P\ F})(\mathrm{F\ Q\ F}) = \{\mathbf{I} + m[\mathrm{F};\mathbf{d}](\mathbf{p};\mathrm{F}^*)\}\{\mathbf{I} + n[\mathrm{F};\mathbf{e}](\mathbf{q};\mathrm{F}^*)\} \quad (12.9)$$

and using Equation 11.6, we see that

$$\begin{aligned}(\mathrm{F\ S\ F})^{-1} &= (\mathrm{F\ Q\ F})^{-1}(\mathrm{F\ P\ F})^{-1}\\ &= \{\mathbf{I} - n[\mathrm{F};\mathbf{e}](\mathbf{q};\mathrm{F}^*)\}\{\mathbf{I} - am[\mathrm{F};\mathbf{d}](\mathbf{p};\mathrm{F}^*)\} \quad (12.10)\end{aligned}$$

where $1/a = \det(\mathrm{F\ P\ F})$ and $\det(\mathrm{F\ Q\ F}) = 1$. Using Equation 12.10, we obtain

$$\begin{aligned}(\mathbf{q};\mathrm{F}^*)(\mathrm{F\ S\ F})^{-1} &= (\mathbf{q};\mathrm{F}^*)\{\mathbf{I} - n[\mathrm{F};\mathbf{e}](\mathbf{q};\mathrm{F}^*)\}\{\mathbf{I} - am[\mathrm{F};\mathbf{d}](\mathbf{p};\mathrm{F}^*)\}\\ &= \{(\mathbf{q};\mathrm{F}^*) - n\,\underbrace{(\mathbf{q};\mathrm{F}^*)[\mathrm{F};\mathbf{e}]}_{=0}(\mathbf{q};\mathrm{F}^*)\}\ \{\mathbf{I} - am[\mathrm{F};\mathbf{d}](\mathbf{p};\mathrm{F}^*)\}\\ &= (\mathbf{q};\mathrm{F}^*)\{\mathbf{I} - am[\mathrm{F};\mathbf{d}](\mathbf{p};\mathrm{F}^*)\}\\ &= (\mathbf{q};\mathrm{F}^*) - b(\mathbf{p};\mathrm{F}^*)\end{aligned}$$

where b is a scalar constant given by $b = am(\mathbf{q};\mathrm{F}^*)[\mathrm{F};\mathbf{d}]$.

Hence,

$$\begin{aligned}b(\mathbf{p};\mathrm{F}^*) &= (\mathbf{q};\mathrm{F}^*) - (\mathbf{q};\mathrm{F}^*)(\mathrm{F\ S\ F})^{-1}\\ &= \begin{pmatrix}\frac{1}{\sqrt{2}} & 0.000 & \frac{1}{\sqrt{2}}\end{pmatrix}\\ &\quad - \begin{pmatrix}\frac{1}{\sqrt{2}} & 0.000 & \frac{1}{\sqrt{2}}\end{pmatrix} \times \begin{pmatrix}0.871896 & 0.018574 & -0.119388\\ 0.030894 & 0.875120 & -0.089504\\ 0.165189 & 0.131307 & 1.226811\end{pmatrix}\\ &= \begin{pmatrix}-0.026223 & -0.105982 & -0.075960\end{pmatrix}. \quad (12.11)\end{aligned}$$

This can be normalized to give **p** as a unit vector:

$$(\mathbf{p};\mathrm{F}^*) \parallel (0.197162\ 0.796841\ 0.571115)$$

$(\mathbf{p};\mathrm{F}^*)$ of course represents the indices of the habit plane of the martensite plate. As expected, the habit plane is irrational. To completely determine the shape deformation matrix (F P F) we also need to know m and **d**. Using Equation 12.9, we see that

$$(\mathrm{F\ S\ F})[\mathrm{F};\mathbf{e}] = [\mathrm{F};\mathbf{e}] + m[\mathrm{F};\mathbf{d}](\mathbf{p};\mathrm{F}^*)[\mathrm{F};\mathbf{e}]$$

Writing c as the scalar constant $c = (\mathbf{p}; \mathrm{F}^*)[\mathrm{F};\mathbf{e}]$, we get

$$\begin{aligned} cm[\mathrm{F};\mathbf{d}] &= (\mathrm{F\ S\ F})[\mathrm{F};\mathbf{e}] - [\mathrm{F};\mathbf{e}] \\ &= \begin{pmatrix} 1.125317 & -0.039880 & 0.106601 \\ 0.023973 & 1.129478 & 0.084736 \\ -0.154089 & -0.115519 & 0.791698 \end{pmatrix} \begin{pmatrix} \frac{1}{\sqrt{2}} \\ 0 \\ -\frac{1}{\sqrt{2}} \end{pmatrix} - \begin{pmatrix} \frac{1}{\sqrt{2}} \\ 0 \\ -\frac{1}{\sqrt{2}} \end{pmatrix} \\ &= [0.013234\ -0.042966\ 0.038334]. \end{aligned} \tag{12.12}$$

Now, $c = (0.197162\ 0.796841\ 0.571115)[\frac{1}{\sqrt{2}}\ 0\ -\frac{1}{\sqrt{2}}] = -0.26442478$ so that $m[\mathrm{F};\mathbf{d}] = [-0.050041\ 0.162489\ -0.144971]$. Since **d** is a unit vector, it can be obtained by normalizing $m\mathbf{d}$ to give

$$[\mathrm{F};\mathbf{d}] = [-0.223961\ 0.727229\ -0.648829]$$

and

$$m = |m\mathbf{d}| = 0.223435$$

The magnitude m of the displacements involved can be factorized into a shear component s parallel to the habit plane and a dilatational component δ normal to the habit plane. Hence, $\delta = m\mathbf{d}.\mathbf{p} = 0.0368161$ and $s = (m^2 - \delta^2)^{\frac{1}{2}} = 0.220381$. These are typical values of the dilatational and shear components of the shape strain found in ferrous martensites.

Using these data, the shape deformation matrix is given by

$$(\mathrm{F\ P\ F}) = \begin{pmatrix} 0.990134 & -0.039875 & -0.028579 \\ 0.032037 & 1.129478 & 0.092800 \\ -0.028583 & -0.115519 & 0.917205 \end{pmatrix}. \tag{12.13}$$

12.8 Stage 4: Nature of the lattice-invariant shear

We have already seen that the shape deformation (F P F) cannot account for the overall conversion of the fcc lattice to that of bcc martensite. An additional homogeneous lattice varying shear (F Q F) is necessary, which in combination with (F P F) completes the required change in structure. However, the macroscopic shape change observed experimentally is only due to (F P F); the effect of (F Q F) on the macroscopic shape change must thus be offset by a system of inhomogeneously applied lattice-invariant shears. Clearly, to macroscopically cancel the shape change due to (F Q F), the lattice-invariant shear must be the inverse of (F Q F). Hence, the lattice invariant shear operates on the plane with unit normal **q** but in the direction −**e**, its magnitude on average being the same as that of (F Q F).

Example 12.4: Lattice–invariant shear

For the martensite reaction discussed in the preceding examples, determine the nature of the homogeneous shear (F Q F) and hence deduce the magnitude of the lattice-invariant shear. Assuming that the lattice-invariant shear is a slip deformation, determine the spacing of the intrinsic dislocations in the habit plane, which are responsible for this inhomogeneous deformation.

Since $(\mathrm{F\ S\ F}) = (\mathrm{F\ P\ F})(\mathrm{F\ Q\ F})$, it follows that $(\mathrm{F\ Q\ F}) = (\mathrm{F\ P\ F})^{-1}(\mathrm{F\ S\ F})$, so that

$$(\mathrm{F\ Q\ F}) = \begin{pmatrix} 1.009516 & 0.038459 & 0.027564 \\ -0.030899 & 0.875120 & -0.089505 \\ 0.027568 & 0.111417 & 1.079855 \end{pmatrix}$$
$$\times \begin{pmatrix} 1.125317 & -0.039880 & 0.106601 \\ 0.023973 & 1.129478 & 0.084736 \\ -0.154089 & -0.115519 & 0.791698 \end{pmatrix}$$
$$= \begin{pmatrix} 1.132700 & 0.000000 & 0.132700 \\ 0.000000 & 1.000000 & 0.000000 \\ -0.132700 & 0.000000 & 0.867299 \end{pmatrix}.$$

Comparison with Equation 11.3 shows that this is a homogeneous shear on the system $(1\ 0\ 1)[1\ 0\ \bar{1}]_\mathrm{F}$ with a magnitude $n = 0.2654$.

The lattice-invariant shear is thus determined since it is the inverse of (F Q F), having the same average magnitude but occurring inhomogeneously on the system $(1\ 0\ 1)[\bar{1}\ 0\ 1]_\mathrm{F}$. If the intrinsic interface dislocations which cause this shear have a Burgers vector $\mathbf{b} = (a_\gamma/2)[\bar{1}\ 0\ 1]_\gamma$, and if they occur on every K'th slip plane, then if the spacing of the $(1\ 0\ 1)_\gamma$ planes is given by d, it follows that

$$n = |\mathbf{b}|/Kd = 1/K \qquad \text{so that} \qquad K = 1/0.2654 = 3.7679$$

Of course, K must be an integral number, and the non-integral result must be taken to mean that there will on average be a dislocation located on every 3.7679th slip plane; in reality, the dislocations will be non-uniformly placed, either 3 or 4 (1 0 1) planes apart.

The line vector of the dislocations is the invariant-line **u** and the spacing of the intrinsic dislocations, as measured on the habit plane is $Kd/(\mathbf{u} \wedge \mathbf{p}.\mathbf{q})$ where all the vectors are unit vectors. Hence, the average spacing would be

$$3.7679(a_\gamma 2^{-\frac{1}{2}})/0.8395675 = 3.1734 a_\gamma$$

and if $a_\gamma = 3.56$Å, then the spacing is 11.3Å on average.

If on the other hand, the lattice-invariant shear is a twinning deformation (rather than slip), then the martensite plate will contain very finely spaced transformation twins, the structure of the interface being radically different from that deduced above, since it will no longer contain any intrinsic dislocations. The mismatch between the parent and product lattices was in the slip case accommodated with the help of intrinsic dislocations, whereas for the internally twinned martensite there are no such dislocations. Each twin terminates in the interface to give a facet between the parent and product lattices, a facet which is forced into coherency. The width of the twin and the size of the facet is sufficiently small to enable this forced coherency to exist. The alternating twin related regions thus prevent misfit from accumulating over large distances along the habit plane.

If the (fixed) magnitude of the twinning shear is denoted S, then the volume fraction V of the twin orientation, necessary to cancel the effect of (F Q F), is given by $V = n/S$, assuming that $n < S$. In the above example, the lattice-invariant shear occurs on $(1\ 0\ 1)[\bar{1}\ 0\ 1]_\gamma$ which corresponds to $(1\ 1\ 2)_\alpha[\bar{1}\ \bar{1}\ 1]_\alpha$, and twinning on this latter system involves a shear $S = \frac{1}{\sqrt{2}}$, giving $V = 0.2654/0.707107 = 0.375$.

It is important to note that the twin plane in the martensite corresponds to a mirror plane in the austenite; this is a necessary condition when the lattice-invariant shear involves twinning. The condition arises because the twinned and untwinned regions of the martensite must undergo Bain strain along different though crystallographically equivalent principal axes [1, 28].

The above theory clearly predicts a certain volume fraction of twins in each martensite plate, when the lattice-invariant shear is twinning as opposed to slip. However, the factors governing the spacing of the twins are less quantitatively established; the finer the spacing of the twins, the lower will be the strain energy associated with the matching of each twin variant with the parent lattice at the interface. On the other hand, the amount of coherent twin boundary within the martensite increases as the spacing of the twins decreases.

A factor to bear in mind is that the lattice-invariant shear is an integral part of the transformation; it does not happen as a separate event after the lattice change has occurred. The transformation and the lattice-invariant shear all occur simultaneously at the interface, as the latter migrates. It is well known that in ordinary plastic deformation, twinning rather than slip tends to be the favoured deformation mode at low temperatures or when high strain rates are involved. It is therefore often suggested that martensite with low M_S temperatures will tend to be twinned rather than slipped, but this cannot be formally justified because the lattice-invariant shear is an integral part of the transformation and not a physical deformation mode on its own. Indeed, it is possible to find lattice-invariant deformation modes in martensite which do not occur in ordinary plastic deformation experiments. The reasons why some martensites are internally twinned and others slipped are not clearly understood [29]. When the spacing of the transformation twins is roughly comparable to that of the dislocations in slipped martensite, the interface energies are roughly equal. The interface energy increases with twin thickness and at the observed thicknesses is very large compared with the corresponding interface in slipped martensite. The combination of the relatively large interface energy and the twin boundaries left in the martensite plate means that internally twinned martensite is never thermodynamically favoured relative to slipped martensite. It is possible that kinetic factors such as interface mobility actually determine the type of martensite that occurs.

12.9 Texture due to displacive transformations

There is enormous utility in treating displacive transformations such as martensite, bainite, Widmanstätten ferrite [30] and in some cases, cementite [31] as physical deformations in which the crystallography is known or can be calculated using the theory described in this chapter, with great rigor. Displacive transformations are confined to individual grains of the parent phase and have a reproducible crystallography with respect to that phase. If there are circumstances in which particular crystallographic variants are favoured, the resulting texture can be estimated from a knowledge of the orientations of the parent crystals and their orientation relationship with the product phase. We shall consider this in the context of the transformation of austenite into martensite in steels.

Given the irrationality of the macroscopic crystallographic features of martensite, cubic symmetry dictates that there always will be 24 variants of the transformation product per austenite grain. If all 24 variants are allowed to form in each austenite grain, then the transformation texture becomes essentially random for any sizeable sample of material. However, circumstances may force variant selection, i.e., where certain crystals which are more compliant to an external influence, such as an externally applied stress.

It is the interaction of the applied stress with the shape deformation of individual martensite plates **P** (page 198 which determines variant selection. The interaction energy which provides the mechanical driving force for transformation is given by [32]:

$$U = \sigma_N \zeta + \tau s \tag{12.14}$$

where σ_N is the stress component normal to the habit plane, τ is the shear stress resolved on the habit plane in the direction of shear and ζ and s are the respective normal and shear strains associated with transformation. The energy U can be used as a rigorous variant selection criterion when the role of any plastic strain is unimportant [12].

With displacive transformations it is possible to calculate the macroscopic plastic-strains as a function of texture; they can also be used to characterize the texture because they are in general anisotropic [12, 33, 34]. An arbitrary vector **u** traversing a grain of austenite prior to transformation (Figure 12.5a) makes an intercept $\Delta\mathbf{u}$ with a domain of austenite that eventually transforms, after which it becomes a new vector **v** given by:

$$\mathbf{v} = \mathbf{P}\Delta\mathbf{u} + (\mathbf{u} - \Delta\mathbf{u}) \tag{12.15}$$

When many plates form in many austenite grains, **u** traverses a polycrystalline sample of austenite so this equation is generalized as follows [12]:

$$\mathbf{v} = \sum_{k=1}^{n}\sum_{j=1}^{24} \mathbf{P}_j^k \Delta\mathbf{u}_j^k + \left(\mathbf{u} - \sum_{k=1}^{n}\sum_{j=1}^{24} \Delta\mathbf{u}_j^k\right) \tag{12.16}$$

where $j = 1 \ldots 24$ represents the 24 crystallographic variants possible in each austenite grain, and $k = 1 \ldots n$ represents the n austenite grains traversed by the vector **u**. In this scenario of a large number of bainite plates, the intercepts $\Delta\mathbf{u}_j^k$ can be approximated by $f_j^k\mathbf{u}$ where f_j^k is the fraction of sample transformed by variant j in austenite grain k.

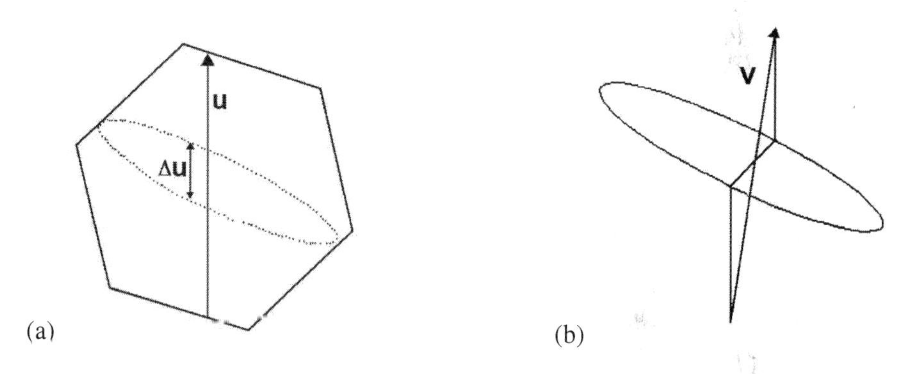

FIGURE 12.5
The deformation of an initial vector **u** by the formation of bainite. (a) An austenite grain prior to transformation, with the ultimate location of a plate of bainite marked. (b) Following displacive transformation.

The deformation due to a particular plate j in austenite grain k, i.e., $(\gamma_k \text{ P}_\text{j} \ \gamma_k) \equiv \mathbf{P}_j^k$. The remaining 23 such matrices for grain 1 of austenite are deduced from this using symmetry operations. They can then be expressed in the reference frame of the sample using a similarity transformation (page 154) as follows:

$$(\text{S P}_j^k \text{ S}) = (\text{S R } \gamma_k)(\gamma_k \text{ P}_j \ \gamma_k)(\gamma_k \text{ R S}) \tag{12.17}$$

where $(\text{S R } \gamma_k)$ is the rotation matrix relating the basis vectors of the kth austenite grain to the sample axes, and $(\gamma_k \text{ R S})$ is the inverse of that rotation matrix. In this way, the calculation described in Equation 12.15 can be conducted in the sample frame of reference.

Calculations illustrating the anisotropy of strains as a function of the number of crystallographic variants of martensite allowed are illustrated in Figure 12.6a for uniaxial tension when transformation occurs from a randomly oriented set of austenite grains. That displacive transformations produce highly anisotropic strains when variant selection is significant has been demonstrated experimentally [33–35].

Figure 12.6b shows that transformation texture is absent when 24 variants form in each austenite grain, for typical intensities of texture in the austenite; the only strain visible in these circumstances is the an averaged isotropic volume expansion. The strength of the transformation texture increases as the number of variants per austenite grain decreases.

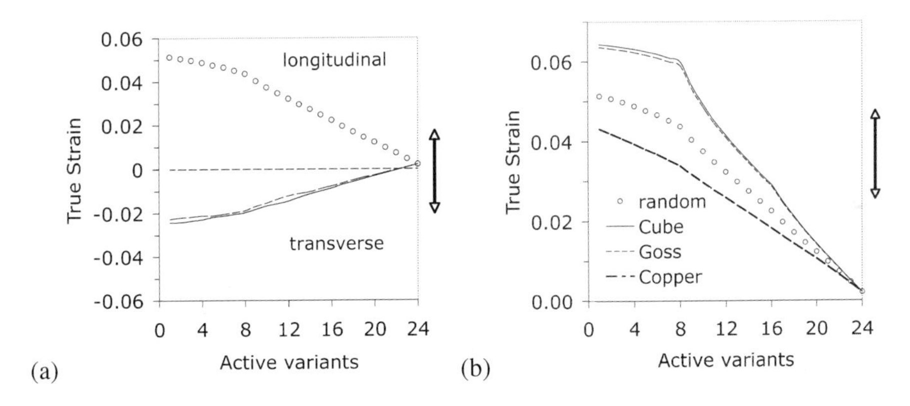

FIGURE 12.6
These diagrams show the plastic strain that develops in a sample of 500 austenite grains placed under tension along the longitudinal axis. The differences in the orthogonal strains correlate with the intensity of transformation texture, assuming that each variant that forms contributes equally to the fraction of transformation. (a) Strains developed due to transformation along the $[1\ 0\ 0]_S$ direction (labelled longitudinal, along the stress axis), and the transverse directions $[0\ 1\ 0]_S$ and $[0\ 0\ 1]_S$. (b) Tensile stress, but transformation beginning from a variety of starting austenite textures and illustrating only the longitudinal stress.

It is emphasised here that the calculation of texture as presented here neglects the fact that favoured variants will in fact have greater volume fractions in the microstructure. The calculations refer simply to the orientations of the martensite plates, not to the intensities as might be observed in X-ray experiments where dominant variants will contribute most. To deal with this problem, it is necessary to couple thermodynamics and crystallography as described elsewhere [36, 37].

12.10 Summary

The crystallographic theory described is one of the most complete methods of predicting observable features of martensitic transformations, including aspects of the lattice-invariant deformation (slip or twinning), the orientation relationship, the shape deformation and even the lenticular shape of the martensite plates. The irrational nature of some of the characteristics of martensite are also predicted to a high precision.

The inputs required are the lattice parameters of the parent and produce unit cells and possible modes of lattice-invariant deformation.

It is worth emphasising that the theory is based on an elastically accommodated plate that is not interacting with any other martensite plates. In particular, plastic deformation driven by the shape deformation, and the impingement of the strain fields of other plates can alter the morphology and crystallography [38, 39].

References

1. J. S. Bowles, and J. K. Mackenzie: The crystallography of martensite transformations, part I, *Acta Metallurgica*, 1954, **2**, 129–137.
2. J. K. Mackenzie, and J. S. Bowles: The crystallography of martensite transformations II, *Acta Metallurgica*, 1954, **2**, 138–147.
3. J. K. Mackenzie, and J. S. Bowles: The crystallography of martensite transformations III FCC to BCT transformations, *Acta Metallurgica*, 1954, **2**, 224–234.
4. M. S. Wechsler, D. S. Lieberman, and T. A. Read: On the theory of the formation of martensite, *Trans. AIME Journal of Metals*, 1953, **197**, 1503–1515.
5. C. M. Wayman: The growth of martensite since E C Bain (1924) - some milestones, *Materials Forum*, 1990, **56-58**, 1–32.
6. T. S. Kuhn: *The structure of scientific revolutions*: Chicago, IL: University of Chicago Press, 1996.
7. D. K. Simonton: Scientific genius is extinct, *Nature*, 2013, **493**, 602.
8. R. Bullough, and B. A. Bilby: Continuous distributions of dislocations: Surface dislocations and the crystallography of martensitic transformations, *Proc. Physics Society B*, 1956, **69**, 1276–1286.
9. W. Bollmann: *Crystal defects and crystalline interfaces*: Berlin, Germany: Springer Verlag, 1970.
10. J. W. Christian: *Theory of transformations in metals and alloys*, Part I: 3 ed., Oxford, U. K.: Pergamon Press, 2003.
11. J. W. Christian: *Theory of transformations in metals and alloys*, Part II: 3 ed., Oxford, U. K.: Pergamon Press, 2003.
12. S. Kundu, K. Hase, and H. K. D. H. Bhadeshia: Crystallographic texture of stress-affected bainite, *Proceedings of the Royal Society A*, 2007, **463**, 2309–2328.

13. H. K. D. H. Bhadeshia: Problems in the calculation of transformation texture in steels, *ISIJ International*, 2010, **50**, 1517–1522.

14. H. K. D. H. Bhadeshia: Multiple, simultaneous, martensitic transformations: implications on transformation texture intensities, *Materials Science Forum*, 2013, **762**, 9–13.

15. J. W. Christian: The origin of surface relief effects in phase transformations, In: V. F. Zackay, and H. I. Aaronson, eds. *Decomposition of austenite by diffusional processes*. New York: Interscience, 1962:371–386.

16. J. D. Eshelby: The determination of the elastic field of an ellipsoidal inclusion and related problem, *Proceedings fo the Royal Society A*, 1957, **241**, 376–396.

17. J. W. Christian: Accommodation strains in martensite formation, the use of the dilatation parameter, *Acta Metallurgica*, 1958, **6**, 377–379.

18. J. W. Christian: Thermodynamics and kinetics of martensite, In: G. B. Olson, and M. Cohen, eds. *International Conference on Martensitic Transformations ICOMAT '79*. Massachusetts: Alpine Press, 1979:220–234.

19. P. Ehrenfest: Phasenumwandlungen im ueblichen und erweiterten sinn, classifiziert nach dem entsprechenden sigularitaeten des thermodynamischen potentiales, *Verhandelingen der Koninklijke Akademie van Wetenschappen (Amsterdam)*, 1933, **36**, 153–157.

20. L. Kaufman, and M. Cohen: Thermodynamics and kinetics of martensitic transformation, *Progress in Metal Physics*, 1958, **7**, 165–246.

21. J. W. Christian, and K. M. Knowles: Interface structures and growth mechanisms, In: H. I. Aaronson, D. E. Laughlin, M. Sekerka, and C. M. Wayman, eds. *Solid-solid phase transformations*. Warrendale, PA: TMS-AIME, 1982:1185–1208.

22. J. W. Christian: *Theory of transformations in metals and alloys*, Part I: 2 ed., Oxford, U. K.: Pergamon Press, 1975.

23. G. F. Bolling, and R. H. Richman: Continual mechanical twinning Parts III, IV, *Acta Metallurgica*, 1965, **13**, 745–757.

24. J. W. Christian, and A. G. Crocker: *Dislocations in solids*, ed. F. R. N. Nabarro: Amsterdam, Holland: North Holland, 1980.

25. A. B. Greninger, and A. R. Troiano: The mechanism of martensite formation, *Trans. A.I.M.E.*, 1949, **185**, 590–598.

26. J. W. Christian, and A. G. Crocker: *Dislocations in solids*, ed. F. R. N. Nabarro, vol. 3, chap. 11: Amsterdam, Holland: North Holland, 1980:165–252.

27. G. B. Olson, and M. Cohen: Interphase boundary dislocations and the concept of coherency, *Acta Metallurgica*, 1979, **27**, 1907–1918.

28. J. W. Christian: *Theory of transformations in metals and alloys*: Oxford, U. K.: Pergamon Press, 1965.

29. G. B. Olson, and M. Cohen: Theory of martensitic nucleation: a current assessment, In: M. S. H. I. Aaronson, D. E. Laughlin:, ed. *Proc. Int. Conf. Solid→Solid Phase Transformations*. Warrendale, Pennsylvania, USA: TMS-AIME, 1981:1209–1213.

30. H. K. D. H. Bhadeshia, and R. W. K. Honeycombe: *Steels: microstructure and properties*: 4th ed., Elsevier, 2017.

31. J. W. Stewart, R. C. Thomson, and H. K. D. H. Bhadeshia: Cementite precipitation during tempering of martensite under the influence of an externally applied stress, *Journal of Materials Science*, 1994, **29**, 6079–6084.

32. J. R. Patel, and M. Cohen: Criterion for the action of applied stress in the martensitic transformation, *Acta Metallurgica*, 1953, **1**, 531–538.

33. H. K. D. H. Bhadeshia, S. A. David, J. M. Vitek, and R. W. Reed: Stress induced transformation to bainite in a Fe-Cr-Mo-C pressure vessel steel, *Materials Science and Technology*, 1991, **7**, 686–698.

34. A. Matsuzaki, H. K. D. H. Bhadeshia, and H. Harada: Stress-affected bainitic transformation in a Fe-C-Si-Mn alloy, *Acta Metallurgica and Materialia*, 1994, **42**, 1081–1090.

35. H. K. D. H. Bhadeshia: Possible effects of stress on steel weld microstructures, In: H. Cerjak, and H. K. D. H. Bhadeshia, eds. *Mathematical modelling of weld phenomena – II*. London, U.K.: Institute of Materials, 1995:71–118.

36. H. N. Han, C. G. Lee, C.-S. Oh, T.-O. Lee, and S.-J. Kim: A model for deformation behavior and mechanically induced martensitic transformation of metastable austenitic steel, *Acta Materialia*, 2004, **52**, 5203–5214.

37. H. K. D. H. Bhadeshia, A. Chintha, and S. Kundu: Model for multiple stress affected martensitic transformations, microstructural entropy and consequences on scatter in properties, *Materials Science and Technology*, 2014, **30**, 160–165.

38. G. Miyamoto, A. Shibata, T. Maki, and T. Furuhara: Precise measurement of strain accommodation in austenite matrix surrounding martensite in ferrous alloys by electron backscatter diffraction analysis, *Acta Materialia*, 2009, **57**, 1120–1131.

39. J. H. Yang, and C. M. Wayman: Self-accommodation and shape memory mechanism of epsilon martensite - I. experimental observations, *Materials Characterization*, 1992, **28**, 23–35.

13

Interfaces

Abstract

That interfaces have structure, some of which may be periodic, is now established and observable using techniques that have sufficient spatial resolution. Small misorientations between like-crystals that are connected at a boundary, can be related to arrays of dislocations, which in turn lead to an estimation of the boundary energy per unit area. In this chapter, we describe a method for estimating the total Burgers vector content of an arbitrary interface that in general connects crystals with different structures, as a function of the degrees of freedom that specify the interface. The concept of the coincidence site lattice is then generalised as Bollmann's O-lattice. The "secondary" dislocations that describe how a boundary may move and yet preserve the level of fit add to the impressive array of tools available in understanding interfaces. There remain some difficulties, which are highlighted toward the end of the discussion.

13.1 Introduction

Atoms located at the boundary between crystals must in general be displaced from positions they would occupy in the undisturbed crystal. Nevertheless, it is now well established that many interfaces have a periodic structure. In such cases, the misfit between the crystals connected by the boundary is not distributed uniformly over every element of the interface; it is localized periodically into discontinuities which separate patches of the boundary where the fit between the two crystals is good or perfect. When these discontinuities are well separated, they may individually be recognized as interface dislocations which separate coherent patches in the boundary, which is macroscopically said to be semi-coherent.

Stress-free coherent interfaces can exist only between crystals which can be related by a transformation strain which is an invariant-plane strain. This transformation strain may be real or notional as far as the calculation of the interface structure is concerned, but a real strain implies the existence of an atomic correspondence and an associated macroscopic shape change of the transformed region between the two crystals, which a notional strain does not.

Incoherency presumably sets in when the misfit between adjacent crystals is so high that it cannot satisfactorily be localised into identifiable interface dislocations, giving a boundary structure which is difficult to physically interpret, other than to say that the motion of such an interface must always occur by the uncoordinated and haphazard transfer of atoms across the interface. This could be regarded as a definition of incoherency. As will become clear later, the intuitive feeling that all "high-angle" boundaries are incoherent is not correct.

13.2 Misfit

The misfit across an interface can formally be described in terms of the net Burgers vector $\mathbf{b}_{\mathrm{t}}$ crossing a vector $\mathbf{p}$ in the interface [1–3]. If this misfit is sufficiently small then the boundary structure may relax into a set of discrete interfacial dislocations where the misfit is concentrated)(which are separated by patches of good fit.

In any case, $\mathbf{b}_{\mathrm{t}}$ may be deduced by constructing a Burgers circuit across the interface, and examining the closure failure when a corresponding circuit is constructed in a perfect reference lattice. The procedure is illustrated in Figure 13.1, where crystal A is taken to be the reference lattice. An initial right-handed Burgers circuit OAPBO is constructed such that it straddles the interface across any vector **p**=OP in the interface; the corresponding circuit in the perfect reference lattice is constructed by deforming crystal B (of the bi-crystal A-B) in such a way that it is converted into the lattice of A, thereby eliminating the interface. If the deformation (A S A) converts the reference lattice into the B lattice, then the inverse deformation $(\mathrm{A\ S\ A})^{-1}$ converts the bi-crystal into a single A crystal, and the Burgers circuit in the perfect reference lattice becomes OAPP′B, with a closure failure PP′, which is identified as $\mathbf{b}_{\mathrm{t}}$. Inspection of the vectors forming the triangle OPP′ of Figure 13.1b shows that:

$$[\mathrm{A};\mathbf{b}_{\mathrm{t}}] = \{\mathbf{I} - (\mathrm{A\ S\ A})^{-1}\}[\mathrm{A};\mathbf{p}]. \tag{13.1}$$

Hence, the net Burgers vector content $\mathbf{b}_{\mathrm{t}}$ crossing an arbitrary vector $\mathbf{p}$ in the interface is given formally by Equation 13.1. The misfit in any interface can in general be accommodated with three arrays of interfacial dislocations, whose Burgers vectors $\mathbf{b}_{\mathrm{i}}$ $(\mathrm{i} = 1, 2, 3)$ form a non-coplanar set. Hence, $\mathbf{b}_{\mathrm{t}}$ can in general be factorised into three arrays of interfacial dislocations, each array with Burgers vector $\mathbf{b}_{\mathrm{i}}$, unit line vector $\mathbf{l}_{\mathrm{i}}$ and array spacing d_i, the latter being measured in the interface plane. If the unit interface normal is $\mathbf{n}$, then a vector $\mathbf{m}_{\mathrm{i}}$ may be defined as (the treatment that follows is due to Knowles [4] and Read [5]):

$$\mathbf{m}_{\mathrm{i}} = \mathbf{n} \wedge \mathbf{l}_{\mathrm{i}}/d_i \tag{13.2}$$

We note the $|\mathbf{m}_{\mathrm{i}}| = 1/d_i$, and that any vector $\mathbf{p}$ in the interface crosses $(\mathbf{m}_{\mathrm{i}}.\mathbf{p})$ dislocations of type I (see Figure 13.1c). Hence, for the three kinds of dislocations we have

$$\mathbf{b}_{\mathrm{t}} = (\mathbf{m}_1.\mathbf{p})\mathbf{b}_1 + (\mathbf{m}_2.\mathbf{p})\mathbf{b}_2 + (\mathbf{m}_3.\mathbf{p})\mathbf{b}_3$$

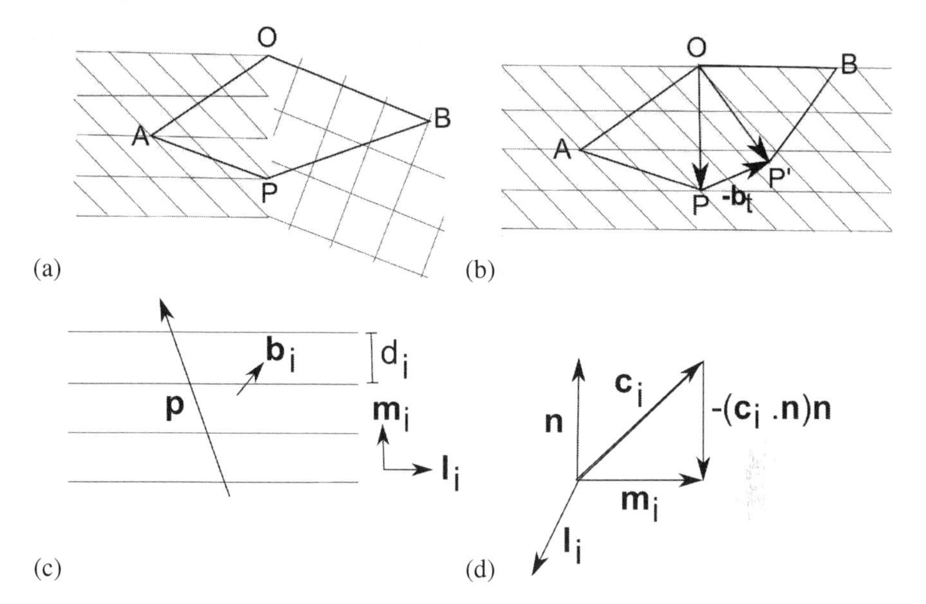

FIGURE 13.1
(a,b) Burgers circuit used to define the formal dislocation content of an interface [3], (c) the vector **p** in the interface, (d) relationship between $\mathbf{l}_i$, $\mathbf{m}_i$, $\mathbf{c}_i$, and **n**.

We note that the Burgers vectors $\mathbf{b}_i$ of interface dislocations are generally lattice translation vectors of the reference lattice in which they are defined. This makes them perfect in the sense that the displacement of one of the crystals through $\mathbf{b}_i$ relative to the other does not change the structure of the boundary. On the basis of elastic strain energy arguments, $\mathbf{b}_i$ should be as small as possible. On substituting this into Equation 13.1, we get:

$$\begin{aligned}(\mathbf{m}_1;\mathrm{A}^*)[\mathrm{A};\mathbf{p}][\mathrm{A};\mathbf{b}_1] \;&+\; (\mathbf{m}_2;\mathrm{A}^*)[\mathrm{A};\mathbf{p}][\mathrm{A};\mathbf{b}_2] \\ &+\; (\mathbf{m}_3;\mathrm{A}^*)[\mathrm{A};\mathbf{p}][\mathrm{A};\mathbf{b}_3] = (\mathrm{A\ T\ A})[\mathrm{A};\mathbf{p}] \end{aligned} \tag{13.3}$$

where $(\mathrm{A\ T\ A}) = \mathbf{I} - (\mathrm{A\ S\ A})^{-1}$.

If a scalar dot product is taken on both sides of Equation 13.3 with the vector $\mathbf{b}_1^*$, where $\mathbf{b}_1^*$ is

$$\mathbf{b}_1^* = \frac{\mathbf{b}_2 \wedge \mathbf{b}_3}{\mathbf{b}_1.\mathbf{b}_2 \wedge \mathbf{b}_3} \tag{13.4}$$

then we obtain

$$(\mathbf{b}_1^*;\mathrm{A}^*)(\mathrm{A\ T\ A})[\mathrm{A};\mathbf{p}] = (\mathbf{m}_1;\mathrm{A}^*)[\mathrm{A};\mathbf{p}] \tag{13.5}$$

If **p** is now taken to be equal to $\mathbf{l}_1$ in Equation 13.5, we find that

$$(\mathbf{b}_1^*;\mathrm{A}^*)(\mathrm{A\ T\ A})[\mathrm{A};\mathbf{l}_1] = 0$$

and

$$(\mathbf{l}_1;\mathrm{A})\underbrace{(\mathrm{A\ T'\ A})[\mathrm{A}^*;\mathbf{b}_1^*]}_{[\mathrm{A}^*;\mathbf{c}_1]}=0 \tag{13.6}$$

where the vector defined as $\mathbf{c}_1$ is normal to $\mathbf{l}_1$. If $\mathbf{m}_1$ is now substituted for $\mathbf{p}$ in Equation 13.5, then we find:

$$\begin{aligned}(\mathbf{b}_1^*;\mathrm{A}^*)(\mathrm{A\ T\ A})[\mathrm{A};\mathbf{m}_1] &= (\mathbf{m}_1;\mathrm{A}^*)[\mathrm{A};\mathbf{m}_1]=|\mathbf{m}_1|^2\\ \therefore\qquad (\mathbf{m}_1;\mathrm{A})(\mathrm{A\ T'\ A})[\mathrm{A}^*;\mathbf{b}_1^*] &= |\mathbf{m}_1|^2\end{aligned}$$

and

$$(\mathbf{m}_1;\mathrm{A})[\mathrm{A}^*;\mathbf{c}_1]=|\mathbf{m}_1|^2 \qquad \text{i.e.,} \qquad \mathbf{m}_1.\mathbf{c}_1=|\mathbf{m}_1|^2$$

These equations indicate that the projection of $\mathbf{c}_1$ along $\mathbf{m}_1$ is equal to the magnitude of $\mathbf{m}_1$. Armed with this and the earlier result that $\mathbf{c}_1$ is normal to $\mathbf{l}_1$, the vector diagram illustrated in Figure 13.1d can be constructed to illustrate the relations between $\mathbf{m}_1$, $\mathbf{l}_1$, $\mathbf{n}$ and $\mathbf{c}_1$. From this diagram, it is evident that

$$\begin{aligned}\mathbf{m}_1 &= \mathbf{c}_1-(\mathbf{c}_1.\mathbf{n})\mathbf{n}\\ \text{so that}\qquad |\mathbf{m}_1\wedge\mathbf{n}| &= |\mathbf{c}_1\wedge\mathbf{n}|=1/d_1\end{aligned} \tag{13.7}$$

and in addition, Figure 13.1d shows that

$$\mathbf{l}_1 \parallel \mathbf{c}_1\wedge\mathbf{n}$$

alternatively,

$$(1/d_1)\mathbf{l}_1=\mathbf{c}_1\wedge\mathbf{n} \tag{13.8}$$

Relations of this type are, after appropriately changing indices, applicable also to the other two arrays of interfacial dislocations, so that a method has been achieved of deducing the line vectors and array spacings to be found in an interface (of unit normal $\mathbf{n}$) connecting two arbitrary crystals A and B, related by the deformation (A S A) which transforms the crystal A to B.

Example 13.1: Symmetrical tilt boundary

Given that the Burgers vectors of interface dislocations in low-angle boundaries are of the form $\langle 1\ 0\ 0\rangle$, calculate the dislocation structure of a symmetrical tilt boundary formed between two grains, A and B, related by a rotation of 2θ about the $[1\ 0\ 0]$ axis. The crystal structure of the grains is simple cubic with a unit lattice parameter.

The definition of a tilt boundary is that the unit boundary normal $\mathbf{n}$ is perpendicular to the axis of rotation which generates one crystal from the other; a symmetrical tilt boundary has the additional property that the lattice of one crystal can be generated from the other by reflection across the boundary plane.

If we choose the orthonormal bases A and B (with basis vectors parallel to the cubic unit cell edges) to represent crystals A and B, respectively, and also arbitrarily choose B to be the reference crystal, then the deformation which generates the A crystal from B is a rigid body rotation (B J B) consisting of a rotation of 2θ about $[1\ 0\ 0]_B$. Hence, (B J B) is given by (see Equation 7.9):

$$(\mathrm{B\ J\ B}) = \begin{pmatrix} 1 & 0 & 0 \\ 0 & \cos 2\theta & -\sin 2\theta \\ 0 & \sin 2\theta & \cos 2\theta \end{pmatrix}$$

and $(\mathrm{B\ T\ B}) = \mathbf{I} - (\mathrm{B\ J\ B})^{-1}$ is given by

$$(\mathrm{B\ T\ B}) = \begin{pmatrix} 0 & 0 & 0 \\ 0 & 1-\cos 2\theta & \sin 2\theta \\ 0 & -\sin 2\theta & 1-\cos 2\theta \end{pmatrix}$$

so that

$$(\mathrm{B\ T'\ B}) = \begin{pmatrix} 0 & 0 & 0 \\ 0 & 1-\cos 2\theta & -\sin 2\theta \\ 0 & \sin 2\theta & 1-\cos 2\theta \end{pmatrix}.$$

Taking $[\mathrm{B};\mathbf{b}_1] = [1\ 0\ 0]$, $[\mathrm{B};\mathbf{b}_2] = [0\ 1\ 0]$ and $[\mathrm{B};\mathbf{b}_3] = [0\ 0\ 1]$, from Equation 13.4 we note that $[\mathrm{B}^*;\mathbf{b}_1^*] = [1\ 0\ 0]$, $[\mathrm{B}^*;\mathbf{b}_2^*] = [0\ 1\ 0]$ and $[\mathrm{B}^*;\mathbf{b}_3^*] = [0\ 0\ 1]$. Equation 13.6 can now be used to obtain the vectors $\mathbf{c}_i$:

$$\begin{aligned}
[\mathrm{B}^*;\mathbf{c}_1] &= (\mathrm{B\ T'\ B})[\mathrm{B}^*;\mathbf{b}_1^*] &=& \begin{bmatrix} 0 & 0 & 0 \end{bmatrix} \\
[\mathrm{B}^*;\mathbf{c}_2] &= (\mathrm{B\ T'\ B})[\mathrm{B}^*;\mathbf{b}_2^*] &=& \begin{bmatrix} 0 & 1-\cos 2\theta & \sin 2\theta \end{bmatrix} \\
[\mathrm{B}^*;\mathbf{c}_3] &= (\mathrm{B\ T'\ B})[\mathrm{B}^*;\mathbf{b}_3^*] &=& \begin{bmatrix} 0 & -\sin 2\theta & 1-\cos 2\theta \end{bmatrix}
\end{aligned}$$

Since $\mathbf{c}_1$ is a null vector, dislocations with Burgers vector $\mathbf{b}_1$ do not exist in the interface; furthermore, since $\mathbf{c}_1$ is always a null vector, irrespective of the boundary orientation $\mathbf{n}$, this conclusion remains valid for any $\mathbf{n}$. This situation arises because $\mathbf{b}_1$ happens to be parallel to the rotation axis, and because (B J B) is an invariant-line strain, the invariant line being the rotation axis; since any two crystals of identical structure can always be related by a transformation which is a rigid body rotation, it follows that all grain boundaries (as opposed to phase boundaries) *need* only contain two sets of interface dislocations. If $\mathbf{b}_1$ had not been parallel to the rotation axis, then $\mathbf{c}_1$ would be finite and three sets of dislocations would be necessary to accommodate the misfit in the boundary.

To calculate the array spacings d_i it is necessary to express $\mathbf{n}$ in the B^* basis. A symmetrical tilt boundary always contains the axis of rotation and has the same indices in both bases. It follows that

$$(\mathbf{n};\mathrm{B}^*) = (0\ \ \cos\theta\ \ -\sin\theta)$$

and from Equation 13.8,

$$(1/d_2)\mathbf{l}_2 = \mathbf{c}_2 \wedge \mathbf{n} = [-2\sin\theta \;\; 0 \;\; 0]_{\mathrm{B}}$$

so that

$$[\mathrm{B}; \mathbf{l}_2] = [\bar{1}\,0\,0]$$

and

$$d_2 = 1/2\sin\theta$$

similarly,

$$(1/d_3)\mathbf{l}_3 = \mathbf{c}_3 \wedge \mathbf{n} = [0\,0\,0]$$

so that dislocations with Burgers vector $\mathbf{b}_3$ have an infinite spacing in the interface, which therefore consists of just one set of interface dislocations with Burgers vector $\mathbf{b}_2$.

These results are of course identical to those obtained from a simple geometrical construction of the symmetrical tilt boundary [3]. We note that the boundary is glissile (i.e., its motion does not require the creation or destruction of lattice sites) because $\mathbf{b}_2$ lies outside the boundary plane, so that the dislocations can glide conservatively as the interface moves. In the absence of diffusion, the movement of the boundary leads to a change in shape of the "transformed" region, a shape change described by (B J B) when the boundary motion is towards the crystal A.

If on the other hand, the interface departs from its symmetrical orientation (without changing the orientation relationship between the two grains), then the boundary ceases to be glissile, since the dislocations with Burgers vector $\mathbf{b}_3$ acquire finite spacings in the interface. Such a boundary is called an asymmetrical tilt boundary. For example, if **n** is taken to be $(\mathbf{n}; \mathrm{B}^*) = (0\;1\;0)$, then

$$[\mathrm{B}; \mathbf{l}_2] = [\bar{1}\,0\,0] \quad \text{and} \quad d_2 = 1/\sin 2\theta$$
$$[\mathrm{B}; \mathbf{l}_3] = [\bar{1}\,0\,0] \quad \text{and} \quad d_3 = 1/(1-\cos 2\theta)$$

The edge dislocations with Burgers vector $\mathbf{b}_3$ lie in the interface plane and therefore have to climb as the interface moves. This renders the interface sessile.

Finally, we consider the structure of a twist boundary, a boundary where the axis of rotation is parallel to **n**. Taking $(\mathbf{n}; \mathrm{B}^*) = (1\;0\;0)$, we find:

$$[\mathrm{B}; \mathbf{l}_2] \parallel [0 \;\; \sin 2\theta \;\; \cos 2\theta - 1],$$

$$[\mathrm{B}; \mathbf{l}_3] \parallel [0 \;\; 1-\cos 2\theta \;\; \sin 2\theta],$$

and $d_2 = d_3 = [2 - 2\cos 2\theta]^{\frac{1}{2}}$.

It is noteworthy that dislocations with Burgers vector $\mathbf{b}_1$ do not exist in the interface since $\mathbf{b}_1$ is parallel to the rotation axis. The interface thus consists of a cross grid of two arrays of dislocations with Burgers vectors $\mathbf{b}_2$ and $\mathbf{b}_3$, respectively, the array spacings being identical. Although it usually is stated that pure twist boundaries contain grids of pure screw dislocations, we see that both sets of dislocations actually have a small edge component. This is because the dislocations, where they mutually intersect, introduce jogs into each other so that the line vector in the region between the points of intersection does correspond to a pure screw orientation, but the jogs make the macroscopic line vector deviate from this screw orientation.

Example 13.2: Interface between alpha and beta brass

The lattice parameter of the fcc alpha phase of a 60 wt% Cu-Zn alloy is 3.6925 Å, and that of the bcc beta phase is 2.944 Å [6]. Two adjacent grains A and B (alpha phase and beta phase respectively) are orientated in such a way that

$$[1\ 1\ 1]_{\mathrm{A}} \parallel [1\ 1\ 0]_{\mathrm{B}}$$

$$[1\ 1\ \overline{2}]_{\mathrm{A}} \parallel [\overline{1}\ 1\ 0]_{\mathrm{B}}$$

$$[1\ \overline{1}\ 0]_{\mathrm{A}} \parallel [0\ 0\ \overline{1}]_{\mathrm{B}}$$

and the grains are joined by a boundary which is parallel to $(\overline{1}\ 1\ 0)_{\mathrm{A}}$. Assuming that the misfit in this interface can be fully accommodated by interface dislocations which have Burgers vectors $[\mathrm{A}; \mathbf{b}_1] = \frac{1}{2}[1\ 0\ \overline{1}]$, $[\mathrm{A}; \mathbf{b}_2] = \frac{1}{2}[0\ 1\ \overline{1}]$ and $[\mathrm{A}; \mathbf{b}_3] = \frac{1}{2}[1\ 1\ 0]$, calculate the misfit dislocation structure of the interface. Also assume that the smallest pure deformation which relates fcc and bcc crystals is the Bain strain.

The orientation relations provided are first used to calculate the coordinate transformation matrix (B J A), using the procedure given in Example 9.3. (B J A) is thus found to be:

$$(\mathrm{B\ J\ A}) = \begin{pmatrix} 0.149974 & 0.149974 & 1.236185 \\ 0.874109 & 0.874109 & -0.212095 \\ 0.874109 & 0.874109 & -0.212095 \end{pmatrix}.$$

The smallest pure deformation relating the two lattices is stated to be the Bain strain, but the total transformation (A S A), which carries the A lattice to that of B, may include an additional rigid body rotation. Determination of the interface structure requires a knowledge of (A S A), which may be calculated from the equation $(\mathrm{A\ S\ A})^{-1} = (\mathrm{A\ C\ B})(\mathrm{B\ J\ A})$, where (A C B) is the correspondence matrix. Since (B J A) is known, the problem reduces to the determination of the correspondence matrix; if a vector equal to a basis vector of B, due to transformation becomes a new vector **u**, then the components of **u** in the basis A form one column of the correspondence matrix.

For the present example, the correspondence matrix must be a variant of the Bain correspondence. The vector that $[1\ 0\ 0]_{\mathrm{B}}$ must become, as a result of transformation, either a vector of the form $\langle 1\ 0\ 0\rangle_{\mathrm{A}}$, or a vector of the form $\frac{1}{2}\langle 1\ 1\ 0\rangle_{\mathrm{A}}$, although we do not know its final *specific* indices in the austenite lattice. However, from the matrix (B J A) we note that $[1\ 0\ 0]_{\mathrm{B}}$ is close to $[0\ 0\ 1]_{\mathrm{A}}$, making it is reasonable to assume that $[1\ 0\ 0]_{\mathrm{B}}$ corresponds to $[0\ 0\ 1]_{\mathrm{A}}$, so that the first column of (A C B) is [0 0 1]. Similarly, using (B J A) we find that $[0\ 1\ 0]_{\mathrm{B}}$ is close to

$[1\,1\,0]_{\mathrm{A}}$ and $[0\,0\,1]_{\mathrm{B}}$ is close to $[\bar{1}\,1\,0]_{\mathrm{A}}$, so that the other two columns of (A C B) are $[\frac{1}{2}\,\frac{1}{2}\,0]$ and $[-\frac{1}{2}\,\frac{1}{2}\,0]$. Hence, (A C B) is found to be:

$$(\mathrm{A\ C\ B}) = \frac{1}{2}\begin{pmatrix} 0 & 1 & \bar{1} \\ 0 & 1 & 1 \\ 2 & 0 & 0 \end{pmatrix}$$

so that $(\mathrm{A\ S\ A})^{-1} = (\mathrm{A\ C\ B})(\mathrm{B\ J\ A})$ is given by:

$$(\mathrm{A\ S\ A})^{-1} = \begin{pmatrix} 0.880500 & -0.006386 & -0.106048 \\ -0.006386 & 0.880500 & -0.106048 \\ 0.149974 & 0.149974 & 1.236183 \end{pmatrix}$$

and since

$$(\mathrm{A\ T\ A}) = \mathbf{I} - (\mathrm{A\ S\ A})^{-1},$$

so that $(\mathrm{A\ T'\ A})$ becomes:

$$(\mathrm{A\ T'\ A}) = \begin{pmatrix} 0.119500 & 0.006386 & -0.149974 \\ 0.006386 & 0.119500 & -0.149974 \\ 0.016048 & 0.016048 & -0.236183 \end{pmatrix}.$$

The vectors $\mathbf{b}_{\mathrm{i}}^*$, as defined by Equation 13.4, are given by:

$$[\mathrm{A}^*; \mathbf{b}_1^*] = \begin{bmatrix} 1 & \bar{1} & \bar{1} \end{bmatrix}$$
$$[\mathrm{A}^*; \mathbf{b}_2^*] = \begin{bmatrix} \bar{1} & 1 & \bar{1} \end{bmatrix}$$
$$[\mathrm{A}^*; \mathbf{b}_3^*] = \begin{bmatrix} 1 & 1 & 1 \end{bmatrix}$$

and applying Equation 13.6 results in

$$\begin{aligned} [\mathrm{A}^*; \mathbf{c}_1] &= [0.263088 \quad 0.036860 \quad 0.236183] \\ [\mathrm{A}^*; \mathbf{c}_2] &= [0.036860 \quad 0.263088 \quad 0.236083] \\ [\mathrm{A}^*; \mathbf{c}_3] &= [-0.024088 \quad -0.024088 \quad -0.024087] \end{aligned}$$

and from Equation 13.8 and the fact that $(\mathbf{n}; \mathrm{A}^*) = 3.6925(-0.707107\ 0.707107\ 0)$,

$$\begin{aligned} (1/d_1)[\mathrm{A}; \mathbf{l}_1] &= [-0.012249 \quad -0.012249 \quad 0.015556] \\ (1/d_2)[\mathrm{A}; \mathbf{l}_2] &= [-0.012249 \quad -0.012249 \quad 0.015556] \\ (1/d_3)[\mathrm{A}; \mathbf{l}_3] &= [\,0.001249 \quad 0.001249 \quad -0.002498] \end{aligned}$$

so that $d_1 = d_2 = 11.63$ Å, and $d_3 = 88.52$ Å.

13.3 Coincidence site lattices

It has been emphasized that Equation 13.1 defines the formal Burgers vector content of an interface; it sometimes is possible to interpret this in terms of physically meaningful interfacial dislocations, if the misfit across the interface is sufficiently small.

For large misorientation boundaries, the predicted spacings of dislocations may turn out to be so small that the misfit is highly localised with respect to the boundary, and the dislocation model of the interface has only formal significance – it often is said that the dislocations get so close to each other that their cores overlap. The arrangement of atoms in such incoherent boundaries may be haphazard, with little correlation of atomic positions across the boundary.

On the other hand, it is unreasonable to assume that all high-angle boundaries have the disordered structure suggested above. There is clear experimental evidence which shows that certain high-angle boundaries exhibit the characteristics of low-energy coherent or semi-coherent interfaces; for example, they exhibit strong faceting, have very low mobility in pure materials and the boundary diffusion coefficient may be abnormally low. These observations imply that at certain relative crystal orientations, which would usually be classified as large misorientations, it is possible to obtain boundaries which have a distinct structure – they contain regions of good fit, which occur at regular intervals in the boundary plane, giving a pattern of good fit points in the interface. It is along these points that the two crystals connected by the boundary match exactly.

If we consider the good fit points to correspond to lattice points in the interface that are common to both crystals, then the following procedure allows us to deduce the pattern and frequency of these points for any given orientation relationship.

If the two lattices with a common origin are notionally allowed to interpenetrate and fill all space, then there may exist lattice points (other than the origin) which are common to both the crystals. The set of these coincidence points forms a coincidence site lattice (CSL) [7, 8]; the fraction of lattice points which are also coincidence sites is a rational fraction $1/\Sigma$, where the denominator is the reciprocal density of coincidence sites relative to ordinary lattice sites. The value of Σ is a function of the relative orientation of the two grains and not of the orientation of the boundary plane. The boundary simply intersects the CSL and will contain regions of good fit which have a periodicity of a planar net of the CSL along which the intersection occurs. Boundaries parallel to low-index planes of the CSL are two dimensionally periodic with relatively small repeat cells, and those boundaries with a high planar coincidence site density should have a relatively low energy per unit area.

Example 13.3: Coincidence site lattices

The axis-angle pair describing the orientation relationship between the two grains (A and B) of austenite is given by:

$$\text{axis of rotation parallel to} \qquad [1\ 1\ 2]_A$$

right-handed angle of rotation 180°

Show that a CSL with $\Sigma = 3$ can be formed by allowing the two lattices sharing an origin to notionally interpenetrate. Show also that 1 in 3 of the sites in an interface parallel to $(0\ \overline{2}\ 1)_A$ are coincidence sites.

Because of the centrosymmetric nature of the austenite lattice, a rotation of 180° about $[1\ 1\ 2]_A$, which generates the B grain from A, is equivalent to a reflection of the A lattice about $(1\ 1\ 2)_A$. We may therefore imagine that B is generated by reflection of the A lattice about $(1\ 1\ 2)_A$ as the mirror plane. The stacking sequence of $(1\ 1\ 2)_A$ is $\ldots ABCDEFABCDEFABCDEF \ldots$ and grain A is represented below as a stack of $(1\ 1\ 2)$ planes, B being generated by reflecting the A lattice about one of the $(1\ 1\ 2)_A$ planes.

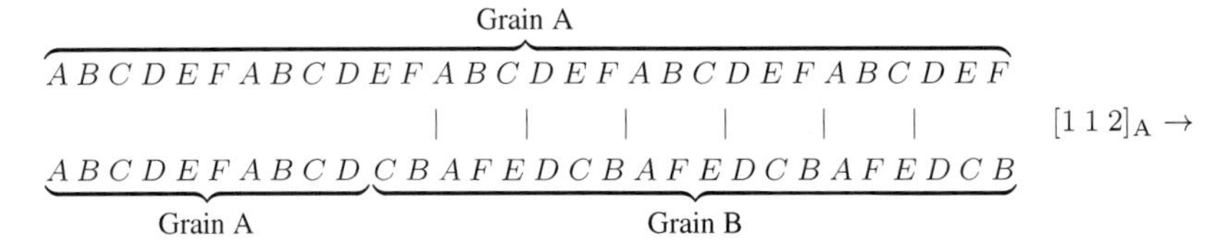

The first sequence represents a stack of $(1\ 1\ 2)_A$ planes, as do the first 9 layers of the second sequence. The remainder of the second sequence represents a stack of $(1\ 1\ 2)_B$ planes (note that B lattice is obtained by reflection of A about the 9th layer of the second sequence). Since the two sequences have the same origin, comparison of the first sequence with the B part of the second sequence amounts to allowing the two crystals to interpenetrate in order to identify coincidences. Clearly, every 3rd layer of grain B coincides exactly with a layer from the A lattice (dashed vertical lines), giving $\Sigma = 3$.

A boundary parallel to $(1\ 1\ 2)_A$ will be fully coherent; at least 1 in 3 of the sites in any other boundary, such as $(0\ \overline{2}\ 1)_A$ will be coincidence sites.

We now consider a mathematical method [3, 9] of determining the CSL formed by allowing the lattices of crystals A and B to notionally interpenetrate. A and B are assumed to be related by a transformation (A S A) which deforms the A lattice into that of B; A and B need not have the same crystal structure or lattice parameters, so that (A S A) need not be a rigid body rotation. Consider a vector **u** which is a lattice vector whose integral components do not have a common factor. As a result of the transformation (A S A), **u** becomes a new vector **x** such that

$$[\mathrm{A}; \mathbf{x}] = (\mathrm{A\ S\ A})[\mathrm{A}; \mathbf{u}] \tag{13.9}$$

x does not necessarily have integral components in the A basis (i.e., it need not be a lattice vector of A). CSL vectors, on the other hand, identify lattice points which are common to both A and B, and therefore are lattice vectors of both crystals. It follows that CSL vectors have integral indices when referred to either crystal. Hence, **x** is only a CSL vector if it has integral components in the basis A. We note that **x** always has integral components in B, because a lattice vector of A (such as **u**) always deforms into a lattice vector of B.

The meaning of Σ is that $1/\Sigma$ of the lattice sites of A or B are common to both A and B. It follows that any primitive lattice vector of A or B, when multiplied by Σ, must give a CSL vector. $\Sigma\mathbf{x}$ must therefore always be a CSL vector and if

Equation 13.9 is multiplied by Σ, then we obtain an equation in which the vector $\mathbf{u}$ always transforms into a CSL vector:

$$\Sigma[\mathrm{A};\mathbf{x}] = \Sigma(\mathrm{A\ S\ A})[\mathrm{A};\mathbf{u}] \qquad (13.10)$$

i.e., given that $\mathbf{u}$ is a lattice vector of A, whose components have no common factor, $\Sigma\mathbf{x}$ is a CSL vector with integral components in either basis. This can only be true if the matrix $\Sigma(\mathrm{A\ S\ A})$ has elements which are all integral since it is only then that $\Sigma[\mathrm{A};\mathbf{x}]$ has elements which are all integral.

It follows that if an integer H can be found such that all the elements of the matrix $H(\mathrm{A\ S\ A})$ are integers (without a common factor), then H is the Σ value relating A and B.

The rotation matrix corresponding to the rotation 180° about $[1\ 1\ 2]_\mathrm{A}$ is given by (Equation 7.9)

$$(\mathrm{A\ J\ A}) = \frac{1}{3}\begin{pmatrix} \bar{2} & 1 & \bar{2} \\ 1 & \bar{2} & 2 \\ 2 & 2 & 1 \end{pmatrix}$$

and since 3 is the integer which when multiplied with (A J A) gives a matrix of integral elements (without a common factor), the Σ value for this orientation is given by $\Sigma = 3$. For reasons of symmetry (see Chapter 9), the above rotation is crystallographically equivalent to a rotation of 60° about $[1\ 1\ 1]_\mathrm{A}$, with the rotation matrix given by

$$(\mathrm{A\ J\ A}) = \frac{1}{3}\begin{pmatrix} 2 & 2 & \bar{1} \\ \bar{1} & 2 & 2 \\ 2 & \bar{1} & 2 \end{pmatrix}$$

so it is not surprising that a rotation of 60° about $[1\ 1\ 1]_\mathrm{A}$ also corresponds to a $\Sigma = 3$ value.

Finally, we see from Equation 13.10 that if the integer H (defined such that $H(\mathrm{A\ S\ A})$ has integral elements with no common factor) turns out to be even, then the Σ value is obtained by successively dividing H by 2 until the result H' is an odd integer. H' then represents the true Σ value. This is because (Equation 13.10) if $H[\mathrm{A};\mathbf{x}]$ is a CSL vector and if H is even, then $H\mathbf{x}$ has integral even components in A, but $H'\mathbf{x}$ would also have integral components in A and would therefore represent a smaller CSL vector. From page 172, the transformation strain relating fcc-austenite and hcp-martensite is given by:

$$(\mathrm{Z\ P\ Z}) = \begin{pmatrix} 1.083333 & 0.083333 & 0.083333 \\ 0.083333 & 1.083333 & 0.083333 \\ -0.166667 & -0.166667 & 0.833333 \end{pmatrix} = \frac{1}{12}\begin{pmatrix} 13 & 1 & 1 \\ 1 & 13 & 1 \\ -2 & -2 & 9 \end{pmatrix}$$

so that $H = 12$, but $H' = \Sigma = 3$. This can be illustrated by considering the stacking sequence of the close-packed planes. The first sequence below represents a stack of the (1 1 1) planes of the fcc lattice, as do the first 9 planes of the second stacking sequence. The other planes of the second sequence represent the basal planes of the hcp lattice. Since two out of every 6 layers are in exact coincidence, $\Sigma = 3$, as shown earlier.

$$\overbrace{A\,B\,C\,A\,B\,C\,A\,B\,C\,A\,B\,C\,A\,B\,C\,A\,B\,C\,A\,B\,C\,A\,B}^{\text{fcc}}$$

$$\qquad\qquad | \; | \qquad\qquad | \; | \qquad\qquad | \; | \qquad [1\;1\;1]_\gamma \parallel [0\;0\;0\;1]_{\text{hcp}} \rightarrow$$

$$\underbrace{A\,B\,C\,A\,B\,C\,A\,B\,C}_{\text{fcc}}\underbrace{A\,C\,A\,C\,A\,C\,A\,C\,A\,C\,A\,C\,A\,C}_{\text{hcp}}$$

Example 13.4: Symmetry and the axis-angle representations of CSL's

Show that the coincidence site lattice associated with two cubic crystals related by a rotation of 50.5° about $\langle 1\;1\;0\rangle$ has $\Sigma = 11$. Using the symmetry operations of the cubic lattice, generate all possible axis-angle pair representations which correspond to this Σ value.

The rotation matrix corresponding to the orientation relation 50.5° about $\langle 1\;1\;0\rangle$ is given by (Equation 7.9):

$$(\text{A J A}) = \begin{pmatrix} 0.545621 & -0.545621 & 0.636079 \\ 0.181961 & 0.818039 & 0.545621 \\ 0.818039 & 0.181961 & -0.545621 \end{pmatrix} = \frac{1}{11}\begin{pmatrix} 6 & -6 & 7 \\ 2 & 9 & 6 \\ 9 & 2 & -6 \end{pmatrix}$$

so that $\Sigma = 11$ (A is the basis symbol representing one of the cubic crystals). The 24 symmetry operations of the cubic lattice are given by:

Angle (degrees)	Axis
0,90,180,270	$\langle 1\;0\;0\rangle$
90,180,270	$\langle 0\;1\;0\rangle$
90,180,270	$\langle 0\;0\;1\rangle$
180	$\langle 1\;1\;0\rangle$
180	$\langle 1\;0\;1\rangle$
180	$\langle 0\;1\;1\rangle$
180	$\langle 1\;\bar{1}\;0\rangle$
180	$\langle 1\;0\;\bar{1}\rangle$
180	$\langle 0\;1\;\bar{1}\rangle$
120,240	$\langle 1\;1\;1\rangle$
120,240	$\langle \bar{1}\;\bar{1}\;1\rangle$
120,240	$\langle 1\;\bar{1}\;\bar{1}\rangle$
120,240	$\langle \bar{1}\;1\;\bar{1}\rangle$

If any symmetry operation is represented as a rotation matrix which then premultiplies (A J A), then the resulting new rotation matrix gives another axis-angle pair representation. Hence, the alternative axis-angle pair representations of $\Sigma = 11$ are found to be:

Angle (degrees)	Axis
82.15	$\langle 1\ 3\ 3 \rangle$
162.68	$\langle 3\ 3\ 5 \rangle$
155.37	$\langle 1\ 2\ 4 \rangle$
180.00	$\langle 2\ 3\ 3 \rangle$
62.96	$\langle 1\ 1\ 2 \rangle$
129.54	$\langle 1\ 1\ 4 \rangle$
129.52	$\langle 0\ 1\ 1 \rangle$
126.53	$\langle 1\ 3\ 5 \rangle$
180.00	$\langle 1\ 1\ 3 \rangle$
100.48	$\langle 0\ 2\ 3 \rangle$
144.89	$\langle 0\ 1\ 3 \rangle$

This procedure can be used to derive all the axis-angle pair representations of any Σ value, and the table below gives some of the CSL relations for cubic crystals, quoting the axis-angle pair representations which have the minimum angle of rotation, and also those corresponding to twin axes.

We note the following further points about CSL relations:

(i) All of the above CSL relations can be represented by a rotation of 180° about some rational axis which is not an even axis of symmetry. Any such operation corresponds to a twinning orientation (for centrosymmetric crystals), the lattices being reflected about the plane normal to the 180° rotation axis. It follows that a twin orientation always implies the existence of a CSL, but the reverse is not always true [3, 10]; for example, in the case of $\Sigma = 39$, there is no axis-angle pair representation with an angle of rotation of 180°, so that it is not possible to find a coherent interface between crystals related by a $\Sigma = 39$ orientation relation.

(ii) Boundaries containing a high absolute density of coincidence sites (i.e., a large value of number of coincidence sites per unit boundary area) can in general be expected to have the lowest energy.

(iii) Calculations of atomic positions [11–13] in the boundary region, using inter-atomic force laws, suggest that in materials where the atoms are hard (strong repulsive interaction at short range), coincidence site lattices *may* not exist. For example, in the case of a $\Sigma 3$ twin in a "hard" bcc material, with a $\{1\ 1\ 2\}$ coherent

TABLE 13.1
Some CSL relations for cubic crystals [3]

Σ	Angle	Axis	Twin axes
3	60.0	$\langle 1\,1\,1\rangle$	$\langle 1\,1\,1\rangle$, $\langle 1\,1\,2\rangle$
5	36.9	$\langle 1\,0\,0\rangle$	$\langle 0\,1\,2\rangle$, $\langle 0\,1\,3\rangle$
7	38.2	$\langle 1\,1\,1\rangle$	$\langle 1\,2\,3\rangle$
9	38.9	$\langle 1\,1\,0\rangle$	$\langle 1\,2\,2\rangle$, $\langle 1\,1\,4\rangle$
11	50.5	$\langle 1\,1\,0\rangle$	$\langle 1\,1\,3\rangle$, $\langle 2\,3\,3\rangle$
13a	22.6	$\langle 1\,0\,0\rangle$	$\langle 0\,2\,3\rangle$, $\langle 0\,1\,5\rangle$
13b	27.8	$\langle 1\,1\,1\rangle$	$\langle 1\,3\,4\rangle$
15	48.2	$\langle 2\,1\,0\rangle$	$\langle 1\,2\,5\rangle$
17a	28.1	$\langle 1\,0\,0\rangle$	$\langle 0\,1\,4\rangle$, $\langle 0\,3\,5\rangle$
17b	61.9	$\langle 2\,2\,1\rangle$	$\langle 2\,2\,3\rangle$, $\langle 3\,3\,4\rangle$
19a	26.5	$\langle 1\,1\,0\rangle$	$\langle 1\,3\,3\rangle$, $\langle 1\,1\,6\rangle$
19b	46.8	$\langle 1\,1\,1\rangle$	$\langle 2\,3\,5\rangle$
21a	21.8	$\langle 1\,1\,1\rangle$	$\langle 2\,3\,5\rangle$, $\langle 1\,4\,5\rangle$
21b	44.4	$\langle 2\,1\,1\rangle$	$\langle 1\,2\,4\rangle$

twin plane, it is found that a small rigid translation (by a vector $\frac{a}{12}\langle\bar{1}\,\bar{1}\,1\rangle$) of the twin lattice lowers the energy of the interface (Figure 13.2) [3, 14]. Because of this relaxation, the lattices no longer have a common origin and so the coincidences vanish. Nevertheless, boundaries which contain high densities of coincidence sites before relaxation may be expected to represent better fit between the lattices, and thus have low energies relative to other boundaries. This is because the periodic nature and the actual repeat period of the structure of the interface implied by the CSL concept is not destroyed by the small translation. The rigid body translations mentioned above have been experimentally established in the case of aluminium; although not conclusively established, the experiments suggest that the translation may have a component outside the interface plane, but the atomistic calculations cannot predict this since they always seem to be carried out at constant volume [14].

(iv) The physical significance of CSL's must diminish as the Σ value increases, because only a very small fraction of atoms in an interface can then be common to both the adjacent crystals.

(v) The $\Sigma = 3$ value is independent [3] of symmetry considerations. This is convenient since such matrices do not uniquely relate the grains; it is usually necessary to impose criteria to allow physically reasonable choices to be made.

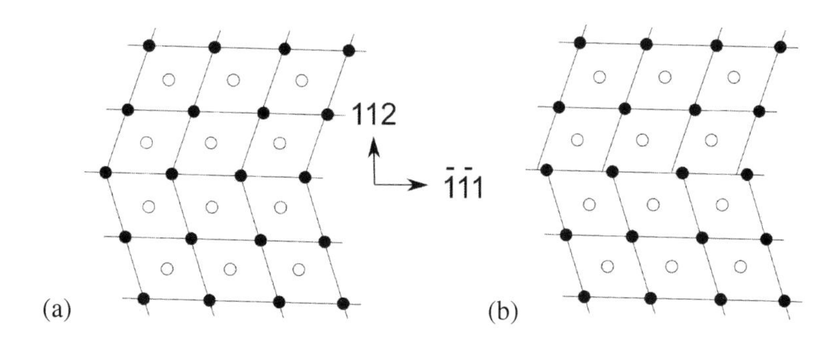

FIGURE 13.2
$\{112\}$ coherent twin boundary in a bcc material, initially with an exact $\Sigma 3$ CSL. The figure on the right illustrates the structure after a rigid body translation is included [3, 14].

13.4 The O-lattice

We have seen that coincidence site lattices can be generated by identifying lattice points which are common to both of the adjacent crystals, when the lattices of these crystals are notionally allowed to interpenetrate and fill all space. The totality of the common lattice points then forms the coincidence site lattice. This procedure is only a formal way of identifying coincidence points; for a real interface connecting the two crystals, only atoms from crystal 1 will exist on one side of the interface and those from crystal 2 on the other. There will, however, be good fit regions in the interface, corresponding to the lattice points that are common to both crystals; the pattern formed by these points consists of a two dimensional section of the CSL. The coincidence site lattice thus concentrates on just lattice points, and this is not the most general case to consider. When the transformation carrying one grain into the other is a rigid body rotation, all points along the rotation axis will represent points of perfect fit between the two crystals, not just the lattice points which may lie on the rotation axis. When the transformation relating two grains is an invariant-plane strain, all points in the invariant-plane represent points of perfect fit between the two lattices and not just the lattice points which lie in the invariant-plane.

Any point $\mathbf{x}$ within a crystal may be represented as the sum of a lattice vector $\mathbf{u}$ (which has integral components) and a small vector β whose components are fractional and less than unity. The internal co-ordinates of the point $\mathbf{x}$ are then defined to be the components of the vector β. The O-lattice method takes account of all coincidences, between non-lattice sites of identical internal co-ordinates as well as the coincidence lattice sites.

All lattice points in a crystal are crystallographically equivalent and any lattice point may be used as an origin to generate the three dimensional crystal lattice. To identify points of the CSL we specify that any lattice vector $\mathbf{u}$ of crystal A must, as a result of the transformation to crystal B, become another lattice vector $\mathbf{x}$ of A - i.e., $\mathbf{x} = \mathbf{u} + \mathbf{v}$ where $\mathbf{v}$ is a lattice vector of A. The CSL point $\mathbf{x}$ can then be considered to be a perfect fit point in an interface between A and B, because it corresponds to a lattice point in both crystals.

Non-lattice points in a crystal are crystallographically equivalent when they have the same internal coordinates. To identify O-points [15, 16], we specify that any non-lattice vector $\mathbf{x}$ of crystal A must, as a result of transformation to crystal B, become another non-lattice vector $\mathbf{y}$ of A such that $\mathbf{y} = \mathbf{x} + \mathbf{v}$ where $\mathbf{v}$ is a lattice vector of A; the points $\mathbf{x}$ and $\mathbf{y}$ thus have the same internal co-ordinates in A. The O-point $\mathbf{y}$ can then be considered to be a perfect fit point in the interface between A and B. Note that when $\mathbf{x}$ becomes a lattice vector, $\mathbf{y}$ becomes a CSL point. The totality of O-points obtained by allowing crystals A and B to notionally interpenetrate forms the O-lattice [15, 16], which may contain the CSL as a sub-lattice if A and B are suitably orientated at an exact CSL orientation. Any boundary between A and B cuts the O-lattice and will contain regions of good fit which have the periodicity corresponding to the periodicity of a planar net of the O-lattice along which the intersection occurs. Boundaries parallel to low-index planes of the O-lattice are in general two dimensionally periodic with relatively small repeat cells, and those with a high planar O-point density should have a relatively low energy per unit area.

Consider two crystals A and B which are related by the deformation (A S A) which converts the reference lattice A to that of B; an arbitrary non-lattice point $\mathbf{x}$ in crystal A thus becomes a point $\mathbf{y}$ in crystal B, where

$$[\mathrm{A}; \mathbf{y}] = (\mathrm{A\ S\ A})[\mathrm{A}; \mathbf{x}]$$

If the point $\mathbf{y}$ is crystallographically equivalent to the point $\mathbf{x}$, in the sense that it has the same internal co-ordinates as $\mathbf{x}$, then $\mathbf{y}$ is also a point of the O-lattice, designated $\mathbf{O}$. This means that $\mathbf{y} = \mathbf{x} + \mathbf{u}$, where $\mathbf{u}$ is a lattice vector of A. Since $\mathbf{y}$ is only an O-point when $\mathbf{y} = \mathbf{x} + \mathbf{u}$, we may write that $\mathbf{y}$ is an O-lattice vector $\mathbf{O}$ if [15, 16]

$$[\mathrm{A}; \mathbf{y}] = [\mathrm{A}; \mathbf{O}] = [\mathrm{A}; \mathbf{x}] + [\mathrm{A}; \mathbf{u}] = (\mathrm{A\ S\ A})[\mathrm{A}; \mathbf{x}]$$

or in other words,

$$[\mathrm{A}; \mathbf{u}] = (\mathrm{A\ T\ A})[\mathrm{A}; \mathbf{O}] \tag{13.11}$$

where $(\mathrm{A\ T\ A}) = \mathbf{I} - (\mathrm{A\ S\ A})^{-1}$. It follows that

$$[\mathrm{A}; \mathbf{O}] = (\mathrm{A\ T\ A})^{-1}[\mathrm{A}; \mathbf{u}]. \tag{13.12}$$

By substituting for $\mathbf{u}$ the three basis vectors of A in turn [3], we see that the columns of $(\mathrm{A\ T\ A})^{-1}$ define the corresponding base vectors of the O-lattice.

Since O-points are points of perfect fit, mismatch must be at a maximum in between neighboring O-points. When **O** is a primitive O-lattice vector, Equation 13.12 states that the amount of misfit in between the two O-points connected by **O** is given by **u**. A dislocation with Burgers vector **u** would thus accommodate this misfit and localise it at a position between the O-points, and these ideas allow us to consider a dislocation model of the interface in terms of the O-lattice theory.

Three sets of dislocations with Burgers vectors $\mathbf{b}_\mathrm{i}$ (which form a non-coplanar set) are in general required to accommodate the misfit in any interface. If the Burgers vectors $\mathbf{b}_1$, $\mathbf{b}_2$, and $\mathbf{b}_3$ are chosen to serve this purpose, and each in turn substituted into Equation 13.12, then the corresponding O-lattice vectors $\mathbf{O}_1$, $\mathbf{O}_2$ and $\mathbf{O}_3$ are obtained. These vectors $\mathbf{O}_i$ thus define the basis vectors of an O-lattice unit cell appropriate to the choice of $\mathbf{b}_\mathrm{i}$. If the O-points of this O-lattice are separated by "cell walls" which bisect the lines connecting neighboring O-points, then the accumulating misfit in any direction can be considered to be concentrated at these cell walls [15, 16]. When a real interface (unit normal **n**) is introduced into the O-lattice, its line intersections with the cell walls become the interface dislocations with Burgers vectors $\mathbf{b}_\mathrm{i}$ and unit line vectors $\mathbf{l}_\mathrm{i}$ parallel to the line of intersection of the interface with the cell walls.

The three O-lattice cell walls (with normals $\mathbf{O}_i^*$) have normals parallel to $\mathbf{O}_1^* = \mathbf{O}_2 \wedge \mathbf{O}_3$, $\mathbf{O}_2^* = \mathbf{O}_3 \wedge \mathbf{O}_1$ and $\mathbf{O}_3^* = \mathbf{O}_1 \wedge \mathbf{O}_2$, so that the line vectors of the dislocations are given by $\mathbf{l}_1 \parallel \mathbf{O}_1^* \wedge \mathbf{n}$, $\mathbf{l}_2 \parallel \mathbf{O}_2^* \wedge \mathbf{n}$ and $\mathbf{l}_3 \parallel \mathbf{O}_3^* \wedge \mathbf{n}$. Similarly, $1/d_i = |\mathbf{O}_i^* \wedge \mathbf{n}|$. These results are exactly equivalent to the theory developed at the beginning of this chapter (Equation 13.1–13.8) but the O-lattice theory perhaps gives a better physical picture of the interface, and follows naturally from the CSL approach [3]. The equivalence of the two approaches arises because Equation 13.11 is identical to Equation 13.1 since **u** and **O** are equivalent to $\mathbf{b}_\mathrm{t}$ and **p**, respectively.

Example 13.5: alpha/beta brass interface using O-lattice theory

Derive the structure of the alpha/beta brass interface described on page 217 using O-lattice theory.

From page 217, the matrix (A T A) is given by:

$$(\mathrm{A\ T\ A}) = \begin{pmatrix} 0.119500 & 0.006386 & 0.016048 \\ 0.006386 & 0.119500 & 0.016048 \\ -0.149974 & -0.149974 & -0.236183 \end{pmatrix}$$

so that

$$(\mathrm{A\ T\ A})^{-1} = \begin{pmatrix} -52.448 & -61.289 & -51.069 \\ -61.289 & -52.448 & -51.064 \\ 72.222 & 72.222 & 60.622 \end{pmatrix}$$

If the vectors $\mathbf{b}_1$, $\mathbf{b}_2$, and $\mathbf{b}_3$ are each in turn substituted for **u** in Equation 13.11, then the corresponding **O** lattice vectors are found to be:

$$[\mathrm{A};\mathbf{O}_1] = [-0.689770 \quad -5.110092 \ 5.799861]$$
$$[\mathrm{A};\mathbf{O}_2] = [-5.110092 \quad -0.689770 \ 5.799861]$$
$$[\mathrm{A};\mathbf{O}_3] = [-56.86861 \quad -56.86861 \ 72.22212]$$

$$[\mathrm{A}^*;\mathbf{O}_1^*] = [\ 0.263088 \ 0.036860 \ 0.236183]$$
$$[\mathrm{A}^*;\mathbf{O}_2^*] = [\ 0.036860 \ 0.263088 \ 0.236183]$$
$$[\mathrm{A}^*;\mathbf{O}_3^*] = [-0.024088 \ -0.024088 \ -0.024088]$$

$$\mathbf{O}_1^* \wedge \mathbf{n} \parallel \mathbf{l}_1 = [-0.012249 \quad -0.012249 \ 0.015556]_\mathrm{A}$$
$$\mathbf{O}_2^* \wedge \mathbf{n} \parallel \mathbf{l}_2 = [-0.012249 \quad -0.012249 \ 0.015556]_\mathrm{A}$$
$$\mathbf{O}_3^* \wedge \mathbf{n} \parallel \mathbf{l}_3 = [\ 0.001249 \ 0.001249 \quad -0.002498]_\mathrm{A}$$

$$d_1 = d_2 = 11.63\ \text{Å} \qquad \text{and} \qquad d_3 = 88.52\ \text{Å}$$

The results obtained therefore are identical to those described beginning page 217.

13.5 Secondary dislocations

Low-angle boundaries contain interface dislocations whose role is to localize and accommodate the misfit between adjacent grains, such that large areas of the boundary consist of low-energy coherent patches without mismatch. These intrinsic interface dislocations are called *primary* dislocations because they accommodate the misfit relative to an ideal single crystal as the reference lattice. Boundaries between grains which are at an exact CSL orientation have relatively low-energy. It then seems reasonable to assume that any small deviation from the CSL orientation should be accommodated by a set of interface dislocations which localise the misfit due to this deviation, and hence allow the perfect CSL to exist over most of the boundary area. These intrinsic interface dislocations [17–20] are called *secondary* dislocations because they accommodate the misfit relative to a CSL as the reference lattice. High-angle boundaries between crystals which are not at an exact CSL orientation may therefore consist of dense arrays of primary dislocations and also relatively widely spaced arrays of secondary dislocations. The primary dislocations may be so closely spaced, that their strain fields virtually cancel each other and in these circumstances only secondary dislocations would be visible using conventional transmission electron microscopy.

Example 13.6: Intrinsic secondary dislocations

The axis-angle pair describing the orientation relationship between the two grains (A and B) of austenite is given by:

$$\text{axis of rotation parallel to } [1\ 1\ 2]_A$$

$$\text{right-handed angle of rotation equal to } 175^\circ$$

Calculate the secondary dislocation structure of an interface lying normal to the axis of rotation, given that the Burgers vectors of the interface dislocations are $[\mathrm{A};\mathbf{b}_1] = \frac{1}{2}[1\ 0\ \bar{1}]$, $[\mathrm{A};\mathbf{b}_2] = \frac{1}{2}[0\ 1\ \bar{1}]$ and $[\mathrm{A};\mathbf{b}_3] = \frac{1}{2}[1\ 1\ 0]$, where the basis A corresponds to the conventional fcc unit cell of austenite, with a lattice parameter $a = 3.56$ Å.

The secondary dislocation structure can be calculated with respect to the nearest CSL, which is a $\Sigma 3$ CSL obtained by a 180° rotation about $[1\ 1\ 2]_A$. The rigid body rotation matrix corresponding to this exact CSL orientation is thus:

$$(\mathrm{A\ J\ A}) = \frac{1}{3}\begin{pmatrix} \bar{2} & 1 & 2 \\ 1 & \bar{2} & 2 \\ 2 & 2 & 1 \end{pmatrix}$$

The rotation matrix (A J_2 A) describing the actual transformation of A to B, corresponding to a rotation of 175° about $[1\ 1\ 2]_A$ is given by:

$$(\mathrm{A\ J_2\ A}) = \begin{pmatrix} -0.663496 & 0.403861 & 0.629817 \\ 0.261537 & -0.663496 & 0.700979 \\ 0.700979 & 0.629817 & 0.334602 \end{pmatrix}.$$

The matrix (A J_3 A) describing the deviation from the exact CSL is given by [16] $(\mathrm{A\ J_3\ A}) = (\mathrm{A\ J_2\ A})^{-1}(\mathrm{A\ J\ A})$, so that

$$(\mathrm{A\ J_3\ A}) = \begin{pmatrix} 0.996829 & 0.071797 & -0.034313 \\ -0.070528 & 0.996829 & 0.036850 \\ 0.036850 & -0.034313 & 0.998732 \end{pmatrix}.$$

If $(\mathrm{A\ T\ A}) = \mathbf{I} - (\mathrm{A\ J_3\ A})^{-1}$, then (A T' A) becomes:

$$(\mathrm{A\ T'\ A}) = \begin{pmatrix} 0.003171 & -0.071797 & 0.034313 \\ 0.070528 & 0.003171 & -0.036850 \\ -0.036850 & 0.034313 & 0.001268 \end{pmatrix}.$$

The secondary dislocation structure can now be calculated using the procedures described on page 213:

$$(\mathbf{n}; \mathrm{A}^*) = a(0.408248\ 0.408248\ 0.816497)$$

$$[\mathrm{A}^*; \mathbf{b}_1^*] = [1\ \bar{1}\ 1]$$
$$[\mathrm{A}^*; \mathbf{b}_2^*] = [\bar{1}\ 1\ \bar{1}]$$
$$[\mathrm{A}^*; \mathbf{b}_3^*] = [1\ 1\ 1]$$

From Equation 13.6,

$$[\mathrm{A}^*; \mathbf{c}_1] = [0.040655\ 0.104207\ -0.072341]$$
$$[\mathrm{A}^*; \mathbf{c}_2] = [-0.0109280\ -0.030507\ 0.069894]$$
$$[\mathrm{A}^*; \mathbf{c}_3] = [-0.034313\ 0.036850\ -0.001268]$$

so that

$$d_1 = 26.71\text{Å}, \qquad d_2 = 26.71\text{Å}, \qquad \text{and} \qquad d_3 = 70.68\ \text{Å}$$

$$\mathbf{l}_1 \parallel [\ 0.860386\ -0.470989\ -0.194695]_\mathrm{A}$$
$$\mathbf{l}_2 \parallel [-0.401046\ 0.883699\ -0.241327]_\mathrm{A}$$
$$\mathbf{l}_3 \parallel [\ 0.607647\ 0.545961\ -0.576794]_\mathrm{A}.$$

13.6 The DSC lattice

In a boundary between two crystals, the Burgers vector **b** of an interface dislocation must be such that the displacement of one of the crystals through **b** relative to the other does not change the structure of the interface. Lattice translation vectors of the reference lattice always satisfy this condition, so that **b** can always equal a lattice translation vector. However, additional possibilities arise in the case of secondary dislocations, whose Burgers vectors are generally DSC lattice translation vectors [16, 21]. Secondary dislocations represent the deviation from a particular coincidence site lattice, and a corresponding DSC lattice may be generated such that the translation vectors of the DSC lattice are possible Burgers vectors of secondary interface dislocations. The interesting point about the DSC lattice is that its lattice vectors need not be crystal lattice translation vectors. Figure 13.3 illustrates a $\Sigma 5$ CSL between two fcc grains (A and B), related by a rotation of 36.87° about $[1\ 0\ 0]_\mathrm{A}$.

Figure 13.3b is obtained by displacing lattice B by $[\mathrm{A}; \mathbf{b}] = \frac{1}{10}[0\ 3\ 1]$ relative to lattice A and it is obvious that the basic pattern of lattice sites and CSL sites remains

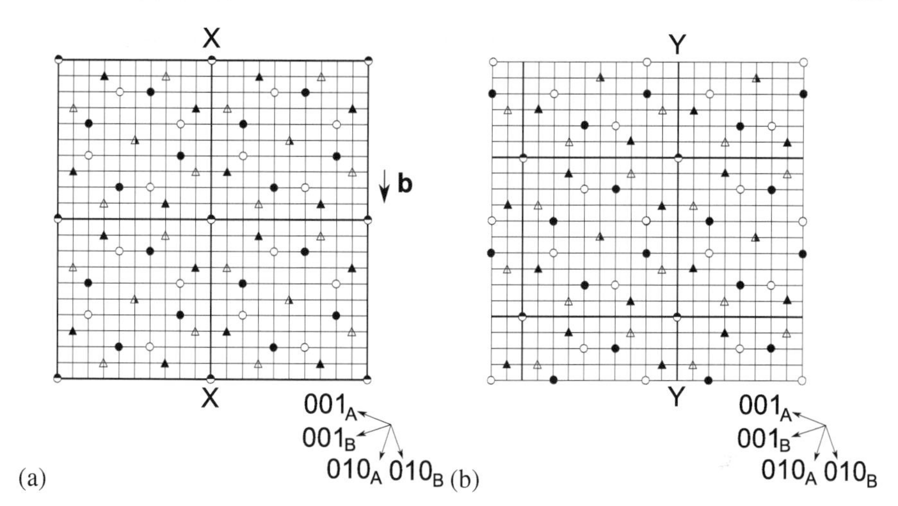

FIGURE 13.3
$\Sigma 5$ coincidence system for fcc crystals [22]. Filled symbols are lattice A, unfilled ones lattice B, and coincidence sites are a mixture of the two. Lattice sites in the plane of the diagram are represented as circles whereas those displaced by $\frac{1}{2}[100]$ are represented as triangles. The [100] axis is normal to the plane of the diagram.

unaffected by this translation, despite the fact that **b** is not a lattice vector of A or B. It is thus possible for secondary dislocations to have Burgers vectors which are not lattice translation vectors, but are vectors of the DSC lattice. The DSC lattice, or the Displacement Shift Complete lattice, is the coarsest lattice which contains the lattice points of both A and B, and any DSC lattice vector is a possible Burgers vector for a perfect secondary dislocation. We note that the displacement **b** causes the original coincidences (Figure 13.3a) to disappear and be replaced by an equivalent set of new coincidences (Figure 13.3b), and this always happens when **b** is not a lattice translation vector. This shift of the origin of the CSL has an important consequence on the topography [22] of any boundary containing secondary dislocations with non-lattice Burgers vectors.

Considering again Figure 13.3, suppose that we introduce a boundary into the CSL, with unit normal $[A; \mathbf{n}] \parallel [0\ 1\ \overline{2}]$, so that its trace is given by XX on Figure 13.3a. The effect of the displacement **b** of crystal B relative to A, due to the presence of a secondary dislocation, is to shift the origin of the CSL; if the boundary originally at XX is to have the same structure after the displacement then it has to shift to the position YY in Figure 13.3b. Because a dislocation separates slipped from unslipped regions, the shift of the boundary occurs at the position of the secondary dislocation so that the boundary is stepped at the core of this dislocation. One such step is illustrated in Figure 13.4.

The following further points about the DSC lattice and its consequences should be noted:

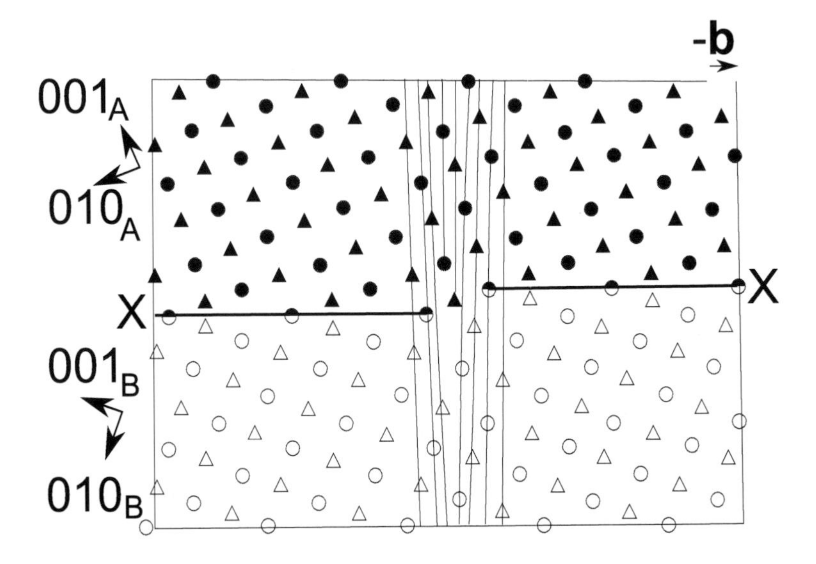

FIGURE 13.4
The presence of a step [23] in a $\Sigma 5$, $(3\,\bar{1}\,0)_A$ boundary of an fcc crystal containing a secondary interface dislocation with $[\mathrm{A};\mathbf{b}] = (a/10)[1\ 3\ 0]$. The symbolism is identical to that of Figure 13.3.

(i) The DSC lattice can be constructed graphically simply by inspection, bearing in mind that it is the coarsest lattice containing lattice sites from both the crystals orientated at an exact CSL orientation. Rather detailed analytical methods for computing the basis vectors of the DSC lattice have been presented elsewhere [24], and tabulations of these DSC lattice calculations as a function of $\Sigma 3$ can also be found for cubic systems [25].

(ii) For the primitive cubic system, the components of the three basis vectors of the DSC lattice are the columns of a 3×3 matrix (**DSC**) obtained by taking the transpose of the inverse of another 3×3 matrix (**CSL**). The columns of **CSL** represent the components of the basis vectors of the coincidence site lattice concerned.

(iii) Crystal lattice vectors (of both the adjacent crystals) form a sub-set of DSC lattice vectors.

(iv) The volume of a CSL unit cell is $\Sigma 3$ times larger than that of the crystal lattice unit cell, whereas the DSC lattice unit cell has a volume $1/\Sigma$ times that of the crystal lattice unit cell [24].

(v) The homogeneous deformation (A S A) of Equation 13.1, describing the transformation from the reference lattice to another crystal is not unique [3], but the CSL and DSC concepts are independent of the choice of (A S A). The O-lattice on the other hand, depends critically on the form of (A S A).

(vi) Primitive DSC lattice vectors can be much smaller than primitive crystal lattice vectors [26]. On the basis of elastic strain energy arguments, smaller interface dislocation Burgers vectors should be favored. This is not confirmed experimentally since secondary dislocations often have Burgers vectors which are crystal lattice vectors.

(vii) Although DSC lattices are defined relative to the unrelaxed CSL, small rigid body translations which destroy exact coincidence (but preserve the CSL periodicity) do not affect the essentials of the DSC concept [14].

13.7 Some difficulties associated with interface theory

Consider two crystals A and B, both of cubic structure, related by a rigid body rotation. Any boundary containing the axis of rotation is a tilt boundary; for the special case of the symmetrical tilt boundary, lattice B can be generated from A by reflection across the boundary plane. By substituting the rigid body rotation for (A S A) in Equation 13.1, the dislocation structure of the symmetrical tilt boundary may be deduced (page 214) to consist of a single array of dislocations with line vectors parallel to the tilt axis.

Symmetry considerations imply that the rigid body rotation has up to 23 further axis-angle representations. If we impose the condition that the physically most significant representation is that which minimizes the Burgers vector content of the interface, then the choice reduces to the axis-angle pair involving the smallest angle of rotation.

On the other hand, (A S A) can also be a lattice-invariant twinning shear on the symmetrical tilt boundary plane [3]; crystal B would then be related to A by reflection across the twin plane so that the resulting bicrystal would be equivalent to the case considered above. The dislocation content of the interface then reduces to zero since the invariant-plane of the twinning shear is fully coherent.

This ambiguity in the choice of (A S A) is a major difficulty in interface theory [3]. The problem is compounded by the fact that interface theory is phenomenological – i.e., the transformation strain (A S A) may be real or notional as far as interface theory is concerned. If it is real then we expect to observe a change in the shape of the transformed region, and this may help in choosing the most reasonable deformation (A S A). For example, in the case of mechanical twinning in fcc crystals, the surface relief observed can be used to deduce that (A S A) is a twinning shear rather than a rigid body rotation. In the case of an fcc annealing twin, which grows from the matrix by a diffusional mechanism (during grain boundary migration), the same twinning shear (A S A) may be used to deduce the interface structure, but the deformation is now notional, since the formation of annealing twins is not accompanied by any surface relief effects. In these circumstances, we cannot be certain that the deduced interface structure for the annealing twin is correct.

The second major problem follows from the fact that the mathematical Burgers vector content $\mathbf{b}_t$ given by Equation 13.1 has to be factorized into arrays of physically realistic dislocations with Burgers vectors which are vectors of the DSC lattice. There is an infinite number of ways in which this can be done, particularly since the interface dislocations do not necessarily have Burgers vectors which minimize their elastic strain energy.

Secondary dislocations are referred to an exact CSL as the reference lattice. Atomistic calculations suggest that boundaries in crystals orientated at exact coincidence contain arrays of primary dislocations (whose cores are called *structural units*) [27]. The nature of the structural units varies with the $\Sigma 3$ value, but some favored CSL's have boundaries with just one type of structural unit, so that the stress field of the boundary is very uniform. Other CSL's have boundaries consisting of a mixture of structural units from various favored CSL's. It has been suggested that it is the favored CSL's which should be used as the reference lattices in the calculation of secondary dislocation structure, but the situation is unsatisfactory because the same calculations suggest that the favored/unfavored status of a boundary also depends on the boundary orientation itself. (We note that the term "favored" does not imply a low interface energy).

Referring now to the rigid body translations which exist in materials with "hard" atoms, it is not clear whether the calculated translations might be different if the relaxation were to be carried out for a three dimensionally enclosed particle.

13.8 Summary

The Burgers vector content of an interface between two crystals that do not necessarily have the same structure, can be obtained by the closure failure of a Burgers circuit constructed across the interface, with the comparison made against the same circuit constructed after transforming the bi-crystal into a single reference-crystal. The Burgers vector content can then be de-convoluted into arrays of interfacial dislocations, with each array defined by a Burgers and line vector. This process requires a prior choice of individual dislocation Burgers vectors but once this is done, the calculations permit a comparison against experimental observations.

There exist special orientations, the coincidence site lattices, where there is a fraction of lattice points that is common to both crystals when their lattices are allowed to interpenetrate and fill all space, assuming they have a common origin. The single crystal may be regarded at $\Sigma = 1$ CSL, and when divided to create a bi-crystal with a small misorientation, leads to the classical dislocation description of the tilt or twist boundary. Other CSL orientations such as $\Sigma = 3$ etc. can also be used as reference lattices, with deviations from the exact CSL described in terms of dislocation arrays. Furthermore, a boundary located on a CSL plane can be displaced so that it maintains the coincidences using secondary dislocations as described in the "complete pattern shift" or DSC lattice.

The idea of coincidences of lattice points can be generalised to coincidences of points that have the same internal coordinates but are not necessarily lattice sites, as in Bollmann O-lattice theory. This again is a useful concept that permits the interface structure to be calculated after making assumptions about the nature of the dislocations in the interface.

All the theory described here refers to pure interfaces that have no segregants and which are not relaxed in the sense of rigid body translations parallel to the interface plane. Furthermore, when the crystals are solid solutions, there will be a chemical component of interfacial energy that would need to be taken into account.

References

1. B. A. Bilby: Types of dislocation source, In: *Bristol Conference Report on Defects in Crystalline Solids*. London, U.K.: The Physical Society, 1955:124–133.
2. F. C. Frank: Report of the pittsburgh symposium, In: *Symposium on the Plastic Deformation of Crystalline Solids*. Pittsburgh, PA: Office of Naval Research, 1950:150.
3. J. W. Christian: *Theory of transformations in metals and alloys*, Part I: 2nd ed., Oxford, U. K.: Pergamon Press, 1975.
4. K. M. Knowles, D. A. Smith, and W. A. T. Clark: On the use of geometric parameters in the theory of interphase boundaries, *Scripta Metallurgica*, 1982, **16**, 413–416.
5. W. T. Read: *Dislocations in crystals*: New York: McGraw-Hill, 1953.
6. G. Bäro, and H. Gleiter: On the structure of incoherent interphase boundaries between FCC/BCC crystals, *Acta Metallurgica*, 1973, **21**, 1405–1408.
7. M. L. Kronberg, and F. H. Wilson: Secondary recrystallization in copper, *Trans. A.I.M.E.*, 1949, **185**, 501–514.
8. K. T. Aust, and J. W. Rutter: Temperature dependence of grain migration in high-purity lead containing small additions of tin, *Trans. A.I.M.M.E.*, 1959, **215**, 820–831.
9. D. H. Warrington, and P. Bufalini: The coincidence site lattice and grain boundaries, *Scripta Metallurgica*, 1971, **5**, 771–776.
10. M. A. Fortes: *Revista de fisica Quimica e Engenharia*, 1972, **4a**, 7.
11. V. Vitek: Multilayer stacking faults and twins on {211} planes in bcc metals, *Scripta Metallurgica*, 1970, **4**, 725–732.

12. D. A. Smith, V. Vitek, and R. C. Pond: Computer simulation of symmetrical high angle boundaries in aluminium, *Acta Metallurgica*, 1977, **25**, 475–483.

13. R. C. Pond: Periodic grain boundary structures in aluminium. II. a geometrical method for analysing periodic grain boundary structure and some related transmission electron microscope observations, *Proceedings fo the Royal Society A*, 1977, **357**, 471–483.

14. J. W. Christian, and A. G. Crocker: *Dislocations in solids*, ed. F. R. N. Nabarro: Amsterdam, Holland: North Holland, 1980.

15. W. Bollmann: On the geometry of grain and phase boundaries: I. general theory, *Philosophical Magazine*, 1967, **16**, 363–381.

16. W. Bollmann: *Crystal defects and crystalline interfaces*: Berlin, Germany: Springer Verlag, 1970.

17. W. T. Read, and W. Schockley: Dislocation models of crystal grain boundaries, *Physical Review*, 1950, **78**, 275–289.

18. D. G. Brandon, B. Ralph, S. T. Ranganathan, and M. S. Wald: A field ion microscope study of atomic configuration at grain boundaries, *Acta Metallurgica*, 1964, **12**, 813–821.

19. T. Schober, and R. W. Balluffi: Extraneous grain boundary dislocations in low and high angle (001) twist boundaries in gold, *Philosophical Magazine*, 1971, **24**, 165–180.

20. J. P. Hirth, and R. W. Balluffi: On grain boundary dislocations and ledges, *Acta Metallurgica*, 1973, **21**, 929–942.

21. D. G. Brandon: The structure of high-angle grain boundaries, *Acta Metallurgica*, 1966, **14**, 1479–1484.

22. A. H. King, and D. A. Smith: The effects on grain-boundary processes of the steps in the boundary plane associated with the cores of grain-boundary dislocations, *Acta Crystallographica A*, 1980, **36**, 335–343.

23. R. C. Pond, and D. A. Smith: On the role of grain boundary dislocations in plastic deformation, In: J. L. Walter, J. H. Westbrook, and D. A. Woodford., eds. *4th Bolton Landing Conference on Grain Boundaries in Eng. Materials*. Baton Rouge, LA: Claitors Publishing Division, 1975:309–315.

24. H. Grimmer, W. T. Bollmann, and D. H. Warrington: Coincidence-site lattices and complete pattern-shift in cubic crystals, *Acta Crystallographica A*, 1974, **30**, 197–207.

25. D. H. Warrington, and H. Grimmer: Dislocation burgers vectors for cubic metal grain boundaries, *Philosophical Magazine*, 1974, **30**, 461–468.

26. W. Bollmann, B. Michaut, and G. Sainfort: Pseudo-subgrain-boundaries in stainless steel, *Physica Status Solidi A*, 1972, **13**, 637–649.

27. A. P. Sutton, and V. Vitek: On the structure of tilt grain boundaries in cubic metals I. symmetrical tilt boundaries, *Philosophical Transactions of the Royal Society A*, 1983, **309**, 1–36.

Appendices

A

Matrix methods

A.1 Vectors

Quantities (such as force, displacement) which are characterised by both magnitude and direction are called vectors; scalar quantities (such as time) only have magnitude. A vector is represented by an arrow pointing in a particular direction, and can be identified by underlining the lower–case vector symbol (e.g., $\mathbf{u}$). The magnitude of $\mathbf{u}$ (or $|\mathbf{u}|$) is given by its length, a scalar quantity. Vectors $\mathbf{u}$ and $\mathbf{v}$ are only equal if they both point in the same direction, and if $|\mathbf{u}| = |\mathbf{v}|$. The parallelism of $\mathbf{u}$ and $\mathbf{v}$ is indicated by writing $\mathbf{u} \parallel \mathbf{v}$. If $\mathbf{x} = -\mathbf{u}$, then $\mathbf{x}$ points in the opposite direction to $\mathbf{u}$, although $|\mathbf{x}| = |\mathbf{u}|$.

Vectors can be added or removed to give new vectors, and the order in which these operations are carried out is not important. Vectors $\mathbf{u}$ and $\mathbf{x}$ can be added by placing the initial point of $\mathbf{x}$ in contact with the final point of $\mathbf{u}$; the initial point of the resultant vector $\mathbf{u} + \mathbf{x}$ is then the initial point of $\mathbf{u}$ and its final point corresponds to the final point of $\mathbf{x}$. The vector m$\mathbf{u}$ points in the direction of $\mathbf{u}$, but $|m\mathbf{u}|/|\mathbf{u}| = m$, m being a scalar quantity. A unit vector has a magnitude of unity; dividing a vector $\mathbf{u}$ by its own magnitude u gives a unit vector parallel to $\mathbf{u}$.

It is useful to refer vectors to a fixed frame of reference; an arbitrary reference frame would consist of three non-coplanar basis vectors $\mathbf{a}_1$, $\mathbf{a}_2$, and $\mathbf{a}_3$. The vector $\mathbf{u}$ could then be described by means of its components u_1, u_2 and u_3 along these basis vectors, respectively, such that

$$u_i = |\mathbf{u}| \cos\theta_i / |\mathbf{a}_i|$$

where I $= 1, 2, 3$ and $\theta_i =$ angle between $\mathbf{u}$ and $\mathbf{a}_i$.

If the basis vectors $\mathbf{a}_i$ form an orthonormal set (i.e. they are mutually perpendicular and each of unit magnitude), then the magnitude of $\mathbf{u}$ is:

$$|\mathbf{u}|^2 = u_1^2 + u_2^2 + u_3^2$$

If the basis vectors $\mathbf{a}_i$ form an orthogonal set (i.e. they are mutually perpendicular) then the magnitude of $\mathbf{u}$ is:

$$|\mathbf{u}|^2 = (u_1|\mathbf{a}_1|)^2 + (u_2|\mathbf{a}_2|)^2 + (u_3|\mathbf{a}_3|)^2$$

A *dot* or *scalar* product between two vectors $\mathbf{u}$ and $\mathbf{x}$ (order of multiplication not important) is given by $\mathbf{u}.\mathbf{x} = |\mathbf{u}| \times |\mathbf{x}| \cos\theta$, θ being the angle between **u** and **x**. If **x** is a unit vector then $\mathbf{u}.\mathbf{x}$ gives the projection of **u** in the direction **x**.

The *cross* or *vector* product is written $\mathbf{u} \wedge \mathbf{x} = |\mathbf{u}| \times |\mathbf{x}| \sin\theta\mathbf{y}$, where **y** is a unit vector perpendicular to both **u** and **x**, with **u**,**x** and **y** forming a right–handed set. A right–handed set **u**, **x**, **y** implies that a right–handed screw rotated through an angle less than 180° from **u** to **x** advances in the direction **y**. The magnitude of $\mathbf{u} \wedge \mathbf{x}$ gives the area enclosed by a parallelogram whose sides are the vectors **u** and **x**; the vector **y** is normal to this parallelogram. Clearly, $\mathbf{u} \wedge \mathbf{x} \neq \mathbf{x} \wedge \mathbf{u}$.

If **u**, **x** and **z** form a right-handed set of three non-coplanar vectors then $\mathbf{u} \wedge \mathbf{x}.\mathbf{z}$ gives the volume of the parallelepiped formed by **u**, **x** and **z**. It follows that $\mathbf{u} \wedge \mathbf{x}.\mathbf{z} = \mathbf{u}.\mathbf{x} \wedge \mathbf{z} = \mathbf{z} \wedge \mathbf{u}.\mathbf{x}$.

The following relations should be noted:

$$\begin{aligned}
&\mathbf{u} \wedge \mathbf{x} = -\mathbf{x} \wedge \mathbf{u} \\
&\mathbf{u}.(\mathbf{x} \wedge \mathbf{y}) = \mathbf{x}.(\mathbf{y} \wedge \mathbf{u}) = \mathbf{y}.(\mathbf{u} \wedge \mathbf{x}) \\
&\mathbf{u} \wedge (\mathbf{x} \wedge \mathbf{y}) \neq (\mathbf{u} \wedge \mathbf{x}) \wedge \mathbf{y} \\
&\mathbf{u} \wedge (\mathbf{x} \wedge \mathbf{y}) = (\mathbf{u}.\mathbf{y})\mathbf{x} - (\mathbf{u}.\mathbf{x})\mathbf{y}
\end{aligned} \tag{A.1}$$

A.2 Matrices

A.2.1 Definition, addition, scalar multiplication

A matrix is a rectangular array of numbers, having m rows and n columns, and is said to have an order m by n. A square matrix **J** of order 3 by 3 may be written as

$$\mathbf{J} = \begin{pmatrix} J_{11} & J_{12} & J_{13} \\ J_{21} & J_{22} & J_{23} \\ J_{31} & J_{32} & J_{33} \end{pmatrix} \quad \text{and its transpose} \quad \mathbf{J}' = \begin{pmatrix} J_{11} & J_{21} & J_{31} \\ J_{12} & J_{22} & J_{32} \\ J_{13} & J_{23} & J_{33} \end{pmatrix} \tag{A.2}$$

where each number J_{ij} ($i = 1, 2, 3$ and $j = 1, 2, 3$) is an element of **J**. **J'** is called the transpose of the matrix **J**. An identity matrix (**I**) has the diagonal elements J_{11}, J_{22}, and J_{33} equal to unity, all the other elements being zero. The trace of a matrix is the sum of all its diagonal elements $J_{11} + J_{22} + J_{33}$. If matrices **J** and **K** are of the same order, they are said to be equal when $J_{ij} = K_{ij}$ for all i, j. Multiplying a matrix by a constant involves the multiplication of every element of that matrix by that constant. Matrices of the same order may be added or subtracted, so that if $\mathbf{L} = \mathbf{J} + \mathbf{K}$, it follows that $L_{ij} = J_{ij} + K_{ij}$.

A.2.2 Einstein summation convention

In order to simplify more complex matrix operations we now introduce the summation convention. The expression

$$u_1 a_1 + u_2 a_2 + ...u_n a_n$$

can be shortened by writing

$$\sum_{i=1}^{3} u_i a_i$$

A further economy in writing is achieved by adopting the convention that the repetition of a subscript or superscript index in a given term implies summation over that index from 1 to n. Using this summation convention, the above sum can be written $u_i a_i$. Similarly,

$$x_i = y_j z_{ij} \quad \text{for} \quad i = 1, 2 \quad j = 1, 2 \qquad \text{implies that}$$
$$x_1 = y_1 z_{11} + y_2 z_{12}$$
$$x_2 = y_1 z_{21} + y_2 z_{22}$$

A.2.3 Multiplication and Inversion

The matrices **J** and **K** can be multiplied in that order to give a third matrix **L** if the number of columns (m) of **J** equals the number of rows of **K** (**J** is said to be conformable to **K**). **L** is given by

$$L_{st} = J_{sr} K_{rt}$$

where s ranges from 1 to the total number of rows in **J** and t ranges from 1 to the total number of columns in **K**. If **J** and **K** are both of order 3×3 then, for example,

$$L_{11} = J_{11} K_{11} + J_{12} K_{21} + J_{13} + K_{31}$$

Note that the product **JK** does not in general equal **KJ**.

Considering a $n \times n$ square matrix **J**, it is possible to define a number Δ which is the determinant (of order n) of **J**. A *minor* of any element J_{ij} is obtained by forming a new determinant of order $(n - 1)$, of the matrix obtained by removing all the elements in the ith row and the jth column of **J**. For example, if **J** is a 2×2 matrix, the minor of J_{11} is simply J_{22}. If **J** is a 3×3 matrix, the minor of J_{11} is:

$$\begin{vmatrix} J_{22} & J_{23} \\ J_{32} & J_{33} \end{vmatrix} = J_{22} J_{33} - J_{23} J_{32} \tag{A.3}$$

where the vertical lines imply a determinant. The cofactor j_{ij} of the element J_{ij} is then given by multiplying the minor of J_{ij} by $(-1)^{i+j}$. The determinant (Δ) of **J** is thus

$$\det \mathbf{J} = \sum_{j=1}^{n} J_{1j}\, j_{1j} \qquad \text{with} j = 1, 2, 3 \tag{A.4}$$

Hence, when $\mathbf{J}$ is a 3×3 matrix, its determinant Δ is given by:

$$\begin{aligned}\Delta &= J_{11}j_{11} + J_{12}j_{12} + J_{13}j_{13} \\ &= J_{11}(J_{22}J_{33} - J_{23}J_{32}) + J_{12}(J_{23}J_{31} - J_{21}J_{33}) + J_{13}(J_{21}J_{32} - J_{22}J_{31})\end{aligned}$$

The inverse of $\mathbf{J}$ is written $\mathbf{J}^{-1}$ and is defined such that

$$\mathbf{J}.\mathbf{J}^{-1} = \mathbf{I}$$

The elements of $\mathbf{J}^{-1}$ are J_{ij}^{-1} such that:

$$J_{ij}^{-1} = j_{ji}/\det \mathbf{J}$$

Hence, if $\mathbf{L}$ is the inverse of $\mathbf{J}$, and if $\det\mathbf{J} = \Delta$, then:

$$\begin{aligned}L_{11} &= (J_{22}J_{33} - J_{23}J_{32})/\Delta \\ L_{12} &= (J_{32}J_{13} - J_{33}J_{12})/\Delta \\ L_{13} &= (J_{12}J_{23} - J_{13}J_{22})/\Delta \\ L_{21} &= (J_{23}J_{31} - J_{21}J_{33})/\Delta \\ L_{22} &= (J_{33}J_{11} - J_{31}J_{13})/\Delta \\ L_{23} &= (J_{13}J_{21} - J_{11}J_{23})/\Delta \\ L_{31} &= (J_{21}J_{32} - J_{22}J_{31})/\Delta \\ L_{32} &= (J_{31}J_{12} - J_{32}J_{11})/\Delta \\ L_{33} &= (J_{11}J_{22} - J_{12}J_{21})/\Delta\end{aligned}$$

If the determinant of a matrix is zero, the matrix is singular and does not have an inverse.

The following matrix equations are noteworthy:

$$\begin{aligned}&(\mathbf{JK})\mathbf{L} = \mathbf{J}(\mathbf{KL}) \\ &\mathbf{J}(\mathbf{K} + \mathbf{L}) = \mathbf{JK} + \mathbf{JL} \\ &(\mathbf{J} + \mathbf{K})' = \mathbf{J}' + \mathbf{K}' \\ &(\mathbf{JK})' = \mathbf{K}'\mathbf{J}' \\ &(\mathbf{JK})^{-1} = \mathbf{K}^{-1}\mathbf{J}^{-1} \\ &(\mathbf{J}^{-1})' = (\mathbf{J}')^{-1}\end{aligned}$$

A.2.4 Orthogonal matrices

A square matrix $\mathbf{J}$ is said to be orthogonal if $\mathbf{J}^{-1} = \mathbf{J}'$. It the columns or rows of a real orthogonal matrix are taken to be components of column or row vectors respectively, then these vectors are all unit vectors. The set of column vectors (or row vectors) form an orthonormal basis; if this basis is right-handed, the determinant of the matrix is unity, but if it is left-handed then $\Delta = -1$.

Orthogonal matrices arise in coordinate transformations between orthonormal bases and where rigid body rotations are represented in a single orthonormal basis.

Example A.1: Simple operations

$$\mathbf{A} = \begin{pmatrix} 2 & 0 & 2 \\ 3 & 4 & 5 \\ 5 & 6 & 7 \end{pmatrix}, \qquad \mathbf{A}' = \begin{pmatrix} 2 & 3 & 5 \\ 0 & 4 & 6 \\ 2 & 5 & 7 \end{pmatrix}, \qquad \det \mathbf{A} = -8$$

$$\mathbf{A}^{-1} = \begin{pmatrix} 0.25 & -1.5 & 1 \\ -0.5 & -0.5 & 0.5 \\ 0.25 & 1.5 & -1 \end{pmatrix}$$

$$\mathbf{B} = \begin{pmatrix} 2 & 3 \\ 1 & 4 \end{pmatrix}, \qquad \mathbf{B}' = \begin{pmatrix} 2 & 1 \\ 3 & 4 \end{pmatrix}, \qquad \det \mathbf{B} = 5, \qquad \mathbf{B}^{-1} = \begin{pmatrix} 0.8 & -0.6 \\ -0.2 & 0.4 \end{pmatrix}$$

$$\mathbf{A} = \begin{pmatrix} 2 & 3 \\ 1 & 4 \end{pmatrix} \qquad \mathbf{A}' = \begin{pmatrix} 2 & 1 \\ 3 & 4 \end{pmatrix}$$

$$\det \mathbf{A} = 5 \qquad \mathbf{A}^{-1} = \begin{pmatrix} 0.8 & -0.6 \\ -0.2 & 0.4 \end{pmatrix}$$

B

General rotation matrix

The simple matrix representing a right-handed rotation of $\theta = 45°$ about [001] was written on page 99 as:

$$\mathbf{R} = \begin{pmatrix} m & n & 0 \\ -n & m & 0 \\ 0 & 0 & 1 \end{pmatrix}$$

where $m = \cos\theta$ and $n = \sin\theta$. The matrix was derived by inspection of Figure 7.5. This was possible because a rotation about one of the basis vectors is easy to visualise. At the same time, it was stated without derivation that a general rotation about an arbitrary unit vector $[u_1\, u_2\, u_3]$ can be described by the matrix ${}^g\mathbf{R} = \mathbf{JRJ}'$, where the rotation matrix $\mathbf{J}$ represents the new coordinate system:

$${}^g\mathbf{R} = \begin{pmatrix} u_1u_1(1-m)+m & u_1u_2(1-m)+u_3n & u_1u_3(1-m)-u_2n \\ u_1u_2(1-m)-u_3n & u_2u_2(1-m)+m & u_2u_3(1-m)+u_1n \\ u_1u_3(1-m)+u_2n & u_2u_3(1-m)-u_1n & u_3u_3(1-m)+m \end{pmatrix} \tag{B.1}$$

The right-handed angle of rotation can be obtained from the fact that

$$J_{11} + J_{22} + J_{33} = 1 + 2\cos\theta \tag{B.2}$$

and the components of the vector $\mathbf{u}$ along the axis of rotation are given by

$$\begin{aligned} u_1 &= (J_{23} - J_{32})/2\sin\theta \\ u_2 &= (J_{31} - J_{13})/2\sin\theta \\ u_3 &= (J_{12} - J_{21})/2\sin\theta \end{aligned} \tag{B.3}$$

We shall now go on to prove all this by doing a similarity transformation of the simple rotation matrix into the more general form, i.e., transforming the rotation into a new coordinate system where the rotation axis is not necessarily parallel to a basis vector. This can be done as follows:

$$
{}^g\mathbf{R} = \begin{pmatrix} J_{11} & J_{12} & J_{13} \\ J_{21} & J_{22} & J_{23} \\ J_{31} & J_{32} & J_{33} \end{pmatrix} \begin{pmatrix} m & n & 0 \\ -n & m & 0 \\ 0 & 0 & 1 \end{pmatrix} \begin{pmatrix} J_{11} & J_{21} & J_{31} \\ J_{12} & J_{22} & J_{32} \\ J_{13} & J_{23} & J_{33} \end{pmatrix}
$$

$$
\equiv \begin{pmatrix} J_{11}m - J_{12}n & J_{11}n + J_{12}m & J_{13} \\ J_{21}m - J_{22}n & J_{21}n + J_{22}m & J_{23} \\ J_{31}m - J_{32}n & J_{31}n + J_{32}m & J_{33} \end{pmatrix} \begin{pmatrix} J_{11} & J_{21} & J_{31} \\ J_{12} & J_{22} & J_{32} \\ J_{13} & J_{23} & J_{33} \end{pmatrix}
$$

$$
\equiv \begin{pmatrix} [m(J_{11}^2 + J_{12}^2) + J_{13}^2] & [m(J_{11}J_{21} + J_{12}J_{22}) + n(J_{11}J_{22} - J_{12}J_{21}) + J_{13}J_{23}] & \ldots \\ \ldots & \ldots & \ldots \\ \ldots & \ldots & \ldots \end{pmatrix} \tag{B.4}
$$

where for clarity, only the first two elements of ${}^g\mathbf{R}$ are listed.

- The components of the rotation axis in the basis **R** are [001], which in the coordinates of **J** become $[J_{13}\, J_{23}\, J_{33}] \equiv [u_1\, u_2\, u_3]$. And since each row and each column of a rotation matrix is a unit vector, $J_{11}^2 + J_{12}^2 = 1 - J_{13}^2$. If follows that the element ${}^gR_{11}$ becomes $u_1u_1(1-m) + m$ as in Equation B.1.
- Since $\mathbf{JJ}' = \mathbf{I}$, the identity matrix, if follows that $J_{11}J_{21} + J_{12}J_{22} + J_{13}J_{23} = 0$ and since $J_{13} = u_1$ and $J_{23} = u_2$, the term $m(J_{11}J_{21} + J_{12}J_{22}) = -u_1u_2m$.
- The determinant of **J** is

$$
J_{33}(J_{11}J_{22} - J_{12}J_{21}) + J_{23}(J_{31}J_{12} - J_{11}J_{32}) + J_{13}(J_{21}J_{32} - J_{22}J_{31}) = 1
$$
$$
u_3(J_{11}J_{22} - J_{12}J_{21}) + u_2(J_{31}J_{12} - J_{11}J_{32}) + u_1(J_{21}J_{32} - J_{22}J_{31}) = 1
$$

Since $u_1^2 + u_2^2 + u_3^2 = 1$, it follows that $J_{11}J_{22} - J_{12}J_{21} = u_3$ etc. so considering the last two listed-items together, ${}^gR_{12} = u_1u_2(1-m) + u_3n$ as in Equation B.1. The remaining terms can be proven similarly.

Finally, the Euler angles described on page 88 can be converted into a rotation matrix. An individual rotation matrix is created for each of the three operations, followed by taking their product in the correct sequence, to give an overall rotation matrix (and therefore an axis-angle pair) for the equivalent orientation relationship.

Index